The Immersive Metaverse Playbook for Business Leaders

A guide to strategic decision-making and implementation in the metaverse for improved products and services

Irena Cronin

Robert Scoble

<packt>

BIRMINGHAM—MUMBAI

The Immersive Metaverse Playbook for Business Leaders

Group Product Manager: Alok Dhuri

Publishing Product Manager: Uzma Sheerin

Book Project Manager: Deeksha Thakkar

Senior Editor: Nithya Sadanandan

Technical Editor: Rajdeep Chakraborty

Copy Editor: Safis Editing

Indexer: Manju Arasan

Production Designer: Vijay Kamble

Senior DevRel Marketing Coordinator: Deepak Kumar

DevRel Marketing Coordinator: Mayank Singh

Business Development Executive: Puneet Kaur

First published: December 2023

Production reference: 1171123

Published by Packt Publishing Ltd.
Grosvenor House
11 St Paul's Square
Birmingham
B3 1RB, UK

ISBN: 978-1-83763-284-8

www.packtpub.com

Endorsements for *The Immersive Metaverse Playbook for Business Leaders*

Just when you think you have the Metaverse figured out, it evolves and shifts. That's just the way it is. Thankfully, Cronin and Scoble are here to help. The duo breaks down trends and hype to help leaders imagine and design the next chapter of hybrid experiences, those that are more immersive, innovative, and unforgettable.

Brian Solis, Head of Global Innovation, ServiceNow

As a business, deciding how and whether to integrate metaverse experiences is an extraordinarily difficult task, given the hype, the wide variety of possible approaches, and the complexity of interfacing with social communities. Irena and Robert are very capable guides, providing both the details and the grand strategy you'll need to survive.

Philip Rosedale, Founder, Second Life

The Metaverse is very alive and well! The Metaverse, together with Spatial Computing, is part of the greatest business transformation of the next few decades. This book is very valuable—I highly recommend it as a business resource!

Sandy Carter, Chief Operating Officer, Unstoppable Domains

Irena and Robert bring practical insights gained as they've explored the vast scope of the Immersive Metaverse. They learned how diverse technical domains such as AR, VR, and AI intersect and leverage each other to drive benefits to business and society, then share how to capitalize on this shift in our relationships with information and other people. Working through the timing of these developments, they clearly explain what is possible now versus what is coming soon and what is still science fiction.

Matt Miesnieks, CEO of LivingCities

Irena Cronin and Robert Scoble have worked hard to be 'the experts' in Spatial Computing and the Metaverse. This book details the trends and provides insights into the paradigm shift we are entering. You need to understand what's about to happen.

Ken Gardner, Founder of DADOS Technology and SOASTA, Seven-time Entrepreneur

Description

"The metaverse" has become a widely known term within a very short time span. The Immersive Metaverse Playbook for Business Leaders explicitly explains what it really refers to and shows you how to plot your business road map using the metaverse.

This book helps you understand the concept of the metaverse, along with the implementation of generative AI in it. You'll not only get to grips with the underlying concepts, but also take a closer look at key technologies that power the metaverse, enabling you to plan your business road map. The chapters include use cases on social interaction, work, entertainment, art, and shopping to help you make better decisions when it comes to metaverse product and service development. You'll also explore the overall societal benefits and dangers related to issues such as privacy encroachment, technology addiction, and sluggishness. The concluding chapters discuss the future of AR and VR roles in the metaverse and the metaverse as a whole to enable you to make long-term business plans.

By the end of this book, you'll be able to successfully invest, build, and market metaverse products and services that set you apart as a progressive technology leader.

What you will learn

- Get to grips with the concept of the metaverse, its origin, and its present state
- Understand how AR and VR strategically fit into the metaverse
- Delve into core technologies that power the metaverse
- Dig into use cases that enable finer strategic decision-making
- Understand the benefits and possible dangers of the metaverse
- Plan further ahead by understanding the future of the metaverse

Foreword

The Metaverse will be everywhere – blending our digital and our physical lives, reflecting the transition from the computer internet to the mobile internet to the upcoming spatial internet. It lives across **virtual reality (VR)**, **augmented reality (AR)**, **mixed reality (MR)** and can be experienced into 2D devices as well. Big tech companies are investing significantly to prepare for the inevitable growth of the Metaverse. Attempting to grasp the Metaverse from foundational technology to business opportunities might feel intimidating. This book, *The Immersive Metaverse Playbook for Business Leaders: A guide to strategic decision-making and implementation in the metaverse for improved products and services*, by Irena Cronin and Robert Scoble of Infinite Retina, can help.

This seminal work bridges the gap between technology and business, between today's early products and the future of our interconnected, digital realities. Irena and Robert's comprehensive guide deciphers the Metaverse's role in our world.

I have seen the emergence of the Metaverse slowly form over the course of last decade, and while the vision is not fully realized yet, its impact can already be seen in various industry segments. It will only accelerate. The concept coalesced suddenly in both my personal and professional life, serving as the Vice President and General Manager of XR at Qualcomm, the market leader of technology platforms in this space.

As a business executive in this space, this playbook serves an ideal reference for others to get started. And while undeniably invaluable for executives attempting to chart the course of technology for billions of people, this book also caters to investors, budding entrepreneurs, and professionals keen to harness the myriad opportunities in this virtual world.

The promise of VR, AR, and MR as fundamental technologies for the Metaverse is broad and boundless. Discerning which products and platforms to invest in, develop, and market requires an in-depth understanding of the relevance and application of these technologies. This playbook provides exactly that. It offers a comprehensive analysis, in layman's terms, backed by powerful and relatable use cases.

This book will transform your perspective. In a world on the cusp of a digital renaissance, Irena and Robert's deft handling of the Metaverse in *The Immersive Metaverse Playbook for Business Leaders* serves as both a compass and a map, beckoning the reader to venture forth, armed with knowledge and clarity, into the brave new world of our future once only imagined, made real today.

Hugo Swart

Vice President and General Manager of XR at Qualcomm

Contributors

About the authors

Irena Cronin is the SVP of product for DADOS technology, which involves making an app for the Apple Vision Pro that offers data analytics and visualization. She is also the CEO of Infinite Retina, which provides research to help companies develop and implement AI, AR, and other new technologies for their businesses. Prior to this, she worked for several years as an equity research analyst and gained extensive experience in evaluating both public and private companies.

Cronin has a joint MBA/MA from the University of Southern California and an MS with distinction in management and systems from New York University. She graduated with a BA from the University of Pennsylvania with a major in economics (*summa cum laude*).

I want to thank my best friend, Carol Cox, who has been my sounding board for many things.

Robert Scoble has coauthored four books on technology innovation – each a decade before the said technology went completely mainstream. He has interviewed thousands of entrepreneurs in the tech industry and has long kept his social media audiences up to date on what is happening inside the world of tech, which is bringing us so many innovations. Robert currently tracks the AI industry and is the host of a new video show, *Unaligned*, where he interviews entrepreneurs from the thousands of AI companies he tracks as head of strategy for Infinite Retina.

I want to thank the 70,000 in AI that I follow on X.com (formerly Twitter). Every day, they teach me what is soon to come and how to best use it.

About the reviewer

Mojtaba Tabatabaie is an augmented and virtual reality specialist. He has been working in this field for the past decade. He is also the CEO and CTO at the Alpha Reality and PendAR companies. He works on cutting-edge technology solutions in his companies, such as **Virtual Positioning Systems** (**VPSs**), and utilizes AI to provide KYC solutions for fintech companies. Additionally, he holds several patents in the aforementioned fields. He is also the host of the AllThingsXR.com podcast, where he interviews the world's leading experts in the AR/VR/AI fields.

I would like to express my heartfelt gratitude to my mother, father, and wife, Saeede, for their unwavering support throughout my entrepreneurial journey.

Table of Contents

Part 2: Key Technologies That Power the Metaverse

5

6

9

Understanding User Experience Design and User Interface   207

Part 3: Consumer and Enterprise Use Cases

10

11

12

3D and 2D Content Forms and Creation 299

13

Retail Experiences   325

Part 4: Why Metaverse Redux?

14

Benefits and Possible Dangers Reframed   355

15

Future Vision   381

Index   401

Other Books You May Enjoy   428

Preface

The Immersive Metaverse Playbook for Business Leaders provides a comprehensive guide to understanding the Metaverse and its underlying technologies, such as AR and VR. It starts by introducing the concept of the Metaverse and why it's important, explaining its evolution and how it improves upon previous technological advancements. The book then moves on to dissect the current state of AR, examining its history, technological developments, and its relevance to the Metaverse.

An in-depth analysis is then conducted on the trajectory of VR, focusing on its initial applications in gaming and its diversification into areas such as training, healthcare, and communications. The book emphasizes how VR will redefine human interaction with technology and within the Metaverse.

Another critical aspect covered is the role of 3D visuals in the Metaverse. It discusses the types of 3D content that can be used and created, as well as the potential for interactivity through apps and digital assistants. This leads to an examination of perception technologies such as artificial intelligence and computer vision, which power the functionalities of AR, VR, and the Metaverse.

Immersive Metaverse Playbook also tackles the need for robust computing technologies, exploring topics such as cloud computing, edge computing, distributed computing, and the significance of each in building a scalable, efficient, and secure Metaverse. Furthermore, the book addresses the importance of APIs in ensuring software interoperability within the Metaverse and discusses the tools required to create 3D models and integrate 2D content.

It also investigates the intricate world of **User Experience (UX)** and **User Interface (UI)** design, revealing how these can be adapted to the unique demands of the Metaverse in both AR and VR contexts. Various use cases highlighting the Metaverse's transformational impact on different work environments are also presented, giving insights into its real-world applications. This is followed by a detailed look into how AR and VR are revolutionizing retail experiences.

Finally, the book covers the societal implications of the Metaverse, discussing both its numerous benefits and potential dangers, such as privacy issues and technological addiction. The book concludes by painting a futuristic vision of the Metaverse, contemplating how it will dramatically alter various facets of life, from the way we work and socialize to the structure of our cities and the role of AI and robots.

Overall, *Immersive Metaverse Playbook* serves as a comprehensive roadmap first and foremost for business leaders, as well as any other professionals looking to understand the complexities of the Metaverse and the technologies that fuel it. It offers comprehensive insights relevant to business needs, spanning from its foundational technologies to its societal implications and future possibilities.

Who this book is for

If you are a C-level suite technology and business executive, this book is ideal for you. Investors, entrepreneurs, and other tech professionals would also benefit.

By the end of this book, C-level suite executives should get a better understanding of what kind of product they should produce and market, business managers should understand the breadth and limitations of the technology that they have been asked to implement, and tech professionals should get a deeper top-down understanding of the area.

What this book covers

Chapter 1, The What and Why of the Metaverse, examines the Metaverse. The term the Metaverse has become widely known within a very short period of time, but what does it really refer to, and why should anyone be interested in it? Part of the answer involves explaining what relevant technologies came before the Metaverse and how the Metaverse improves on them. This is addressed in this chapter, along with a fuller explanation of what the Metaverse is and why we should care about it.

Chapter 2, Augmented Reality Status Quo, reviews the current state of Augmented Reality (AR), providing details of its first iterations with enterprise AR headsets and experiences created for the smartphone. Then, AR glasses are discussed, most notably Apple's upcoming product. Lastly, we will explain AR's significance to the Metaverse.

Chapter 3, Where Is Virtual Reality Heading?, explores VR, which has had somewhat of a bumpy ride, starting with its promise for 3D gaming and then branching out to enterprise uses, such as training and education, health, and communication. Here we will review how VR started, how VR is used, and where it is heading. Most importantly, we will show how VR and the Metaverse will significantly change how we interact with technology and each other.

Chapter 4, The Value of Using 3D Visuals to Interact, dives deep into 3D visuals. AR and VR technologies used in the Metaverse enable significant new ways of seeing and doing things. Existing and professionally made 3D images and video, as well as created ones that are now easily made by consumers with generative AI, will be available to you for display and use in the Metaverse. All the 3D visuals that can be found there will provide you with a gamut of information and entertainment, and they can also be searched for objects using AI, as well as for a particular language used in videos. Interactivity made possible by apps and digital assistants will greatly enhance the way you do things, in addition to introducing new ways of doing them.

Chapter 5, Understanding Perception Technologies, covers **perception technologies**, which are essential to enable AR, VR, and the Metaverse. We will present here what and how the main technologies work, helping you to understand what the Metaverse is capable of and how businesses can benefit. Areas covered in this chapter include artificial intelligence, computer vision, and tracking and capture technologies.

Chapter 6, The Different Types of Computing Technologies, examines the various types of computing power that will play significant roles. The most comprehensive form of the Metaverse will demand more computing power than anything we've seen so far. We will kick things off with cloud computing, detailing how it serves as the Metaverse's key structure for storage and processing capacities. We will then discuss the importance of edge computing in reducing delays, enhancing user experiences, and facilitating real-time engagement. Then, we will review the role of distributed computing in effectively spreading workloads across numerous systems for maximum performance and scalability. Finally, we will delve into the value of decentralized computing, emphasizing its essential part in promoting a secure, user-driven, and democratic Metaverse.

Chapter 7, Where Are APIs Needed, examines application programming interfaces (APIs). Applications built using older versioned software, as well as currently incompatible software, need to have APIs built that allow for those applications to interface with new applications. Additionally, new applications in the Metaverse, for which there are and will be many, need APIs so that they can interconnect to function. Building APIs can be expensive in that many could be needed, so it is important to know where they are necessary. Here, we will discuss the areas where APIs need to be built.

Chapter 8, Making and Using 3D Models and Integrating 2D Content, explores software tools for 3D imaging, avatar creation, and the integration of 2D visuals for the Metaverse, which are vital. These software tools, together with AI software, enable the Metaverse to have content. Avatar creation, especially, is undergoing rapid change from what was originally cartoony to more realistic renders. In this chapter, we will review software functionality for each software area, as well as explain who the main companies are.

Chapter 9, Understanding User Experience Design and User Interface, unpacks the unique user concerns associated with VR compared to AR, highlighting the fundamental reasons behind their UX and UI disparities. In the realm of the Metaverse, the intricacies of UX and UI design manifest distinctively between VR and AR. As we delve deeper, we'll also shed light on the overarching principles of UX and UI design, as they seamlessly integrate and shape the comprehensive user journey within the expansive Metaverse ecosystem.

Chapter 10, New Ways of Social Interaction, provides an in-depth exploration of the Metaverse's transformative impact on office and virtual work, walk-around jobs such as retail and field operations, and factory-related tasks including training and diagnostics. As technology continues to advance, the Metaverse will emerge as a groundbreaking development, poised to revolutionize the way work is conducted in diverse sectors. Leveraging AR and VR, we will present use cases that not only highlight the changing dynamics of these work environments but also offer best practices to optimize efficiency and effectiveness in this new virtual paradigm. Whether you're an organizational leader contemplating how to harness the capabilities of the Metaverse, or an individual keen on future-proofing your skills, the practical insights and competencies gained from this chapter are indispensable to navigate the future of work.

Chapter 11, Virtual and Onsite Work, addresses how the Metaverse emerges as a groundbreaking development, poised to revolutionize the way work is conducted in diverse sectors. This chapter provides an in-depth exploration of the Metaverse's transformative impact on office and virtual work, walk-around jobs like retail and field operations, and factory-related tasks including training and diagnostics. Leveraging AR and VR, we present use cases that not only highlight the changing dynamics of these work environments but also offer best practices for optimizing efficiency and effectiveness in this new virtual paradigm. Whether you're an organizational leader contemplating how to harness the capabilities of the Metaverse, or an individual keen on future-proofing your skills, the practical insights and competencies gained from this chapter are indispensable for navigating the future of work.

Chapter 12, 3D and 2D Content Forms and Creation, explores the multifaceted domains of gaming, streaming entertainment, and art within the Metaverse. As 3D games, immersive streaming content, and creative expression take center stage, we will examine the distinctions between AR and VR in their diverse applications. From the fusion of 3D and 2D content to the democratizing influence of generative AI, we will uncover how accessibility and contribution to this digital realm extend to everyone. Through a series of use cases and real-world examples, we will reveal the best practices that enhance business efficiency and effectiveness, equipping you with valuable skills to navigate this dynamic virtual frontier.

Chapter 13, Retail Experiences, dives deep into shopping using AR and VR in the Metaverse, which is one of its most anticipated features for consumers. The Metaverse will deliver on-demand online retail experiences, from customer query to fulfillment, which includes getting personalized matches and unsolicited (but opted-in) recommendations, based on data from AI digital assistants, digitally trying on goods such as clothes, footwear and cosmetics, viewing furniture and home goods as they would appear in a room, and pulling up 3D images and videos of food dishes when choosing restaurants and even when seated in one. Much of this can be replicated when a person who is wearing glasses is outside walking or driving around. Use cases using AR and VR, exemplifying particular Metaverse retail scenarios, are included in this chapter.

Chapter 14, Benefits and Possible Dangers Reframed, examines the innumerable benefits and a number of possible dangers that the Metaverse provides. The benefits include those discussed in the previous chapters, which included use cases – benefits ranging from socializing, networking, and creating to office and virtual work, entertainment, and shopping, among others. A discussion of overall societal benefits will also be included in this chapter. We will also address the possible dangers of privacy encroachment, technology addiction, and laziness.

Chapter 15, Future Vision, discusses the future of AR and VR roles in the Metaverse, their ramifications, and the Metaverse as a whole. In the Metaverse, 3D images, models and videos will be expected and commonplace. Due to AI digital personalized assistants in the Metaverse, information imparting knowledge will be seemingly instant, making the Metaverse an even better extension of our minds than a smartphone. Working virtually will become quite effortless, making people more productive and giving them more time with their family and friends. Since remote work will become more of the norm, a city's landscape will change and become more decentralized – downtown will no longer be a much-desired location destination. Autonomous cars, which are considered robots by engineers, and future robots that are made to improve our lives will interface with the Metaverse, receiving information and directives to do tasks. All in all, the Metaverse will dramatically change how we see and do things.

To get the most out of this book

Just read. No prior knowledge of immersive technologies is needed to understand this book.

Conventions used

> **Tips or important notes**
> Appear like this.

Get in touch

Feedback from our readers is always welcome.

General feedback: If you have questions about any aspect of this book, email us at customercare@packtpub.com and mention the book title in the subject of your message.

Errata: Although we have taken every care to ensure the accuracy of our content, mistakes do happen. If you have found a mistake in this book, we would be grateful if you would report this to us. Please visit www.packtpub.com/support/errata and fill in the form.

Piracy: If you come across any illegal copies of our works in any form on the internet, we would be grateful if you would provide us with the location address or website name. Please contact us at copyright@packt.com with a link to the material.

If you are interested in becoming an author: If there is a topic that you have expertise in and you are interested in either writing or contributing to a book, please visit authors.packtpub.com.

Share Your Thoughts

Once you've read *The Immersive Metaverse Playbook for Business Leaders*, we'd love to hear your thoughts! Scan the QR code below to go straight to the Amazon review page for this book and share your feedback.

https://packt.link/r/1837632847

Your review is important to us and the tech community and will help us make sure we're delivering excellent quality content.

Download a free PDF copy of this book

Thanks for purchasing this book!

Do you like to read on the go but are unable to carry your print books everywhere? Is your eBook purchase not compatible with the device of your choice?

Don't worry, now with every Packt book you get a DRM-free PDF version of that book at no cost.

Read anywhere, any place, on any device. Search, copy, and paste code from your favorite technical books directly into your application.

The perks don't stop there, you can get exclusive access to discounts, newsletters, and great free content in your inbox daily

Follow these simple steps to get the benefits:

1. Scan the QR code or visit the link below

https://packt.link/free-ebook/9781837632848

2. Submit your proof of purchase
3. That's it! We'll send your free PDF and other benefits to your email directly

Part 1: The Reality of the AR/MR Metaverse

In *Part 1*, the concept of the Metaverse, which is gaining rapid recognition, is explored in depth. It encompasses advanced technologies such as **Augmented Reality** (**AR**) and **Virtual Reality** (**VR**). AR's evolution, from early smartphone applications to upcoming AR glasses, as well as VR, plays a pivotal role in shaping the Metaverse and its experiences in gaming, training, healthcare, and so on.

Within this digital realm, users are exposed to a wealth of 3D content, both professionally generated and user-created, often with the assistance of generative AI. These visual assets serve not only as sources of information and entertainment but also as repositories of searchable objects and specific language within videos, thanks to AI-driven search capabilities.

Furthermore, the integration of apps and digital assistants enhances interaction within the Metaverse, presenting a paradigm shift in how tasks are performed and introducing innovative approaches to daily activities.

This part has the following chapters:

- *Chapter 1, The What and Why of the Metaverse*
- *Chapter 2, Augmented Reality Status Quo*
- *Chapter 3, Where Is Virtual Reality Heading?*
- *Chapter 4, The Value of Using 3D Visuals to Interact*

1

The What and Why of the Metaverse

The term **Metaverse** has become widely known within a very short period of time. But what does it really refer to, and why should anyone be interested in it? Part of the answer is in explaining which relevant technologies came before the Metaverse and how the Metaverse improves on them. This is addressed in this chapter, along with a fuller explanation of what the Metaverse is and why we should care about it.

In this chapter, we're going to cover the following main topics:

- Origins of the Metaverse

- What is the Metaverse?

- Why is the Metaverse so important?

Origins of the Metaverse

All of a sudden, we have something called the Metaverse that's in the media and being talked about by seemingly everyone. Almost everyone's consensus is that it will be revolutionary for business and society. However, the main topic when talking about the Metaverse is to try to pin down what it actually is, because it is not here today. Before beginning to explain what the Metaverse is, it helps to understand its origins and then, what it is and the reasons behind its anticipation.

Human beings have a wish to communicate with each other in person, and when not in person, in a way that they can convey their thoughts and—in many instances—their emotions. Whether it was writing on animal bones, stone, clay, papyrus, or—more recently—paper, or speaking on the telephone, humans always found some way to communicate. There are so many reasons to want to communicate, from the personal— to catch up with close and not-so-close friends and family—to the non-personal and professional—to complain to a co-op board, to appraise a work of art, negotiate to buy a house or building, buy clothes and food, applaud or disparage a certain politician, provide instructions to an employee, and so on. The motivations to communicate are many.

Several major things have changed from several thousands of years ago to today when it comes to communication, apart from the method of communication—speed, ease, precision, and audience reach.

The most obvious change is the increased speed with which communication can be done. Speed has even increased when it comes to communicating in person. From coming together by walking, riding horses, or traveling by chariot, carriage, car, plane, or helicopter, the opportunity for in-person communication has increased manifold, and with it, the speed of getting that communication done. And that is just in-person communication. When communicating not in person, the speed of that kind of communication has increased tremendously—from messengers and postmen delivering items of communication by walking or running, by horse, then by car and carrier plane, to wires being sent, to phone calls by landline, to email, cellphone, and videoconference.

With increased speed comes the ease with which communication is accomplished. That ease has compelled more people to communicate more often. That increase in communication has allowed for more precision of communicated intent. Let's say that when letters were the largest mode of not-in-person communication, if the receiver of a letter misconstrued what the writer of that letter intended, the receiver could decide to not communicate anymore with the writer, or if they did, a return letter would have to be sent, and so on. The ease of faster communications allows for the correcting of any misconstrued messaging. And this is important when communication is able to reach as many people as is digitally possible.

More precise, easier, and faster communication, together with the capability of large audience reach is where we are today. Social media can be counted as a mode of communication, but the current formats don't allow for much flexibility and personalization. Outside of social media, increased speed, ease, precision, and audience reach of current communications have made people more efficient and productive in both their personal and professional lives. Although it's commonly thought that the main original motivation behind the Metaverse is to foster interoperability among different computer games, the origins of the current manifestation of the imagined Metaverse come out of a need for more improved and enhanced communication capabilities. And this improved and enhanced communication has the benefit of bringing about multiplying business opportunities. Use cases that exemplify what can be done in the Metaverse come later in this book, in *Part 3*.

To better understand how the Metaverse came about and its place in technology, it's helpful to think of it as part of a paradigm shift; in this case, the fourth paradigm shift— spatial computing.

The Metaverse – part of the fourth paradigm

A technological paradigm shift is a change in the underlying principles that shape the development and use of technology in a society. A classic technological paradigm shift is the shift from horse and buggy to the automobile. Four technological paradigm shifts have been recognized in computing.

The first paradigm – the personal computer arrives

The shift from mainframe computers to **personal computers** (**PCs**) is considered the first technological paradigm shift. Mainframe computers were large, expensive, and complex machines that were primarily used by businesses, government agencies, and other organizations. They were typically housed in dedicated computer rooms and operated by trained technicians.

The IAS machine (also known as the **Institute for Advanced Study** computer) was an early computer built between 1946 and 1951 at the **Institute for Advanced Study** (**IAS**) in Princeton, New Jersey. It was one of the first electronic computers to be built and was used continually and productively until 1960 for a wide range of research projects, including the development of the first high-level programming language, called FORTRAN.

In the 1970s, computer rooms were typically large, specialized spaces that housed mainframe computers and their associated equipment such as IBM's very large **Access Client Solutions** (**ACS**) chip arrays that are shown in *Figure 1.1*. These computers were much larger and more expensive than the PCs that became popular in the 1980s, and they required dedicated space with specialized cooling and electrical systems to operate. The computer room was usually a secure area that was only accessible to authorized personnel, and it was often monitored by technicians who were responsible for maintaining the computer equipment:

Figure 1.1 – A section of IBM's 1968-era very large ACS circuit board with a 10 x 10 array
of chip packages that were used to power one computer (source: Robert Scoble)

The Apple 1 was a PC released in 1976 by Apple Computer, Inc. It was a small, relatively inexpensive PC that could be used by an individual or small group and was designed to be assembled by the user to be used in a home or small office setting. It was one of the first PCs on the market and was designed to be a kit that users could assemble themselves. The Apple 1 was powered by a MOS Technology 6502 microprocessor and had 4 KB of RAM, which could be expanded to 8 or 48 KB. It used a cassette tape to store data and programs, and it had a simple command-line interface for users to input commands.

Figure 1.2 – Steve Wozniak, co-founder of Apple Computer, stands with the Apple II that he helped develop and is now in the Computer History Museum (source: Robert Scoble)

The development and widespread adoption of PCs represented a paradigm shift in the way that people used computers. Before the development of PCs, computers were large, expensive machines that were used primarily by large organizations, such as businesses, universities, and government agencies, to support the computing needs of hundreds or thousands of users. These computers were operated by specialized personnel and were typically accessed remotely through terminals or other devices.

In contrast, PCs are smaller, more affordable, and easier to use than mainframe computers. They can be used by individuals and small businesses and do not require specialized training to operate. The development of the microprocessor and the PC revolutionized the way people interacted with computers and made it possible for people to use computers for a wide range of tasks, from word processing and spreadsheet creation to internet browsing and gaming. The development of PCs was a key factor in the growth of the digital economy.

The second paradigm – graphical user interfaces

A **graphical user interface** (**GUI**) is a type of UI that allows users to interact with electronic devices through graphical icons and visual indicators, rather than text-based commands. GUIs are designed to make it easier and more intuitive for users to access and use computer programs and other electronic devices. They use visual elements, such as icons, menus, and buttons, to represent different options and functions, which users can access using a pointing device, such as a mouse or a touchpad.

The concept of a GUI was first introduced in the 1970s, but it was not until the 1980s that GUIs became widely adopted. The first GUI was developed at Xerox **Palo Alto Research Center** (**PARC**) in the 1970s, and it was used on the Xerox Alto, one of the first PCs. The Xerox Alto was the first computer to use a mouse-based input system, which made it possible to use a GUI to navigate and interact with the computer.

The first widely available PC to use a GUI was the Apple Macintosh, which was introduced in 1984 and it helped to popularize the use of GUIs in PCs. In the following years, other companies, such as Microsoft, introduced their own GUI-based operating systems, and the use of GUIs became widespread in the PC market. Today, GUIs are the standard interface for most PCs and are widely used in a variety of electronic devices.

GUIs represented a paradigm shift in the way that people interact with computers because they made it much easier and more intuitive for users to access and use computer programs. Prior to the development of GUIs, computers used command-line interfaces, which required users to input commands using a keyboard. This was a time-consuming and error-prone process, and it was difficult for people who were not familiar with computers to learn how to use them.

The adoption of GUIs had a significant impact on the way that people use computers and has contributed to the widespread adoption of PCs. GUIs made it possible for people with little or no computer experience to use computers with ease, which has had a profound impact on many aspects of society, including education, business, and communication.

The third paradigm – mobile

The first **mobile phones** were developed in the late 1940s and 1950s, but they were large and expensive and were only used by a small number of people, such as wealthy individuals and businesses. The first commercially available mobile phone was the Motorola DynaTAC 8000X, which was released in 1983. These early mobile phones were quite large and expensive and were only used by a small number of people. Over time, mobile phones became smaller, less expensive, and more widely available, and their use became more widespread.

The **LG Prada** (also known as the **LG KE850**) was a mobile phone released by LG Electronics in May 2007. It was one of the first phones to feature a touchscreen display and was widely considered to be a fashionable and high-end device.

The first iPhone, on the other hand, was released by Apple in June 2007. It was a revolutionary device that introduced a new type of UI based on a multi-touch screen and established the smartphone as a new category of device. The iPhone also had a number of features that set it apart from other mobile phones at the time, such as a high-resolution display, a digital camera, and the ability to access the internet and run a wide range of apps.

Overall, the LG Prada was an important early touchscreen phone, but the iPhone was a more significant and influential device that set the stage for the modern smartphone market:

Figure 1.3 – The first iPhone versus the Nokia N97; the first iPhone was released in
June 2007 and the Nokia N97 in December 2008 (source: Robert Scoble)

Apple is also widely known for obliterating the importance of Nokia when it comes to mobile phones. Nokia was considered the mobile phone leader before the iPhone came out. Yet, due to its miscalculation of the importance of the iPhone's innovations, Nokia mistakenly thought that it would not need to do too much to stay ahead, which led to its steady downfall in the area.

The mobile phone has become a technological paradigm because it has fundamentally changed the way that people communicate and access information. Before the widespread adoption of mobile phones, people had to be physically present in a specific location to make phone calls or access information. With the advent of mobile phones, people are able to communicate and access information from anywhere at any time. This has had a profound impact on society and has led to the development of new industries and business models. Mobile phones have also had a major impact on the way that people interact with each other and with the world around them, and they have become an essential part of daily life for many people.

The fourth paradigm – spatial computing

Spatial computing refers to the use of technology to create an immersive, 3D digital environment that interacts with the physical world. It is a multidisciplinary field that combines computer science, engineering, design, and other areas to create an interactive experience that goes beyond traditional 2D screens. Spatial computing includes any technology that would be used to move about in a virtual or augmented 3D world. This includes **virtual reality (VR)**, **augmented reality (AR)**, **mixed reality (MR)**, **artificial intelligence (AI)**, **computer vision (CV)**, and sensor technology, among others.

Spatial computing is considered the fourth paradigm because it represents a new way of interacting with technology that goes beyond traditional 2D screens and input methods. Applications of spatial computing include gaming, education, design, and industrial training, and it has emerging uses in many other industries such as healthcare, retail, and entertainment.

In 1987, Jaron Lanier coined the term **VR**. Lanier was a founder of VPL Research, a company that made early commercial VR headsets and wired gloves. There were earlier attempts to make headsets that were either completely experimental or commercially failed, such as Morton Heilig's Telesphere Mask:

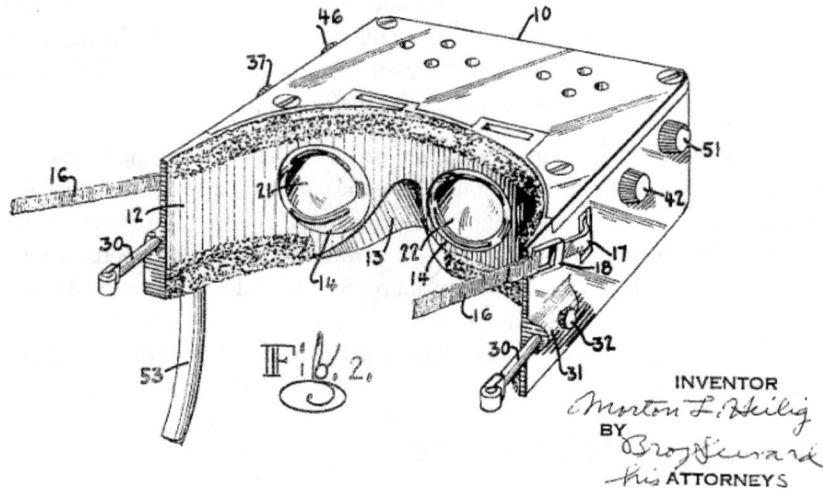

Figure 1.4 – Morton Heilig's Telesphere Mask, a head-mounted display device patented in 1960 that commercially failed (source: United States Patent and Trademark Office (USPTO))

Others, such as a patent filed in 2008 by Apple for a VR headset and a remote controller, portrayed a product that was never produced. In 2012, the company Oculus VR was founded, with a VR headset, the Oculus Rift, becoming commercially available in 2016. A couple of years earlier in 2014, Facebook bought Oculus VR and started on the journey to creating more VR headset models. HTC and a couple of other players joined Oculus in creating competitive VR headsets:

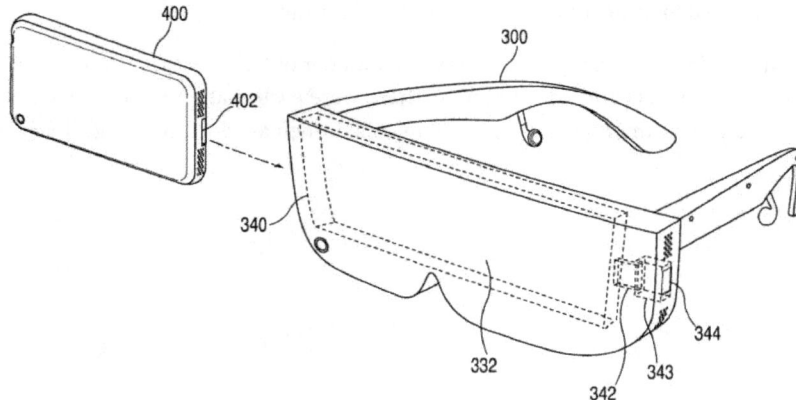

Figure 1.5 – A patent filed in 2008 by Apple for a VR headset and a remote controller that would use an iPhone's screen as the headset's primary display; the headset was never commercially made (source: USPTO)

The first functional AR headset was made in 1980 by Steve Mann and was called the EyeTap, a helmet that displays virtual information in front of the wearer's eye. Early AR headsets were not widely adopted due to limitations in technology and high cost. In the 2010s, advancements in technology, such as the development of smartphones and improved displays, led to the resurgence of interest in AR and the introduction of more advanced and affordable AR headsets, such as the Microsoft HoloLens and the Magic Leap.

Spatial computing has many potential benefits, some of which include the following:

- **Immersive experience**: Spatial computing allows for a more immersive and engaging experience for users, as it creates a 3D digital environment that interacts with the physical world. This allows for a more natural and intuitive way to interact with information and technology.

- **Enhanced productivity**: Spatial computing can be used to create more efficient and effective ways of working, such as VR and AR tools for industrial training, design, and education. It can also improve remote collaboration by creating shared virtual spaces.

- **Improved accessibility**: Spatial computing can be used to create more accessible experiences for users with disabilities, such as those who are visually impaired or have difficulty with fine motor skills.

- **New opportunities in various industries**: Spatial computing has potential use cases in various industries such as healthcare, retail, and entertainment. For example, in healthcare, it can be used for training and surgeries, in retail for virtual shopping, and in entertainment for games and movies.

- **Increased convenience**: Spatial computing can make it more convenient for users to access and interact with information, such as overlaying virtual instructions on real-world objects for repair or assembly.

- **Data visualization**: Spatial computing can be used to create 3D visualizations of complex data, making it easier to understand and analyze.

Spatial computing is a key enabler for the Metaverse, providing technology that allows for the creation of immersive, 3D digital environments that can be used for socializing, entertainment, work, and many other use cases. Now that we have glanced through the history of the technology that led up to the Metaverse, let's understand what the Metaverse actually is.

What is the Metaverse?

What the Metaverse promises to bring to communications would be an improvement in the areas of speed, ease, precision, and audience reach. So, what is this Metaverse?

Here is writer Matthew Ball's definition, which has gained some traction in the public eye:

"The Metaverse is a massively scaled and interoperable network of real-time rendered 3D virtual worlds and environments which can be experienced synchronously and persistently by an effectively unlimited number of users with an individual sense of presence, and with a continuity of data, such as identity, history, entitlements, objects, communications, and payments."

Does this explain why the Metaverse is an improvement on other means of communication? Kind of…let's unpack Ball's definition:

- **Massively scaled and interoperable network**

  This means that the Metaverse can handle a large number of people and their activities at the same time and that people can easily roam and operate between different environments and apps.

- **Real-time rendered 3D virtual worlds and environments**

  Real-time rendered here means that images are dynamically produced and updated in near real time so that a person can interact or move around in a virtual environment without lag.

- **Experienced synchronously and persistently**

  Synchronously means that people's interactions in the Metaverse happen simultaneously. *Persistently* means that in whichever state the environment exists and whatever was updated in it remains in that state if and until that environment is actively changed.

- **Effectively unlimited number of users with an individual sense of presence, and with a continuity of data, such as identity, history, entitlements, objects, communications, and payments**

  An individual sense of presence means that a person in the Metaverse experiences it almost as if they were there in a psychological sense.

My definition

The Metaverse allows many people to simultaneously come together and interact in real-time 3D virtual environments or worlds with the capability of retaining distinct and persistent digital identities, including the trail of each individual's activities.

With my definition, how the Metaverse can provide improvements to communication is much clearer. Improvements to speed, ease, precision, and audience reach are all there.

Here is a breakdown:

- **Speed**: *Simultaneously come together and interact in real time* speaks for itself.

- **Ease**: *Real-time* and *retaining distinct and persistent digital identities* –

    The ease of communication in *real time* is intuitively understood. *Retaining distinct and persistent digital identities* allows a person to return to the Metaverse and interact more easily each time because that person's data is continuously being stored for that person's use.

- **Precision**: *Real-time 3D virtual environments or worlds* and *retaining distinct and persistent digital identities* – *Real-time 3D virtual environments or worlds* bring about precision by providing *real-time* capability and by allowing people to view objects more realistically since they are in 3D. This could be immensely helpful in business scenarios, such as when a product is demoed or shown in 3D in the Metaverse.

- **Audience reach**: *Allows many people* speak for itself.

The definition of the Metaverse is different from its fictional version. The term *The Metaverse* was coined by the author Neal Stephenson in his 1992 novel *Snow Crash*. In that novel, there was only *one* Metaverse and people would use VR headsets to go online and enter it to escape their dystopian reality. More recently, and inherent in the definition, the idea that there will be many available Metaverse manifestations or platforms has gained traction and is seen as the most viable, given business limitations, profit motives, and antitrust concerns. Additionally, the Metaverse is envisioned to be accessed by both AR headsets and glasses and VR headsets.

In addition to what was covered here in this chapter, in this section, and under *The fourth paradigm – spatial computing* subheading, in-depth detail on AR and VR, including relevant companies and their products, can be found in *Chapters 2* and *3*, respectively, and Metaverse-supporting technologies are covered in *Part 2, Key Technologies that Power the Metaverse*. As an introduction, next is a short overview of technologies that are generally needed to create, manage, and/or experience the Metaverse, including blockchain and other decentralized system technologies (more detail on these can be found in *Chapter 6, The Different Types of Computing Technologies*).

Basic technical needs

The following technologies are summarized here: game engines, other design software, AI, CV, payment processing systems, UIs, cloud computing, **application programming interfaces** (**APIs**), tracking and capture technologies, VR headsets, and AR headsets and glasses.

Game engines

Game engines are software frameworks that provide developers with the tools to build video games more efficiently. These tools typically include things such as a rendering engine for graphics, a physics engine for simulating realistic movements, and support for input, audio, and networking. Some game engines also offer additional features such as level editors and animation tools. Game engines are designed to be flexible and reusable so that developers can build a variety of different types of games with them.

In addition to video games, game engines are also used for other types of interactive applications, such as the following:

- VR and AR experiences
- Simulation and training software
- Interactive architectural and product visualizations
- Educational software
- GUI applications

Game engines are well suited to these types of applications because they provide a high-performance environment for rendering 3D graphics and handling user interaction.

Some well-known game engine companies include the following:

- **Unity Technologies**: Unity is a cross-platform game engine that is widely used for building 2D and 3D games, as well as other interactive content. It is known for its ease of use and flexibility, and it has a large community of developers who contribute to its development.
- **Epic Games**: Epic Games is the company behind Unreal Engine, a powerful game engine used for building high-quality 3D games. Unreal Engine is known for its advanced graphics capabilities and is used by many AAA game studios.
- **Crytek**: Crytek developed CryEngine, a game engine that is used for building 3D games. CryEngine is known for its advanced graphics and its support for VR.
- **Open 3D Engine (O3DE)**: O3DE is a free and open source 3D game engine developed by Open 3D Foundation, a subsidiary of the Linux Foundation. The engine's initial version came from an updated version of the Amazon Lumberyard engine, which was developed and contributed by Amazon Games.

Other design software – in-production 3D software and avatar creation

3D modeling software is a type of software that allows users to create three-dimensional digital models of objects or environments. These models can be used for a variety of purposes, such as creating 3D art, visualizing architectural plans, and designing products. Some common features of 3D modeling software include the ability to create and manipulate 3D shapes, apply textures and materials, and add lighting and other effects. There are many different 3D modeling software programs available, ranging from professional tools used by artists and designers to more beginner-friendly options that are accessible to hobbyists and students. Some examples of 3D modeling software include the following:

- **Autodesk 3ds Max**: 3ds Max is a professional 3D modeling, animation, and rendering software that is widely used in the film, television, and gaming industries.
- **Autodesk Maya**: Maya is a professional 3D modeling, animation, and rendering software that is used in a variety of industries, including film, television, and games.
- **Houdini**: Houdini is a 3D animation and visual effects software that is used in the film and television industry to create complex effects and simulations.
- **Blender**: Blender is a free, open source 3D modeling and animation software that is used in a variety of industries, including film, television, and games.
- **Cinema 4D**: Cinema 4D is a professional 3D modeling, animation, and rendering software that is used in the film, television, and games industries.

Avatar creation software is software that allows users to create custom avatars or digital representations of themselves that could be used in the Metaverse. Some avatar creation software allows users to create avatars by selecting from a range of predefined options, such as hairstyles, facial features, and clothing. Other software allows users to create more detailed avatars by importing and manipulating 3D models or by using tools to sculpt and shape the avatar's appearance. Some avatar creation software allows users to create a realistic 3D model of their own face or body, while others offer a more stylized or cartoonish approach.

AI

AI has the potential to be used in a variety of applications within the Metaverse. The type of AI that is most promising for the Metaverse is **generative AI (genAI)**.

GenAI refers to a type of AI that is able to generate new content or ideas based on a set of input data or rules. This can be achieved using techniques such as **machine learning (ML)**, **neural networks (NNs)**, and evolutionary algorithms.

Examples of genAI models are GPT-3 (includes ChatGPT; GPT-4 forthcoming), DALL-E 2, Stable Diffusion, Midjourney, Meta's Make-A-Video, and Google DreamFusion (text-to-3D image generator; in research stage).

In the context of the Metaverse, genAI could be used in a variety of ways, such as the following:

- **Generating virtual objects**: GenAI could be used to create a wide range of virtual objects and assets for use in the Metaverse, such as buildings, furniture, vehicles, or clothing.

- **Designing virtual environments**: GenAI could be used to design and generate virtual environments and landscapes for the Metaverse, such as cities, forests, or planets.

- **Generating virtual events**: GenAI could be used to create and schedule virtual events and experiences within the Metaverse, such as concerts, festivals, or sports games.

- **Creating virtual characters**: GenAI could be used to design and generate virtual characters or avatars for use in the Metaverse, such as **non-player characters** (**NPCs**) or virtual assistants that can converse with users naturally.

- **Generating virtual content**: GenAI could be used to create a variety of virtual content for the Metaverse, such as music, videos, or games.

- **Personalization**: GenAI could be used to create personalized experiences within the Metaverse. For example, an AI system could analyze a user's preferences and generate customized virtual spaces or events that match their interests.

- **Automation**: GenAI could be used to automate tasks and processes within the Metaverse, such as managing virtual assets or handling transactions.

CV

CV is a field of study that focuses on enabling computers to interpret and understand visual information from the world, such as images and videos. It involves a combination of computer science, electrical engineering, and AI to develop algorithms, models, and systems that can recognize and interpret visual data and make decisions based on that data.

Several technologies are used in CV, including the following:

- **Image processing**: Techniques for improving the quality of images, such as noise reduction and image enhancement, which make it easier for CV algorithms to interpret them

- **Feature extraction**: Algorithms for identifying and extracting key features from images, such as edges, corners, and textures, which are used to identify objects and understand the scene

- **Deep learning (DL)**: Using NNs, specifically **convolutional NNs** (**CNNs**) that can learn to recognize patterns in images and videos

- **ML**: Techniques for training models to recognize patterns in data, such as **supervised learning** (**SL**), **unsupervised learning** (**UL**), and **reinforcement learning** (**RL**)

- **Object detection**: Algorithm to detect objects of a certain class in images or videos

- **Semantic segmentation**: Algorithm that assigns a class label to each pixel in images

- **Motion analysis and object tracking**: Algorithms that can track the movement of objects in a scene over time

- **3D CV**: Techniques for creating 3D models of scenes and objects from 2D images, such as structure from motion and stereo vision

Specifically, for the Metaverse, CV can be used in many ways, such as for the following:

- **User identity and tracking**: CV algorithms can be used to identify and track users within the Metaverse, and to recognize and respond to their gestures and movements. This could be used to create more immersive and interactive experiences and to enable new types of social interactions.

- **Environment and object recognition**: CV algorithms can be used to recognize and understand the environment and objects within the Metaverse, and to respond to them in a realistic and believable way. This could be used to create more realistic virtual worlds and to enable new types of interactions with virtual objects.

- **Spatial mapping and navigation**: CV can be used to map and understand the spatial layout of the Metaverse, and to enable users to navigate through it. This could be used to create more intuitive and user-friendly Metaverse experiences and to enable new types of virtual travel.

- **Human-computer interaction (HCI)**: CV can be used to enable more natural and intuitive forms of interaction with virtual worlds, such as hand and body tracking, facial recognition, and speech recognition.

- **AR**: AR technology can be integrated with CV to overlay virtual elements onto the real world, providing more immersive and interactive experiences.

Overall, CV has a lot of potential to enable new and exciting experiences within the Metaverse, by enabling computers to understand and interpret visual information and to respond to it in real time.

Payment processing systems

Traditional **payment systems** can be used in the Metaverse in a few different ways. Here are some examples:

- **In-game currencies and microtransactions**: Many Metaverse platforms allow users to purchase virtual currencies or items that can be used within a virtual world. These transactions can be processed using traditional payment systems such as credit cards, debit cards, and digital wallets.

- **Virtual real estate**: Some Metaverse platforms allow users to purchase virtual real estate or other virtual assets, which can be treated as a form of investment. These transactions can be processed using traditional payment systems, and in some cases, virtual assets may be traded on secondary markets.

- **Subscriptions**: Access to certain areas or services within the Metaverse may require a subscription. Traditional payment systems can be used to process these payments on a recurring basis.

- **Virtual goods and services**: Virtual goods such as clothes, skins, and other accessories for avatars can be sold for real money. Similarly, virtual services such as tutoring or guiding in a virtual world can also be sold with traditional payment systems.

Next is a diagram of how a traditional payment processing system works. When a customer orders an item on a website using a credit or debit card, the payment goes through a gateway and a verification process before it goes from the card-issuing bank to the merchant:

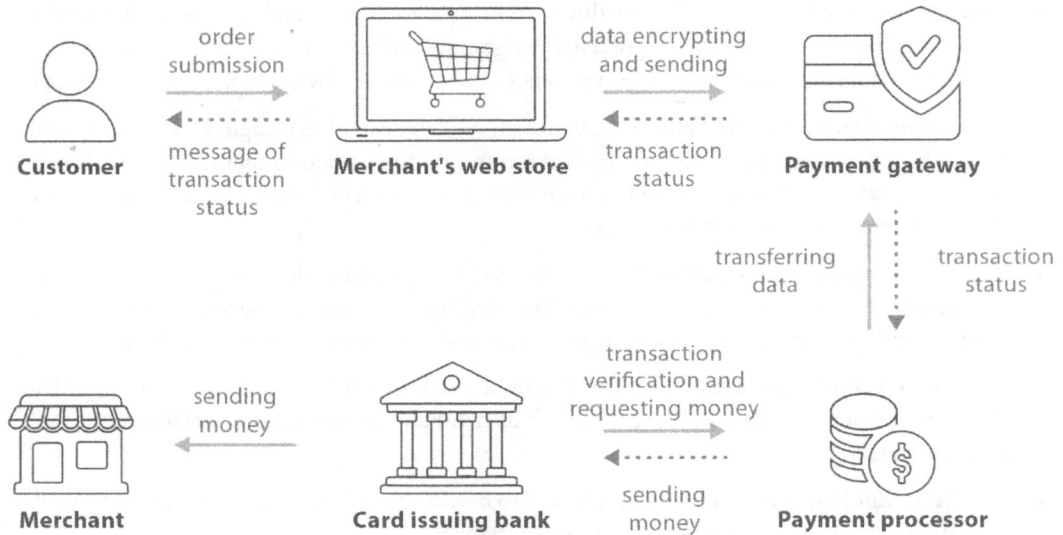

greenice

Figure 1.6 – Traditional payment processing system

While traditional payment systems can be used to facilitate transactions in the Metaverse, the Metaverse itself may introduce new and innovative forms of payment as well. For example, virtual currencies and blockchain technology could become increasingly important in the Metaverse economy.

The next diagram shows how a transaction that is done on a blockchain would typically work, which goes as follows. Someone initiates a transaction, which is then broadcast to computers whose networks are called "nodes." The next step is validation; passing that allows for a transaction to form a data block. This block is then in turn added to a chain of other blocks—a "blockchain." This blockchain is then broadcast to nodes, whereupon the transaction is entered into a decentralized digital ledger:

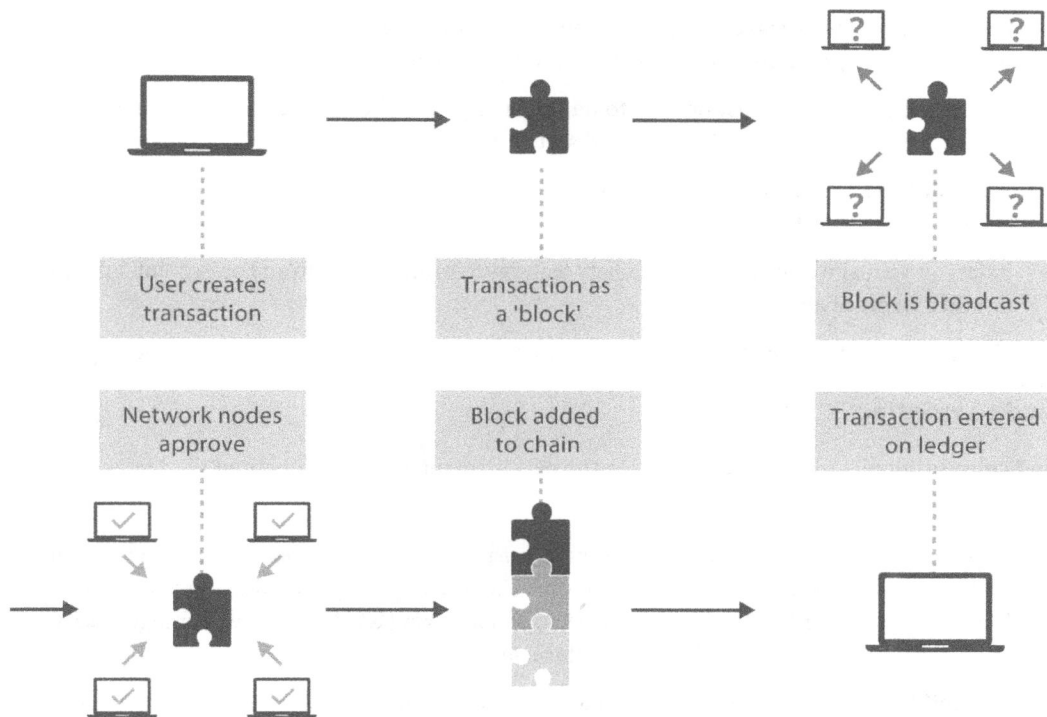

Figure 1.7 – Blockchain payment flow

UI

A **UI** is the point of interaction between a user and a computer or other device. It refers to the way in which a user interacts with and controls the device, and it includes visual elements, such as buttons and icons, as well as non-visual elements, such as audio and haptic feedback. The goal of a UI is to make the device easy to use and understand, by providing a clear and consistent way for the user to interact with the device, and by providing clear and useful feedback to the user.

Some issues related to a UI for the Metaverse include the following:

- **Navigation**: In a virtual world, users must be able to easily move around and explore the environment. This can be challenging to implement in a way that feels intuitive and natural.

- **Interaction**: Users must be able to interact with virtual objects and other users in ways that are familiar and easy to understand.

- **Identity**: Users must be able to express their identity in a way that is unique and meaningful in a virtual world, while also protecting their privacy.

- **Performance**: The Metaverse should be able to perform well on a wide range of devices and networks while providing a smooth and responsive experience.

- **Scale**: The Metaverse must be able to handle a large number of users and a wide range of activities while maintaining stability and security.

- **Accessibility**: The UI for the Metaverse should be accessible to people with disabilities, the elderly, and non-technical users.

- **Safety**: A virtual world should have safety features to protect users from harassment, bullying, or other forms of abuse.

- **Usability**: The interface should be easy to use and understand, with minimal training required.

Cloud computing

Cloud computing plays an important role in the development and deployment of the Metaverse in the following ways:

- **Hosting**: The Metaverse relies on servers to host a virtual world and all of its components, such as 3D models, textures, and animations. Cloud computing providers, such as **Amazon Web Services (AWS)**, Azure, and **Google Cloud Platform (GCP)**, offer a wide range of server resources that can be used to host the Metaverse.

- **Scalability**: The Metaverse is expected to grow rapidly in terms of the number of users and the amount of content that is created. Cloud computing allows the Metaverse to scale up and down as needed, by allocating more or fewer resources as needed.

- **Security**: Cloud computing providers offer a wide range of security features—such as encryption, authentication, and access control—that can be used to protect the Metaverse and its users.

- **Analytics and intelligence**: Cloud computing services can provide advanced analytics and AI capabilities that can be used to understand user behavior, optimize performance, and improve the overall user experience.

- **Infrastructure as a Service (IaaS)**: Instead of building and managing the hardware and software infrastructure for the Metaverse, developers and companies can use a **cloud service provider's (CSP's)** infrastructure to build and deploy their own Metaverse experiences.

- **Platform as a Service (PaaS)**: Some companies offer a ready-made platform for building Metaverse experiences. Developers can build and run their experiences on that platform.

Overall, cloud computing is an important enabler for the development and deployment of the Metaverse. It allows developers and companies to focus on creating the best possible user experience, without having to worry about the underlying infrastructure and scaling concerns.

APIs

An **API** is a set of tools, protocols, and standards for building and integrating software applications. APIs allow different software applications to communicate and exchange data with one another. They provide a way for different pieces of software to talk to each other, and for different systems to share data and functionality.

APIs are a key technology used in the Metaverse for the following reasons:

- **Interoperability**: Metaverse experiences are built by different creators and companies, and APIs allow them to communicate and interact with each other seamlessly. This enables users to move between different parts of the Metaverse without interruption and for different experiences to interact with one another in meaningful ways.

- **Data sharing**: APIs can be used to share data between different parts of the Metaverse, such as user information, inventory, and transactions. This enables a more cohesive and integrated user experience across the Metaverse.

- **Third-party integration**: APIs can be used to integrate external services, such as payments, identity management, and analytics, into the Metaverse. This enables Metaverse developers to leverage the capabilities of these services to enhance their experiences.

- **Development and automation**: APIs can be used to automate certain functions and tasks, such as creating new assets or managing inventory, as well as to create tools for developers to work more efficiently.

- **Extending functionality**: APIs can enable developers to access and extend the functionality of different components of the Metaverse, such as avatars, physics engines, or scripting engines.

- **Analytics and monitoring**: APIs can be used to gather data on the usage, behavior, and performance of the Metaverse, which can be used for monitoring and improving Metaverse experiences.

APIs are a powerful technology that can be used to connect different parts of the Metaverse, enabling a more seamless and integrated user experience and creating opportunities for innovation and growth.

Tracking and capture technologies

Tracking and capture technologies are a group of tools and methods used to measure and record various aspects of the physical world and human behavior. These technologies are used to capture and track different types of data, such as movement, position, and expression, in order to provide a more realistic and immersive experience.

Here are some examples of tracking and capture technologies used in the Metaverse:

- **Head-mounted displays (HMDs), AR glasses, and hand controllers**: These devices are used to track the position and movement of a user's head and hands in a virtual world, allowing for a more natural and intuitive experience.

- **Motion capture**: This technology is used to track the movement and posture of a user's body, using sensors or cameras. The data is then used to control the movement of an avatar in a virtual world, providing a more realistic representation of the user.

- **Facial tracking and expression**: This technology tracks the movements of the user's face and uses this data to control the expression of the avatar in a virtual world.

- **Voice recognition**: This technology is used to capture and interpret a user's voice, allowing for natural speech-based interactions in a virtual world.

- **Eye tracking**: This technology is used to track a user's gaze, allowing for more realistic interactions with virtual objects and other users.

- **Haptic feedback**: This technology is used to provide tactile feedback to the user, allowing them to feel like they are physically interacting with objects in a virtual world.

These technologies help create a more realistic and immersive experience for users in the Metaverse, allowing them to interact with virtual objects and other users in natural and intuitive ways. Additionally, the captured data can be used to improve performance, analyze user behavior, and even personalize the experience for the user. What is not covered here is **brain-computer interfaces** (**BCIs**) or capturing neuromotor impulses for tracking and capture purposes; these technologies will take at least a few decades and, as such, are out of the scope of this book, which is meant to help people within this decade.

VR and MR headsets

VR and MR (combined VR and AR) **headsets** are key technologies used in the Metaverse:

Figure 1.8 – The popular Oculus Quest VR headset being used by a child at school (source: Robert Scoble)

They provide the following:

- **Immersive experience**: VR and MR headsets provide users with a fully immersive experience by blocking out the real world and creating a sense of presence in a virtual one.

- **Interaction**: They can be used in conjunction with hand-held controllers, or even with hand and finger-tracking, to allow users to interact with virtual objects and other users in natural and intuitive ways.

- **Exploration**: They allow users to explore virtual worlds in a more natural and intuitive way, providing a sense of freedom and presence in virtual space.

- **Social interaction**: They can be used to socially interact in the Metaverse, allowing users to communicate with others in real time.

- **Training and education**: They can be used to provide a safe and immersive environment for training and education.

- **Remote collaboration**: They can be used to facilitate remote collaboration in fields such as architecture, design, and engineering.

- **Treatment**: They have been used in the treatment of psychological disorders such as **post-traumatic stress disorder (PTSD)** and phobias, as well as in physical therapy and pain management.

Overall, VR and MR headsets are important technologies for viewing immersive and interactive experiences in the Metaverse, providing users with a sense of presence and realism in a virtual world.

Apple Reality Pro

On June 5, 2023, Apple announced an MR headset that was highly anticipated, taking over 7 years and a purported $40 billion to produce. This MR headset, Apple Reality Pro, is very important as validation that the immersive industry is a very viable one that is here to stay for the long term. By extension, it also validates the Metaverse.

More about Apple Reality Pro and other VR and MR headsets can be found in *Chapter 3, Where is Virtual Reality Heading?*

AR headsets and glasses

AR headsets and glasses are technologies that can be used in the Metaverse, providing the following:

- **Enhanced reality**: AR headsets and glasses superimpose virtual objects and information onto the real world, rather than completely blocking it out, providing a more enriched version of the real world that can be utilized to enhance various activities.

- **Navigation and information**: They can be used to provide navigation and information in the Metaverse, such as showing users where to go, providing information about nearby virtual objects, and displaying notifications or messages.

- **Interaction**: They can be paired with hand controllers or other input devices, to allow users to interact with virtual objects and other users in natural and intuitive ways.

- **Remote collaboration**: They can be used to connect users remotely and collaborate on different tasks and projects, and they can work together in the same virtual space, even if they are in different physical locations.

- **Training and education**: They can be used to create interactive and immersive training simulations that can enhance the learning experience and increase retention of the material.

- **Industrial and commercial uses**: They can be used in a variety of industries and commercial settings, such as providing instructions and guidance for maintenance and repair, or for helping with customer service and sales.

- **Maintenance and repair**: They can be used to assist in maintenance and repair tasks, by superimposing virtual instructions, information, and even guidance over the real world to help technicians, mechanics, and engineers.

- **Gaming**: They can enhance the gaming experience by superimposing virtual characters, objects, and information over the real world, creating a more immersive and interactive gaming experience.

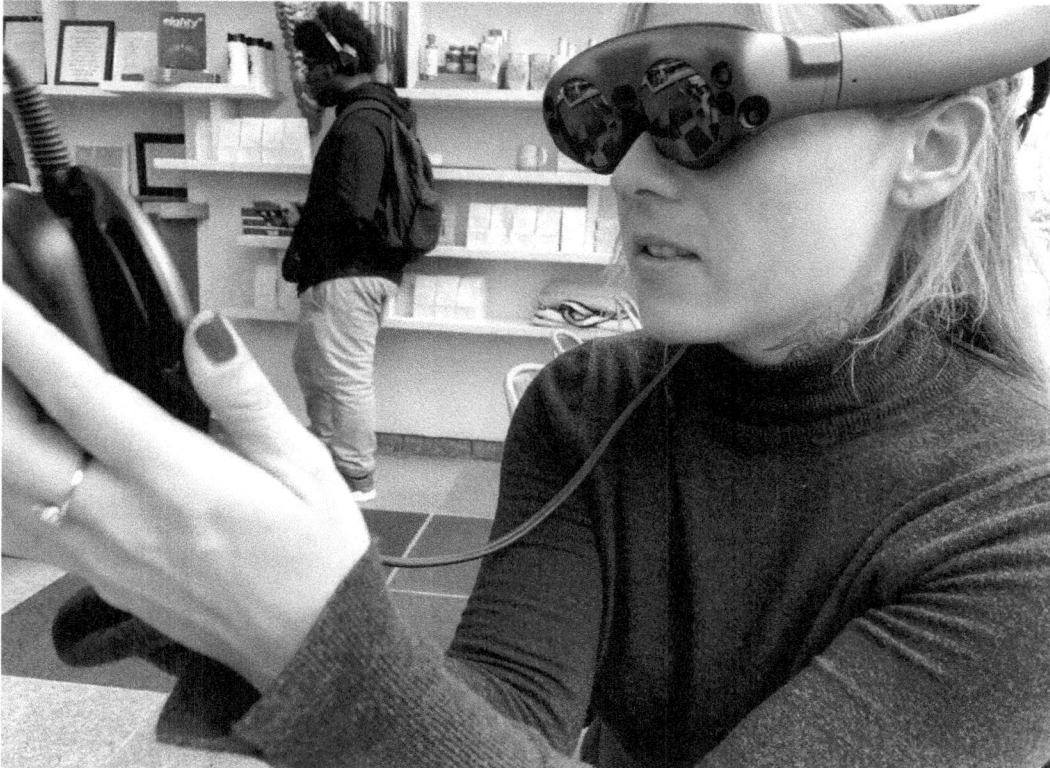

Figure 1.9 – The original Magic Leap AR headset (source: Robert Scoble)

Overall, AR headsets and glasses are technologies that can enhance the Metaverse experience by providing users with virtual objects and information that can interact with the real world, allowing for new possibilities for interaction, navigation, information, collaboration, and more.

More about AR headsets and glasses can be found in *Chapter 2, Augmented Reality Status Quo*.

Why is the Metaverse so important?

The significance of the Metaverse for human communication was discussed at the beginning of this chapter. More detail on this is provided next, together with how the Metaverse is specifically beneficial for businesses and then economies.

For human beings

The Metaverse is capable of simultaneously bringing together many human beings to interact in a 3D virtualized realistic or imaginative space with speed, ease, and precision. What is the significance of this? Why is it so important?

The increase in communication that the Metaverse brings is its biggest benefit. The reason why it is such a big benefit is because humans are social creatures, and the continued existence of societies relies on the efficient communication and interaction of the human beings who live in them.

From an April 2022 survey of US developers, 55% of them thought that it was likely that in 5 years, the Metaverse will replace real-life interactions. This sentiment provides credence that the Metaverse is a desired technology for social interaction:

Will the Metaverse Replace Real-Life, In-Person Social Interactions and Experiences in the Next 5 Years?

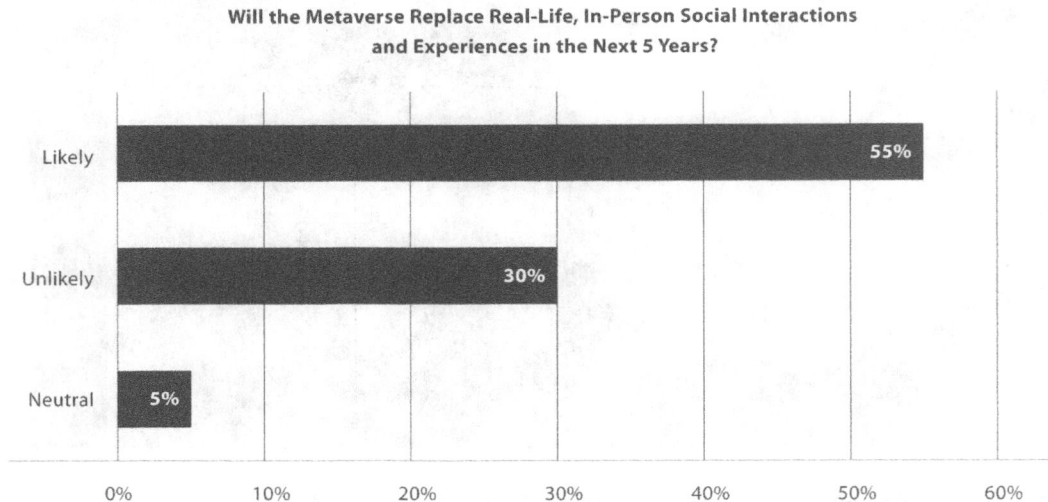

Figure 1.10 – Results from an April 2022 survey of 300 US-based developers (source: Data from Agora)

The Metaverse brings family and friends together, makes new friendships very possible, and allows for connections with people who share the same interests. The argument might be made that Zoom video calls and other companies that offer video calls can do the same thing. After all, many people can get on a video call to *meet*, and a video call is an improvement over having a phone call. With a phone call that is done outside of a typical business setting, only a limited number of people can get on a call together, and they cannot see each other's expressions and physical presence, so it is hard to feel close to anyone when they are on a phone call.

However, a video call is only a slight improvement over a phone call. Moving around is hard to maneuver—a person could take a walk around with their laptop or smartphone while on their video call, but it is really cumbersome, and there are a limited number of things that can be done in a video call besides talking and pulling up documents.

With the Metaverse, a person can enter a 3D virtual location using an avatar that very soon will be extremely photorealistic, looking like that person, and will also be able to move in a seemingly natural human way. Or, if a person wants their avatar to look like anything else than them, they could appear as any kind of animal or mystical creature—there is no limit except a person's imagination here.

As their avatar in the Metaverse, a person can move around—again, in very realistic environments or immensely imaginative ones—and jump instantly from one environment to another, meeting people they've set up ahead of time to meet, or meeting people they don't know that share the same interests they do. Setting privacy options is going to be an extremely important part of setting up Metaverse preferences, including restricting who can see you're online and blocking people, but on the positive side, preferences could be used to hone interests so that like-minded people you don't know can serendipitously come in contact with you. Family and friend circles would also be a very useful feature to set up.

In terms of environments in the Metaverse, they could resemble what has historically been seen in *Second Life*, an online virtual world that peaked at about 1 million users in 2013, but tremendously better. The Metaverse, in 3D with heightened resolution and very close to zero lag, is nirvana for those who have been in previous 2D virtual worlds. The richness in color and extremely realistic objects and forms, if that is what is preferred, will lure countless other people in, just as the internet did several years ago. So, people could meet in a virtual version of New York's Central Park, on top of the Eiffel Tower in Paris, in a house that was digitally created and furnished by an environment's host, or on the northern plains of Mars (modeled on existing photos). Also, journeying into the fantastical will be very popular—whether it is on a planet with two moons in the presence of giant docile bunnies, walking down the yellow brick road to the land of Oz, or traveling back in history to be in Ancient Greece, in Paris taking part in the French Revolution, or meeting with Ernest Hemingway in the 1920s, and so on.

With the alluring environments comes what can be accomplished there. What kinds of activities and actions are possible in the Metaverse? In *Part 3, Consumer and Enterprise Use Cases*, I relay what many of those could be in detail—gaming, socializing, sports, retail, work, and more. Imagine being able to play a really good game of baseball, followed up by buying clothes after virtually trying them on given your actual physical dimensions, then meeting up with a friend to talk and share realistic virtual wine with them—and that's on a day off. So much work can take place in the 3D space of the Metaverse. Businesses will also benefit greatly, both in terms of Metaverse marketplaces and other modes of retail and the ability to have large virtual workspaces and other virtual work capabilities.

Business, which is the focus of this book, will have a large foothold in the Metaverse. I continue my introduction on what the Metaverse can do for business next, with more detail using illustrative use cases to be found in *Part 3*.

For businesses

The Metaverse brings an immense improvement for commerce that could be likened to what the advent of the internet did for it. It is the next evolutionary step following the internet.

The Metaverse has the potential to provide many benefits for businesses, including the following:

- **New revenue streams**: The Metaverse could create new opportunities for businesses to generate revenue through the sale of virtual goods and services, such as virtual real estate and advertising.

- **Increased customer engagement**: It could provide businesses with new ways to interact with and engage their customers, such as through virtual storefronts and interactive experiences.

- **Improved collaboration and productivity**: It could enable businesses to connect and collaborate with other companies and individuals in new ways, such as through virtual workspaces and remote communication tools.

- **Enhanced customer insights**: It could provide businesses with new ways to gather data about their customers, such as through the analysis of user behavior in virtual environments.

- **Expanding reach**: It could help businesses expand their reach beyond the physical constraints of their geographic location.

- **Cost savings**: It could also help businesses save money on physical expenses such as real estate, transportation, and employee benefits.

According to a survey done in 2022, over 50% of the people surveyed can see themselves as partly working in the Metaverse. It is no surprise that these people were from Gen Z, which is defined as those who were born between the late 1990s and early 2010s. This gives an indication of how closely connected the Metaverse is to the notion of business and how a new generation foresees its importance in that area.

For economies

The Metaverse has the potential to be beneficial for countries' economies in several ways. The development and growth of the Metaverse can create new opportunities for businesses and entrepreneurs within a country, leading to increased economic activity and job creation.

The Metaverse can also attract foreign investment and talent to a country, as companies and individuals look to participate in and benefit from the Metaverse's market growth, which is forecasted by the year 2030 to be a total of $10 trillion given average expansion, according to *Grand View Research*.

Furthermore, the Metaverse can open up new export opportunities for countries, as businesses and entrepreneurs within a country develop and sell virtual goods and services to users in other countries.

Additionally, the Metaverse can be a platform for new forms of entertainment and leisure, which can also drive economic growth.

Lastly, the Metaverse can also be a platform for innovation, and countries that foster the development of the Metaverse may be able to leverage this innovation to gain a competitive advantage in other areas of the economy.

Summary

After reading this chapter, it should be clear what the core definition of the Metaverse is, its true origins, which relevant technologies came before it, and why it is important to us as human beings and for businesses and economies.

It is important to take what has been learned in this chapter and use it as a springboard to further comprehend the importance of what's to come, both in terms of practical use cases and large societal impacts.

In the next chapter, the current state of AR, a rapidly growing area, is covered. Its significance to the Metaverse is now understood to be equal, if not more than equal, to that of VR.

Augmented Reality Status Quo

In this chapter, we will review the current state of **augmented reality** (**AR**), providing details of its first iterations with enterprise AR headsets and experiences created for the smartphone. Next, we'll discuss AR glasses, most notably Apple's upcoming product. Lastly, we'll explain AR's significance to the Metaverse.

In this chapter, we're going to cover the following main topics:

- The current state of AR
- What led up to the current state of AR, including which AR headsets and glasses are relevant and which industries are using them
- The significant AR glasses that are anticipated
- Understanding what AR's significance is to the Metaverse

After reading this chapter, you will be much better equipped to understand what AR hardware can do for you.

Smartphone, headset, and glasses endeavors

AR has been several decades in the making, with hardware going through many periods of experimentation. In this section, we'll review smartphone AR and AR headsets and glasses, enabling you to understand AR's current status. In learning this, you will be able to make educated decisions about what AR can do for your business.

The concept of AR can be traced back to the 1960s and 1970s when computer graphics researchers developed the first **head-mounted displays** (**HMDs**), which could overlay computer-generated imagery in the real world. However, it wasn't until the 1990s that AR began to emerge as a distinct technology.

In the 1990s, computer graphics and video processing technology improved, enabling the development of more sophisticated AR systems. One of the first commercial AR applications was a toy called *Virtual Fixtures*, developed by Louis Rosenberg at the US Air Force's Armstrong Laboratory. Virtual Fixtures used an HMD to overlay computer-generated images onto the user's real-world view, enabling them to manipulate virtual objects in a physical space.

In 1992, Tom Caudell and David Mizell, two Boeing researchers, coined the term *augmented reality* to describe a system they had developed to display aircraft wiring diagrams on a head-mounted display. This was one of the earliest examples of an AR system that combined computer graphics with real-world objects.

Another early example of AR included the ARToolKit, an open source software library for creating AR applications that was first released in 1999.

As computer processing power continued to improve, AR applications became more advanced and widespread. In the early 2000s, several companies began developing AR systems for industrial and military use, including helmet-mounted displays for fighter pilots and heads-up displays for drivers. BMW used AR for assembly and quality control of their vehicles and also created an AR app in 2003 that allowed customers to configure and visualize their cars in 3D.

In 2004, Boeing began using AR technology to assist with the assembly of aircraft, allowing workers to view digital instructions and diagrams overlaid on the physical components they were working on. This allowed workers to see exactly where each component should be placed and ensured that everything was assembled correctly. The launch of smartphones and tablets with built-in cameras, advanced sensors, and powerful processors in the late 2000s paved the way for the popularization of AR applications for consumer use.

Next, we'll review the history of smartphone AR, including some relevant apps, and the current status and significant models of AR headsets and glasses.

Smartphone AR

One of the earliest examples of smartphone AR was the ARToolKit, an open source software library for creating AR experiences that was first released in 1999 and made open source in 2001. ARToolKit allowed developers to create interactive AR applications using markers, which were printed images that could be recognized by a camera and used to overlay digital content on top of the real world.

In 2005, Nokia developed AR tennis for both the Nokia 6600 and 6630 models due to their good quality displays, processing power, and camera. With Bluetooth being used to synchronize the movement of the ball between both players' phones, the players were able to hit a virtual ball over the net and play against each other.

The AR app *Wikitude*, which was released in 2008, was a milestone in the history of AR and was followed by the release of many other AR apps and games over the next few years. The Wikitude app brought location-based AR to smartphones, allowing users to see information and points of interest displayed over a live camera view of their surroundings.

The first smartphone with native AR capabilities was the iPhone 3GS, released by Apple in 2009. This device featured a built-in compass and accelerometer, which allowed it to track the user's movements and orientation in real time. Additionally, the iPhone 3GS had a rear-facing camera that could be used to capture video and images, which made it possible to overlay digital content onto the physical world. Over the next few years, other manufacturers, including Samsung and LG, released smartphones with AR capabilities.

Here are the historically significant smartphone AR apps after the launch of the iPhone in 2009:

- **Layar**: One of the first well-known smartphone AR apps was called Layar and it was launched for the iPhone in 2009 and for the Samsung Galaxy S model in 2010. Layar allowed users to point their phone's camera at real-world objects and see information overlays about nearby businesses, landmarks, and other points of interest on top of the camera view.

- **Yelp Monocle**: In 2009, the first AR app for the iPhone, made specifically for the iPhone 3GS model, Yelp Monocle, was released, which allowed users to see restaurant ratings and reviews overlaid onto a live video feed of their surroundings.

- **Nearest Tube**: Also in 2009, acrossair developed Nearest Tube. This app used the iPhone 3GS's GPS and compass to display the location of nearby London Underground stations in real time. Overlaying this information onto the phone's live camera feed allowed users to see the station's location and distance overlaid onto the physical world.

- **Google Goggles**: In 2010, Google released the Android-based Nexus One, which featured a similar set of sensors as the iPhone 3GS and was capable of running AR apps. This was followed in 2011 by Google releasing an AR app called Google Goggles, which allowed users to search the internet by taking a picture of an object. This app was later integrated into the Google Translate app, allowing users to translate foreign text by pointing their phone's camera at it.

- **Vuforia**: In 2013, Qualcomm released the Vuforia AR platform, which allowed developers to create AR experiences using image recognition and tracking technology. Vuforia was used to create several popular AR apps, including the *LEGO AR Playgrounds* app, which allowed users to build virtual LEGO sets and play with them in the real world.

- **Pokémon GO**: In 2016, the release of Pokémon GO, a mobile game made by Niantic that used location-based AR to place virtual creatures in the real world, brought the technology to the attention of an even wider audience. The game became a massive hit, with millions of people around the world downloading and playing it.

- **ARKit and ARCore**: The release of ARKit by Apple in 2017 and ARCore by Google in 2018 further accelerated the development of smartphone AR applications. These **software development kits (SDKs)** provided developers with tools to build AR apps that could take advantage of the advanced sensors and processing capabilities of modern smartphones.

Since ARKit and ARCore, smartphone AR has evolved significantly, with the development of more advanced sensors, more powerful processors, and better software. Additionally, advances in machine learning and computer vision have enabled more sophisticated AR applications, such as object recognition and tracking.

Today, smartphone AR has become ubiquitous, with millions of people using AR apps on their devices for a wide range of applications in fields such as gaming, entertainment, retail, manufacturing, logistics, education, and healthcare. For example, logistics companies use AR to help warehouse workers pick and pack orders more quickly and accurately, while healthcare providers use AR to help medical students and doctors visualize complex medical procedures.

The development of AR devices, such as Microsoft HoloLens and Magic Leap One, started to open new possibilities for AR, enabling users to interact with virtual objects in a more natural and immersive way. In the next section, some current notable AR headsets and glasses will be presented, along with how AR headsets and AR glasses differ and what the desirable specifications are for both.

AR devices

In this section, we will cover how AR headsets are different from AR glasses, the desirable specifications of each, and notable AR headsets and glasses models, as well as the relevant AR glasses that are anticipated from Apple, Meta, and Google.

AR headsets versus AR glasses

Both AR headsets and AR glasses are types of AR devices that display digital information over the user's real-world view. However, there are some differences between the two.

Here are the differences:

AR headsets:

- AR headsets are more immersive than AR glasses as they typically cover the user's entire **field of view (FOV)**.

- They often include additional features such as spatial audio and hand-tracking to enhance the user's experience.

- AR headsets tend to be bulkier and heavier than AR glasses, which can make them less comfortable to wear for extended periods.

AR glasses:

- AR glasses are more lightweight and less obtrusive than AR headsets, making them more suitable for everyday use.

- They can be designed to look like regular glasses or sunglasses, making them more fashionable and appealing to wear in public.

- AR glasses typically have a smaller FOV than AR headsets, which can limit the amount of information that can be displayed.

Ultimately, the choice between AR headsets and AR glasses will depend on the specific use case and personal preferences of the user. AR headsets may be more appropriate for immersive experiences such as gaming or training simulations, while AR glasses may be better suited for daily use cases such as navigation, messaging, or social media.

Desirable specifications

Since AR headsets and AR glasses have different tech specifications, here, we will provide the desirable specifications in the relevant categories for both kinds of devices to make it easier for your purchase decision-making.

Key AR headset specifications

AR technology has advanced significantly in recent years, and there are now a range of AR headsets available on the market.

When looking for an AR headset, here are some important specifications to consider:

- **Display**: Look for a high-resolution display that is capable of producing clear and vivid visuals. A minimum resolution of 1080p per eye is recommended, but higher resolutions in most future headsets such as 2K or 4K will be greatly preferable.

- **Field of view (FOV)**: FOV is the extent of the observable world that is visible through the headset. A larger FOV can provide a more immersive experience. A good consumer AR headset should have a FOV of at least 40-50 degrees. Less is acceptable for an enterprise AR headset due to its combination with specialized features.

- **Tracking**: Look for a headset with accurate and reliable tracking that can be achieved through a combination of sensors and cameras. This includes both positional tracking (tracking the headset's location in 3D space) and hand tracking (tracking hand and finger movements).

- **Comfort**: Comfort is an important factor for any wearable device. Look for a headset that is lightweight, well-balanced, and has comfortable padding around the eyes and nose.

- **Battery life**: AR experiences can be quite demanding on battery life, so it's important to look for a headset with a decent battery life. Aim for a minimum of 2-3 hours of continuous use, but ideally closer to 5-6 hours.

- **Connectivity**: Look for a headset that is compatible with your device (for example, iOS or Android) and has reliable connectivity options such as Bluetooth or Wi-Fi.

- **Price**: AR headsets can vary greatly in price, so consider your budget when selecting a headset. More expensive headsets often have better specifications, but there are also affordable options that can provide a decent AR experience.

The best AR headset for you will depend on your individual needs and preferences. Consider these specifications when you're making your decision and do your research to find the headset that best meets your needs.

Key AR glasses specifications

When looking for AR glasses, there are several important specifications to consider. Here are some key factors to keep in mind:

- **Display quality**: AR glasses should have a FOV of at least 40-50 degrees.
- **FOV**: A larger FOV can provide a more immersive experience. Some AR glasses have a limited FOV, which may make it difficult to experience AR content.
- **Comfort**: Comfort is essential when it comes to AR glasses. Look for glasses that are lightweight and have an adjustable fit. It is also important to consider the materials used and the ergonomics of the design.
- **Battery life**: AR glasses require power to operate and therefore, battery life is an important consideration. Look for glasses with a long battery life of at least 5 hours or the ability to easily swap out batteries.
- **Connectivity**: AR glasses may require connectivity to a smartphone, tablet, or computer. It is important to ensure that the glasses are compatible with the device you plan to use and that the connectivity is stable and reliable.
- **Price**: AR glasses can range from a few hundred to several thousand dollars. Consider your budget when selecting AR glasses and be sure to compare the features and specifications of different models within your price range.

Overall, it is important to carefully consider these factors when selecting AR glasses to ensure that you get the best possible experience.

Notable AR headsets and glasses

There are many brands and makes of AR headsets and glasses, but there are relatively few that either approach or meet the needs of consumers and/or enterprise users. Here are the significant ones that meet those criteria.

HoloLens 2

- **It's an AR headset developed by Microsoft**. It was released in November 2019 as an update to the original HoloLens, which was released in 2016. The HoloLens 2 is a standalone device that does not require any external sensors or computing devices.

- **It's designed for use in a variety of industries**, including manufacturing, construction, healthcare, and retail. It is intended to help workers perform their jobs more efficiently by providing them with contextual information and virtual assistance. It has been used in a variety of settings, including for remote assistance, 3D visualization, and training simulations.

- **It features several improvements over the original HoloLens**. It has a wider FOV at 52 degrees diagonally, which allows users to see more virtual objects in their environment. It also has a higher resolution display at 2,048 x 1,080 per eye, a more comfortable fit, which makes it easier to wear for extended periods. It also has spatial sound and a weight of 1.25 pounds.

- **It is one of the first AR headsets to have a 6-degree-of-freedom (6DoF) tracking system**, which includes hand-tracking technology and allows users to interact with virtual objects using natural hand gestures. It also has eye-tracking technology, which can be used to select objects and scroll through content.

- **It runs on a custom version of Windows 10 called Windows Holographic**, which is designed specifically for mixed-reality applications.

- **It has several sensors**, including four visible light cameras, two infrared cameras, and a depth camera, enabling it to accurately map and track the environment.

The following figure shows the HoloLens 2 headset being worn and tested by two people, with one of them using a hand gesture to choose and manipulate images:

Figure 2.1 – Two people testing out the Microsoft HoloLens 2 AR
headset at a Saab Automobile booth at a conference

The HoloLens 2 seems to point to there being continuing future models. However, there is a big possibility that Microsoft will not be making the updated model, HoloLens 3, or any other models, as had been previously expected. In early 2023, there were large layoffs in the areas connected with the HoloLens, calling into question whether it will be continued in the future.

Magic Leap 2

- **This is a second-generation device that launched in 2022**, following the release of the original Magic Leap One in 2018 by the US company Magic Leap.

- **It's designed to provide a more immersive and realistic AR experience** by integrating computer-generated images and information with the real world. It uses a combination of sensors, cameras, and projectors to create virtual objects and place them. The Magic Leap 2 is equipped with an FOV of 70 degrees diagonally, a 6DoF controller, and voice recognition technology.

- **It is designed to be lightweight and comfortable at 0.57 pounds**, with a slim profile and a sleek, modern design. It features a high-resolution display at 1,536 × 1,856 per eye, an upgraded processor, and improved tracking capabilities, allowing for smoother and more responsive interactions with virtual objects.

- **It can support multiple users in the same physical space**, allowing for shared experiences and collaborative interactions. This is achieved through the use of spatial computing, which enables the device to understand the user's environment and place virtual objects in a way that is consistent with the real world.

- **It's targeted primarily at enterprise customers**, with a focus on applications such as remote collaboration, training and education, and product visualization.

The following figure shows a person at a conference being watched while he tests out a Magic Leap One.

While Magic Leap's focus was initially on consumers, they ran out of cash before having a product generate revenue, so now, it is focused on enterprise. The company has been through a major downsizing and significant management and investment changes during the last few years. The next few years will be very important for them to prove that they are here to stay:

Figure 2.2 – A person testing out the Magic Leap One AR headset at a conference

Epson Moverio BT-45CS

- **It's designed for use in industrial and commercial applications** made by the Japanese company Epson. These glasses are an updated version of the Epson Moverio BT-35E, with improved features and capabilities.

- **It weighs 1.2 pounds**, with a flexible headband and adjustable nose pads. The glasses display has a resolution of 1,920 x 1,080 and an FOV of 34 degrees.

- **It features head-tracking sensors and a built-in 5-megapixel camera** that can capture images and video, as well as a microphone and touchpad for navigation.

- **It's powered by an Android-based operating system**, which allows users to download and run a variety of AR applications. It also has built-in Wi-Fi and Bluetooth connectivity, making it easy to connect to other devices and networks.

- **It can be used for remote assistance**, training and education, field service, and logistics and warehouse operations. It is also popular in the entertainment industry for use in immersive experiences and theme park attractions.

Overall, the Epson Moverio BT-45CS is a powerful and versatile tool for AR applications, with advanced features and capabilities that make it ideal for use in a wide range of industrial and commercial settings.

Lenovo ThinkReality A3

- **It's designed for enterprise use** and comes in at 0.29 pounds and has a 47-degree FOV. It was first introduced by the Chinese company Lenovo in early 2021.

- **It comes in two variants**: a PC-tethered version and a standalone version. The PC-tethered version connects to a laptop or PC via a USB-C cable and is designed for use in industries such as manufacturing, logistics, and healthcare, where workers need to access information while keeping their hands free.

- **The standalone version runs on an Android-based operating system** and is designed for use in field service, retail, and other industries where mobility is essential.

- **It features 1,920 x 1,080 per-eye resolution displays,** which can project up to five virtual monitors at once. The glasses come equipped with a variety of sensors, including an IMU, magnetometer, and ambient light sensor. The glasses also include an 8 MP RGB camera and a 5 MP depth camera for capturing images and videos.

- **It's designed to be comfortable to wear for extended periods**, with adjustable nose pads and a lightweight design. They also feature a built-in microphone and speakers for voice control and audio feedback.

Overall, the Lenovo ThinkReality A3 is a powerful tool for enterprise use, offering hands-free access to digital information and applications in a variety of industries.

RealWear Navigator 500

- **It's designed specifically for use in industrial and field service environments**. Made by the US company RealWear, it's a rugged and durable device that can be worn on the head like a pair of safety glasses or a hard hat, allowing workers to keep their hands free while accessing critical information and communication tools.

- **It features a micro display** that has a 20-degree diagonal FOV and a resolution of 854 x 480 per eye that is positioned just below the user's line of sight, allowing them to view digital content without having to take their eyes off the task at hand. The display is also adjustable and can be moved up and down or tilted to accommodate different user preferences and work environments.

- **It's powered by a Qualcomm Snapdragon processor** and runs on the Android operating system. It includes voice recognition technology and noise-canceling microphones, allowing users to control the device and communicate with others in noisy or busy environments.

- **One of its key features is its robust design**, which includes durable housing that is resistant to dust, water, and shock. This makes it well-suited for use in challenging environments such as construction sites, manufacturing facilities, and oil rigs. It weighs 0.6 pounds.

- **It includes a range of software applications** designed to enhance productivity and collaboration in industrial settings. These include remote expert assistance, training and instruction, and digital work instructions.

- **It can be remotely managed and configured** using RealWear's cloud-based management platform, allowing IT teams to quickly deploy and update devices as needed. Additionally, the device is equipped with multiple layers of security, including biometric authentication and data encryption, to protect sensitive information.

Overall, the RealWear Navigator 500 is a powerful tool for industrial and field service workers, providing them with access to critical information and communication tools while keeping their hands free and maintaining a safe work environment.

Nreal Air

- **It is a lightweight AR headset that weighs 0.17 pounds** and resembles a pair of glasses, designed and manufactured by the Chinese company, Nreal. The glasses are designed to provide an affordable and portable experience to users.

- **It uses a tethered connection** to a compatible smartphone or PC to display augmented and virtual reality content. It comes with a small processing unit that is attached to the user's waistband or pocket, which helps offload some of the computational load from the smartphone.

- **It has a resolution of 1,920 x 1,080 per eye**, a diagonal FOV of 46 degrees, and a 6DoF tracking system, which allows for more precise and accurate tracking of the user's movements and interactions with virtual objects.

- **It is compatible with a wide range of devices**, including smartphones and PCs running Android, Windows, and iOS.

- **It is primarily targeted at consumers**, with a focus on gaming, entertainment, and social experiences. However, the device also has potential applications in fields such as education, healthcare, and remote collaboration.

- **It is currently available in select markets**, including the United States and South Korea, and is expected to be released in additional markets in the future.

Nreal still has a lot left to prove in terms of overall performance as its visuals could be uneven. And even though the glasses are lightweight, their design is currently not the most comfortable.

Vuzix Blade 2

- **It was released in 2021 as an updated version** of the original Vuzix Blade AR glasses and features several improvements and new features. It has a sleeker and more lightweight design at 0.2 pounds, with a smaller factor and improved comfort for extended use. Its resolution, though, is relatively low at 480 x 480 per eye, and it has a relatively low diagonal FOV of 20 degrees.

- **It supports Wi-Fi and Bluetooth connectivity** and uses a variety of sensors, offering gesture recognition, voice control, and video recording with an 8-megapixel camera.

- **Its use cases include industrial applications** such as maintenance and repair, logistics and warehousing, and field service operations, but it is much more of a consumer device that's good for gaming, entertainment, and navigation.

Vuzix still has more to go before it could be considered an all-around consumer or enterprise device due to its relatively low resolution and FOV. However, each new version of a device that Vuzix produces shows improvement.

In summary, there are currently several AR headsets and glasses available on the market, each with its unique features and use cases. From enterprise-focused devices such as the HoloLens 2, Magic Leap 2, Epson Moverio BT-45CS, ThinkReality A3, and RealWear Navigator 500 to more consumer-focused devices such as the Nreal Air and the Vuzix Blade 2, there is an AR device to suit a wide range of needs and preferences. However, much better AR devices will be coming within the next few years.

Anticipated AR glasses

For several years now, Apple, Meta, and Google have been working on developing AR glasses. The most highly anticipated of the three, though, is what Apple is projected to deliver. Before discussing the potential product specs and timing of Apple's future AR glasses (which we'll do in the *Product information and timing* section), it would be helpful to first understand its particular circumstances within Apple's business narrative.

Apple

Apple has significant advantages over other companies when it comes to producing and launching new products, including their future AR glasses. We will discuss the major advantages in this section, as well as their probable learnings as to what the specs for their AR glasses should be. We will also make informed estimates based on current technology of what the specs of their AR glasses will be and the timing of their launch.

iPhone connections

The success of the Apple iPhone has been a defining moment in the history of mobile technology. It is a revolutionary device that has transformed the way we communicate, access information, and interact with technology. The iPhone's success can be attributed to several factors, such as its sleek design, user-friendly interface, and seamless integration of hardware and software. However, the success of the iPhone has also paved the way for Apple to explore other innovative technologies, such as AR glasses.

Apple has been investing in AR technology for several years now, and the company has made significant progress in this area. AR glasses are a natural extension of the iPhone as they allow users to interact with the digital world in a more immersive way.

Apple has also been working to create a seamless integration between the iPhone and AR glasses. For example, users may be able to control their AR glasses using their iPhone, or they may be able to view AR content on their iPhone and then seamlessly transition to viewing it on their AR glasses. This integration will make it easier for users to adopt AR glasses and help ensure that they become a mainstream technology.

The success of the Apple iPhone and the potential success of Apple AR glasses are interconnected in several ways:

- **The success of the Apple iPhone has established Apple as a leader in the technology industry**, particularly in the area of consumer electronics. Apple's reputation for quality and innovation has led to a loyal customer base and a strong brand image. This brand image is likely to extend to any new products that Apple releases, including AR glasses. Customers who have had a positive experience with Apple products in the past are likely to be more receptive to trying new Apple products, such as AR glasses.

- **The Apple iPhone has been instrumental in creating an ecosystem of apps and services that are available exclusively on iOS devices**. This ecosystem has been a key factor in the success of the iPhone as it has provided users with a wide range of tools and services that are tailored to their needs. When Apple releases its AR glasses, it will have also begun to develop an ecosystem of apps and services that are tailored to the unique features of the glasses. This ecosystem could be a key factor in the success of the glasses as it would provide users with a compelling reason to purchase the glasses over competing products.

- **The Apple iPhone has demonstrated Apple's ability to create products that are both technically advanced and user-friendly**. The iPhone's user interface is widely regarded as one of the most intuitive and easy-to-use interfaces on the market. Apple would prioritize creating a user-friendly interface for the glasses that would make it easy for users to interact with the glasses. This focus on user experience could be a key factor in the success of the glasses as it would differentiate them from other AR glasses on the market that may be more technically advanced but more difficult to use.

- **The Apple iPhone has created a strong network effect**. As more people use iPhones, the value of the iPhone ecosystem increases as there are more people to interact with and more apps and services to use. This network effect will also play a role in the success of AR glasses. Apple will release AR glasses that are compatible with the iPhone ecosystem, and it will lead to the value of the ecosystem increasing as there will be more ways to interact with the ecosystem and more services that could be used with the glasses.

The success of the Apple iPhone and the potential success of Apple AR glasses are closely intertwined. Apple's reputation for quality and innovation, the ecosystem of apps and services that have been created around the iPhone, the user-friendly interface of the iPhone, and the strong network effect that has been created by the iPhone all have the potential to impact the success of AR glasses very positively. Leveraging these factors to create a compelling AR glasses product could be a major new revenue stream for Apple and a key driver of future growth.

Needed specs

As Apple designed its move into AR, it did a great amount of human factor research – probably more than it has done on any product combined before.

What are some of the things Apple learned?

- **Glasses must be very lightweight**. Current vision correction glasses are running at 25-30 grams. By comparison, Apple's AirPods Max over-the-ears headphones weigh 386 grams, which are way too heavy to wear all day long.

- **They can't generate much heat on a wearer's head**. Facial skin is very sensitive to heat. VR passthrough headsets can consume 5 watts of electricity. That should not be true for glasses, which should have very small batteries that use much less power (1 watt or less); otherwise, they would generate too much heat. Apple solves these problems by putting most of the computing in your phone and edge computing, leaving the glasses to do very little computing on your face.

- **They must exceed the usability of everyday tasks compared to phones, tablets, or laptops**. If a call comes in while a user is playing a game or watching a concert in the glasses, they must be able to answer the call easily.

- **Super short latency is key, especially in multi-party video games**. Even during concerts, while dancing, any latency will be noticed, which would ruin the experience. Since some users will be using headphones, latency is key, especially since audio is instrumental to making AR objects and beings "talk" and provide immersion.

- **A new smart network is needed**. A *smart network* will split the computing up between glasses, phones, and other devices in the home and determine when cloud computing is needed (for multi-party video games, for instance). Having AI computers on the network will enable the network itself to take over computationally intensive tasks. This network will also let Apple enable very advanced AI inferencing without sending data to the cloud, which strengthens Apple's privacy stance over its competitors, who need to send data to the cloud.

- **Great audio will enable it to differentiate**. This is why Apple has invested so much in Spatial Audio/Dolby Atmos on its music, TV, and movie services. Great audio builds the illusion that augmented beings and items seem "real" and will make its devices stand out from others that have headphones that can't do concert-level music.

- **A new search capability is needed**. 8 years ago, Apple knew Siri would need to be rebuilt from scratch to compete with Google's search (it had built a system to watch how fast each company's AI was learning and knew that Siri was learning slower than Google). There is an expectation that there will be a new Siri associated with the Apple Vision Pro when it becomes available for sale in Spring 2024 (more on this headset in the next chapter, which is on VR). There has also been a leak advanced by Mark Gurman, the Apple pundit, that Apple is working on its own LLM, "Apple GPT," which could be utilized during the search for enhanced capabilities. Once you get people to wear cameras and microphones, a new kind of search is possible – one that can see what you are holding, touching, or looking at. These search capabilities could be a lot more like talking to an assistant than typing into a box at the top of a window, too. A new word to describe this new kind of search, which will give you semantic results that you can select with your voice, eyes, or hands, will probably be coined.

- **Great screens are needed**. It isn't enough to match a computer screen, either. Screens on glasses must be excellent at mixing virtual information, objects, scenes, and beings with the real world. So far, AR screens have been dim and lack the visual quality of any 4K TV, particularly outdoors, and have small fields of view so that AR items live in a small box in front of your eyes.

If Apple, or its competitors, get it right, then these AR glasses will become like a browser for the real world, unlike any browser we have today.

As you walk around, you will see lightweight information constantly appearing when you need it. So, for example, when a call comes in, you'll see a photo of who is calling and you can choose to answer just by saying, "*Please answer.*"

More on particular scenarios and other use cases can be found in *Part 3, Consumer and Enterprise Use Cases.*

Product information and timing

Though specifications for Apple's AR glasses have not yet been disclosed, we can make informed estimates based on current technology. For example, the FOV and resolution will probably be quite a bit greater than those of the HoloLens 2 (52 degrees and 2,048 x 1,080 per eye, respectively).

Apple's focus is to provide users with an immersive experience that goes significantly beyond basic 2D notifications and map displays as was seen on Google Glass. To achieve this, it is expected that their upcoming AR glasses will be connected to the iPhone via Wi-Fi. This will allow the iPhone to handle all the video footage captured by the glasses' cameras and transmit high-quality 3D images back to the glasses at a high frame rate. Bluetooth bandwidth is not sufficient for this purpose.

As for battery life, it is expected to last at least 5 hours on average to remain competitive (but closer to 8 hours). A wireless charging glasses case, like the one for Apple AirPods, could potentially extend the glasses' operating time throughout the day.

A baseline feature for Apple's AR glasses will be to enable users to access phone-related information directly on their faces. Specifically, the glasses are expected to sync with the wearer's iPhone to display various types of content, such as texts, emails, maps, and games, directly within the user's field of vision.

In addition to this functionality, Apple is expected to develop a dedicated app store for third-party AR glasses apps, similar to the app stores for Apple TV and Apple Watch.

Apple's AR glasses may have the potential to enhance their finger and hand tracking accuracy with smart rings that have been patented by Apple. This advancement may eliminate the requirement for external sensors and improve the system's accuracy. Apple's recent patent victories suggest that they will integrate wearables into various features, including finger gesture support.

The following figure from a patent application shows how the rings might be worn:

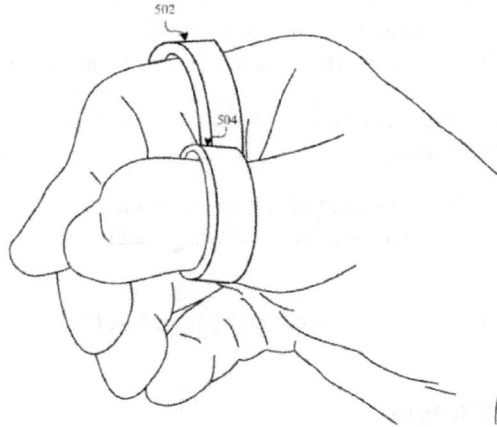

Figure 2.3 – An illustration of Apple's smart ring design

Source: United States Patent and Trademark Office

The smart rings can identify the object that a user is holding in their hand, allowing Apple's AR glasses to respond accordingly. For instance, when holding an Apple Pencil, the glasses can track movements and convert them into handwritten text.

Overall, AR glasses pose greater technical challenges than AR headsets due to their stricter weight, bulk, and aesthetic requirements, as well as their intended use for longer periods. A cumbersome battery may need to be carried separately to achieve a battery life of even 2 hours, which is far less than that of a smartphone. Additionally, current unanswered questions related to processors, software, and manufacturing still exist.

Due to these issues, in January 2023, a leak regarding the timing of Apple's AR glasses indicated that a formerly expected launch in 2024 was not realistic. However, Apple did announce its MR headset in June 2023 (more on the Apple Vision Pro in the next chapter). The leak did not provide a new date for the AR glasses launch; however, there have been other indications that point to a 2025-2026 launch.

Meta

In 2021, Meta announced Project Nazere, Meta's first AR glasses. In 2022, it was reported that the glasses would launch in 2024, that they would weigh 0.22 pounds (around four times normal glasses) and would resemble Clark Kent's thick frames. The indication was that Nazare would be powered by a small wireless puck and that its battery life would be just 4 hours. The targeted FOV was 70 degrees. The first version, at least, is intended to be used mostly indoors. Though Meta first intended to deliver the first generation of AR glasses by 2024, followed by lighter, more advanced designs in 2026 and 2028, the first delivery has been moved up to 2025 due to company cost-cutting measures. Most of the billions of dollars Meta has spent on headset innovation has gone toward AR versus VR efforts. Despite investing billions, Meta has low sales expectations in the low tens of thousands for the first version, which is targeting early adopters and developers.

Mark Zuckerberg has stated that a "killer use case" of the Metaverse is wearing AR glasses when interacting with people in real life, so you can continue to text message others without the people in real life noticing. The glasses will feature holographic communication capabilities that Meta believes will provide a more immersive experience than video calling.

However, AR glasses have privacy concerns as they can be used to invade privacy in public places and can also be used as cameras and microphones in people's homes, potentially exposing private information to companies such as Meta. Meta has a lengthy and comprehensive history of problems related to privacy, so it will need to secure the trust of potential users before its AR glasses can be deemed a success.

Google

In 2022, Google announced that it will begin testing camera-equipped AR glasses in public. Google has had to be very careful about privacy-related issues ever since its debacle over Google Glass.

Google Glass was a wearable smart device prototype developed by Google that resembled a pair of glasses. It was first introduced in 2013 and marketed as a device for personal use and workplace productivity. The device featured a small display screen that was mounted on the frame of the glasses, and it allowed users to view information and interact with the device using voice commands and gestures.

Google Glass utilized a combination of technologies, including a high-resolution display, a camera, a microphone, a touchpad, and wireless connectivity, to provide a variety of features such as voice-activated search, turn-by-turn navigation, and the ability to record photos and videos hands-free. The device also had several third-party apps available, including apps for social media, productivity, and fitness.

While Google Glass was initially met with a lot of excitement and interest, it faced several challenges, including privacy concerns. This included questions about its potential to infringe on people's privacy and safety. People wearing the device became known as "Glassholes." This term was coined due to the behavior of some early adopters of the technology, who were seen as being rude, invasive, and inconsiderate when using the device.

Some people felt uncomfortable around Glass wearers as they were perceived to be constantly recording and taking pictures without consent. Others criticized the technology for its potential to infringe on people's privacy, and for the fact that wearers could easily access information about others without their knowledge.

In 2015, Google announced that it would halt consumer sales of Google Glass, but it continued offering an enterprise version of Google Glass until March 2023, when it was announced that it would be no longer offered.

This legacy has left Google, along with other AR device developers, with a significant privacy hurdle that it must clear before any AR devices come to market.

There has been no other update to the status of Google's new AR glasses since the announcement that was put out regarding public testing.

Others

Samsung has been rumored to be working on an AR headset, though some of the latest rumors about a particular headset, model SM-I120, are pointing to it being a mixed-reality headset.

It is an open secret that Sony has also been developing an AR headset, but it has been pulled from development and reactivated several times over the last few years. It's unknown whether it will be fully produced; Sony has been known to invest heavily in projects and then not come out with marketable products.

Several other US and Asian companies have been working on AR headsets and glasses. For example, Xiaomi, a well-known Chinese company, announced its Xiaomi Wireless AR Smart Glass Explorer Edition in February 2023. But the consensus among tech insiders is that once Apple comes out with its AR glasses, it will dominate, and companies such as Meta and Google will be playing catch-up, not leaving much of a market for other companies.

AR and the Metaverse

Imagine a world where only AR exists and there is no VR – the old concept of the Metaverse as VR-centric disappears, allowing it to take on a new form. At its core, the Metaverse is a virtual space that connects people from around the world in a shared, interactive experience. In an AR-only world, the Metaverse would take the form of a digital layer superimposed on top of the physical world, accessible through mobile devices, smart glasses, and other AR-enabled technologies. This layer would be designed to enhance and augment the real world, providing a new layer of information and interactivity that is seamlessly integrated into our everyday lives.

Imagine walking down a busy street and seeing a layer of digital information overlaid on top of the physical environment. You can see the names and reviews of nearby restaurants, navigate to your destination with virtual arrows and waypoints, and even see the names and faces of people you pass by, thanks to real-time facial recognition technology. This augmented layer would provide a wealth of information and opportunities for interaction, making the real world feel more vibrant and engaging than ever before.

But the potential of the Metaverse goes far beyond simple information and navigation. In an AR-only world, the possibilities for immersive and interactive experiences are virtually limitless. With AR, you could visit a museum and see digital reconstructions of ancient artifacts, interact with virtual creatures in a park, or even join a virtual concert where the performers are projected onto the stage in front of you.

One of the key features of the Metaverse in an AR-only world would be the ability to customize and personalize your experience. Users could create their own virtual spaces, whether it's a private room or a public space, and share them with others in the Metaverse. These spaces could be anything from a virtual art gallery to a full-scale replica of a famous city, allowing users to explore and interact with each other in new and exciting ways.

The Metaverse in an AR-only world could also be used for a variety of practical applications. Imagine a construction worker using AR technology to see digital blueprints overlaid onto a real-world construction site, or a doctor using AR glasses to see medical information in real time during surgery. The possibilities for AR in the workplace are vast, offering new ways to increase productivity and efficiency.

Of course, with any new technology, there are potential downsides and challenges to consider. One of the most significant concerns with AR technology is privacy and security. With facial recognition technology, there is a risk of people being identified and tracked without their knowledge or consent. It would be essential to implement strong privacy protections and security measures to ensure that users' personal information and data are protected.

Another challenge with AR technology is the potential for sensory overload. With so much information and interaction available in the Metaverse, it could be easy to become overwhelmed and disoriented. It would be crucial to design AR experiences that are intuitive, easy to navigate, and don't distract from the real world.

An AR-only world has the potential to be a transformative and exciting new frontier. The Metaverse in this world would offer a wealth of opportunities for exploration, interaction, and customization, enhancing the way we live, work, and play.

Now, let's end this exercise and bring back VR as part of the Metaverse with AR. What is the role of AR in the Metaverse?

Apple has been very vocal that it will not use the term "Metaverse" at any time in connection with any of its technology. The end game for Apple is more like the "Appleverse." Apple did make clear during its presentation on the Apple Vision Pro that the official term that they would be using instead of the "Metaverse" is "Spatial Computing."

With its AR glasses, Apple will be able to capture the majority of people's minds and hearts in its mission to dominate the Metaverse.

Particular use cases that illustrate AR's place in the Metaverse can be found in *Part 3, Consumer and Enterprise Use Cases.*

Currently, AR has some limitations that should be overcome for it to live up to its full potential.

AR's current limitations

AR is a rapidly evolving technology that has the potential to revolutionize how we interact with the world around us. AR devices are an essential tool in this revolution as they enable us to superimpose digital information onto our physical surroundings. However, despite the exciting potential of AR devices, they still face significant limitations that prevent them from becoming widespread and mainstream technology.

Here are some of the current limitations of AR headsets and glasses in detail:

- **Field of view**: One of the most significant limitations of current AR headsets and glasses is their narrow FOV. The FOV is the angle of the visible world that can be seen through the headset's display. In most AR headsets, this is limited to a small portion of the user's FOV, often around 50-60 degrees, and with AR glasses even less, which is not enough to create a truly immersive experience. This limitation is due to technical challenges in developing high-resolution displays with a large FOV that are also compact and lightweight enough to be worn on the head. A limited FOV can make it challenging to interact with the digital content in a natural way and can break the illusion of the augmented environment, which can negatively impact the user experience.

- **Display quality**: Another limitation of AR headsets and glasses is the quality of the display. While many modern AR headsets offer high-resolution displays, the image quality can still be relatively poor compared to other display technologies. This is partly due to the limitations of the optics used, which can introduce distortion and reduce clarity. Additionally, the limited brightness and contrast of the displays can make it difficult to view digital content in bright or outdoor environments. Additionally, lower-resolution displays can lead to eye strain and fatigue, which can be a significant concern, especially for extended use.

- **Battery life**: Another significant limitation of AR headsets and glasses is their battery life. Many require a lot of power to run, which can lead to short battery life, often lasting only a few hours. This can be especially problematic for users who require extended use or those in industries where continuous usage is required. For instance, medical professionals who rely on AR devices for patient treatment, or industrial workers who use them for maintenance or repair work, may face difficulties with frequent battery replacements or the need for a wired power source.

- **Processing power**: AR headsets and glasses require high processing power to provide a seamless and immersive experience. The devices must capture and analyze real-world data, then combine it with digital content, and display it to the user in real time. The current state-of-the-art in processing power is often not enough to provide a smooth experience, especially with complex and dynamic content. This can result in slow or laggy performance, which can break the illusion of the augmented environment and negatively impact the user experience.

- **Size and weight**: AR devices need to be lightweight, compact, and comfortable to wear for extended periods to be viable for everyday use. However, current headsets are often bulky and heavy, making them uncomfortable to wear for extended periods. The size and weight of the headset can also impact the FOV and the overall balance, which can be a significant issue for users who require precise movements, such as gamers or athletes.

- **Cost**: Current AR devices are expensive, with many high-end models costing several thousand dollars. This high cost is often due to the advanced technologies required to create a functional AR headset, such as high-resolution displays, sensors, and processing units. This limits the adoption of the technology, making it inaccessible to the general public.

- **Limited interactivity**: AR headsets and glasses can be used to interact with the digital world, but currently, there are still limitations to the level of interactivity. Some AR headsets use hand gestures or voice commands to interact with the digital world, but these methods can be imprecise and unreliable. Additionally, the lack of physical feedback can make it challenging to interact with the virtual world naturally and intuitively.

- **Accessibility**: AR devices can be challenging to use for people with certain disabilities or impairments, such as visual or hearing impairments. The limited FOV can also be problematic for individuals with restricted mobility or those who require a larger FOV to compensate for visual impairments.

- **Content creation**: Another significant limitation of AR headsets and glasses is the limited amount of content available. Developing AR content is a complex and time-consuming process, requiring specialized skills and tools, which makes it challenging to create a vast library of content. Additionally, creating AR content that is compatible with various AR devices can be difficult as each has different hardware and software specifications.

- **Privacy concerns**: AR devices often rely on sensors and cameras to track the user's movements and environment. While this information is typically only used for AR-related purposes, there are still concerns about privacy and data security. Additionally, the use of AR devices in public spaces can raise concerns about privacy and data collection from bystanders.

Progress is being made. Apple and other companies making AR devices have been very aware of these limitations and have been working on addressing them for several years.

Summary

This chapter was a deep dive into AR – its history, current and anticipated technology and limitations, and its significance to the Metaverse.

AR's role in the Metaverse will be very rich and in many ways more far-reaching than VR's role since AR allows a person to still be connected to the real world. In this way, many everyday practical tasks could be accomplished using AR, as well as the most fantastical entertainment.

Next up is virtual reality – how it came about, where it's going, and its particular relevance to the Metaverse.

3
Where Is Virtual Reality Heading?

Virtual reality (VR) has had somewhat of a bumpy ride, starting with its promise for 3D gaming and then branching out to enterprise uses, such as training and education, health, and communication. Here, we review how VR started, how VR is being used, and where it is heading. Most importantly, we show how VR and the Metaverse will significantly change how we interact with technology and each other.

In this chapter, we're going to cover the following main topics:

- The history of VR, including its gaming start and what the current status is
- Which VR headsets are relevant and which industries have been using them
- Understand what VR's significance is to the Metaverse

History of VR

VR has come a long way since its early beginnings, evolving from the realm of science fiction to an everyday technology that has transformed various industries. Over its history, many researchers, inventors, and enthusiasts have contributed to the development of VR. The following are important points in VR's history.

Early concepts and precursors (1800–1950s):

- **Stereoscopy and the stereoscope (1838):** The concept of virtual reality can be traced back to the early 19th century with the invention of the stereoscope by Charles Wheatstone. Stereoscopy involves the presentation of two slightly offset images to the left and right eye, creating an illusion of depth perception. The stereoscope marked the beginning of 3D imagery and laid the groundwork for future VR technologies.

- **Sensorama (1956)**: In the 1950s, filmmaker Morton Heilig designed and built the Sensorama, an arcade-style machine that offered an immersive multi-sensory experience. The Sensorama used stereoscopic 3D images, motion, sounds, and even smells to transport users to different environments, such as a motorcycle ride through Brooklyn.

Foundations of modern VR (1960s– 1980s):

- **The Ultimate Display (1965)**: Computer scientist Ivan Sutherland proposed the concept of the "Ultimate Display," which he envisioned as a room where a computer could control the existence of matter. Sutherland believed this display could simulate reality to the point where it was indistinguishable from the real world. This concept marked the foundation of modern VR.

- **The first head-mounted display (1968)**: Ivan Sutherland and his student, Bob Sproull, created the first **head-mounted display (HMD)**, known as the Sword of Damocles. The HMD used computer graphics to project virtual images onto the user's field of view, creating a sense of immersion. The Sword of Damocles was large and cumbersome, but it was the first step towards modern VR headsets.

- **Dataglove (1982)**: Thomas Zimmerman invented the first data glove—a wearable input device that allowed users to interact with virtual environments using hand gestures. Zimmerman and Jaron Lanier later founded VPL Research, which became the first company to sell VR products, including the Dataglove and the EyePhone, a head-mounted display.

- **NASA's Virtual Interface Environment Workstation (VIEW) (1985)**: NASA began developing the VIEW system, a combination of HMD, data gloves, and 3D audio, to control remote robotic devices. The VIEW system allowed operators to see and manipulate virtual objects in a 3D environment, paving the way for VR's use in teleoperation and telepresence.

The rise of consumer VR (1990s):

- **Virtuality Group's arcade machines (1991)**: Virtuality Group released a series of VR arcade machines, including the Virtuality 1000CS and the Virtuality 2000, which offered multiplayer gaming experiences using HMDs and data gloves. These arcade machines introduced the public to the concept of VR gaming.

- **Sega VR and Nintendo Virtual Boy (1993, 1995)**: Sega and Nintendo, two major video game companies, attempted to bring VR gaming to the consumer market with the Sega VR headset and the Nintendo Virtual Boy. Both products failed to gain widespread adoption due to technical limitations, high costs, and discomfort during use.

- **The CAVE (1992)**: The **CAVE (Cave Automatic Virtual Environment)** was developed by a team at the University of Illinois in Chicago. The CAVE was a room-sized, immersive VR environment that used projectors to display 3D graphics on the walls and floor. The CAVE allowed multiple users to interact with virtual objects using 3D glasses and a wand, demonstrating the potential for collaborative VR experiences.

Advancements in VR technologies (2000s):

- **Development of Motion Tracking**: In the 2000s, advancements in motion-tracking technology enabled more precise and intuitive user interactions with virtual environments. Devices such as the Wii remote (2006) popularized motion-controlled gaming, while the development of **inertial measurement units (IMUs)** and optical tracking systems provided more accurate and responsive tracking.

- **The emergence of augmented reality (AR)**: The 2000s also saw the emergence of AR, a technology that overlays digital information onto the user's view of the real world. AR devices such as the Microsoft HoloLens (2016) demonstrated the potential for mixed reality experiences that combine elements of both VR and AR.

- **The rise of mobile VR**: The rapid growth of smartphones with powerful processors and high-resolution displays led to the development of mobile VR experiences. Google Cardboard (2014) and Samsung Gear VR (2015) provided affordable, accessible VR experiences by leveraging the capabilities of smartphones.

Figure 3.1 – A person using a developer version of the Oculus Rift, the DK 2 (2014)

Resurgence and mainstream adoption of VR (2010s –present):

- **Oculus Rift (2016)**: In 2014, Facebook (now called Meta) acquired Oculus VR for $2 billion, further solidifying the potential of VR gaming. The consumer version of the Oculus Rift headset, the CV1, was released in 2016, and it set the standard for VR gaming headsets with its high-resolution displays, wide field of view, and comfortable design. The Rift's release sparked a renewed interest in VR, with developers and content creators flocking to create new experiences for the platform. Despite its high price and the requirement of a powerful gaming PC, the Rift became a symbol of the potential of VR.

- **HTC Vive (2016)**: Following the Rift's release, other tech giants entered the VR market with their offerings. HTC and Valve Corporation introduced the HTC Vive in 2016, which featured room-scale tracking, allowing users to move freely within a predefined space. The Vive also included motion controllers, enabling a more interactive experience.

- **PlayStation VR (PSVR) (2016)**: Sony launched PSVR in 2016, designed for use with the PlayStation 4 gaming console. The PSVR's lower price point and compatibility with an existing console made VR accessible to a wider audience.

- **Microsoft Windows Mixed Reality Headsets (2017)**: Microsoft released **Windows Mixed Reality (WMR)** headsets in 2017, developed in partnership with various manufacturers such as Samsung, Dell, and HP. These devices offered inside-out tracking, eliminating the need for external sensors.

Figure 3.2 – A person using the Samsung Gear VR headset paired up with a Samsung smartphone

Mobile VR and standalone headsets:

- **Google Cardboard (2014) and Daydream View (2016)**: Google introduced the Cardboard in 2014 and the Daydream View in 2016. Both are low-cost VR solutions for smartphones.

- **Samsung Gear (2015)**: Mobile VR became increasingly popular with the release of the Samsung Gear VR in 2015, utilizing compatible Samsung smartphones as the display and processing unit.

- **Oculus Go (2018)**: The Oculus Go was released in 2018 as a standalone VR headset, not requiring a smartphone or PC. The Go was designed for casual VR experiences, and it had a lower price point and ease of use.

The quest for the perfect VR Headset:

- **Oculus (now Meta) Quest (2019)**: Oculus released the Quest in 2019, a standalone VR headset with room-scale tracking and motion controllers. The Quest marked a significant milestone in the industry, offering a high-quality VR experience without the need for external hardware.

- **Valve Index (2019)**: Valve released the Index in 2019, featuring high-quality optics, a wider field of view than the Meta Quest, and a high refresh rate.

- **HTC Vive Cosmos (2019)**: HTC released the Vive Cosmos in 2019 with modular faceplates for customization and inside-out tracking.

- **Meta Quest 2 (2020)**: The Quest 2, released in 2020, improved on its predecessor with a more powerful processor, higher resolution, and a lower price point.

Our next section goes into more detail on how VR outgrew the need to have 3D gaming and how it's grown in other ways.

Gaming start and its current growth

As a driving force in VR adoption, VR gaming has revolutionized the way we interact with digital environments, allowing players to immerse themselves in realistic 3D worlds. Over the past few decades, VR gaming has evolved from a niche concept to a widely accessible and increasingly sophisticated technology. In this section, we review the history of VR and gaming, including location-based VR and social VR, and explore the key developments, challenges, and breakthroughs that have shaped this industry.

Early beginnings: 1960s–1980

As previously noted, the concept of VR dates back to the 1960s when Ivan Sutherland and Bob Sproull developed the **Sword of Damocles** HMD. Although rudimentary, this device laid the foundation for future VR technologies.

In the 1980s, the term VR was coined by Jaron Lanier, founder of VPL Research. VPL Research developed the DataGlove—an early VR input device that allowed users to interact with digital environments using hand gestures. The company also produced the EyePhone, one of the first VR headsets designed for gaming.

During this period, arcade gaming was booming, and several companies attempted to incorporate VR into their offerings. One notable example was the Virtuality Group, which introduced a series of VR arcade machines in the early 1990s. These systems, like the Virtuality 1000CS, provided players with a fully immersive experience that included a head-mounted display, 3D graphics, and motion-tracking technology.

Nintendo's Virtual Boy: 1995

In 1995, Nintendo released the Virtual Boy, a standalone VR gaming console. While the Virtual Boy was the first mass-market VR gaming device, it was not without its limitations. The monochromatic red display and lack of head-tracking caused discomfort for users, and the high price tag and limited game library led to poor sales. Despite its shortcomings, the Virtual Boy was a significant milestone in the history of VR gaming.

VR in-home gaming systems: late 1990s

During the late 1990s, several companies attempted to integrate VR technology into their gaming systems. Sega developed a VR headset for the Sega Genesis, but it never reached the market due to concerns about motion sickness. Similarly, Sony experimented with a VR headset for the PlayStation, but this project, called the **Glasstron**, was ultimately canceled.

Advances in VR hardware: the 2000s

In the 2000s, technology advancements led to significant improvements in VR hardware. The development of powerful **graphics processing units** (**GPUs**), accelerometers, and gyroscopes enabled more sophisticated VR experiences. Furthermore, the proliferation of smartphones equipped with high-resolution displays and motion sensors made it possible to create affordable VR headsets, such as the Google Cardboard, which utilized a smartphone as the display and processor.

During this time, researchers and developers continued to experiment with VR gaming. Before 2010, notable examples include the CAVE—a room-sized VR environment that used projectors to create immersive experiences—and the VirtuSphere—a large, walkable sphere that allowed users to navigate virtual worlds.

Early innovations and Kickstarter campaigns: 2010–2012

VR gaming has come a long way since its inception in the late 20th century, but the 2010s have been particularly transformative in terms of technology, adoption, and content. This decade saw VR gaming shift from a niche hobby to a mainstream entertainment medium, with significant advancements in hardware, software, and the gaming experiences it offered.

The beginning of the decade was marked by early innovations and experimentation. In 2010, Palmer Luckey, a 19-year-old VR enthusiast and the founder of Oculus VR, started working on the first prototype of the Oculus Rift. The Rift was a virtual reality headset designed to provide an immersive gaming experience and be an affordable and high-quality VR headset, unlike other devices at the time. Luckey's idea gained traction after an online demo attracted the attention of renowned video game programmer John Carmack, who showcased the device at the **Electronic Entertainment Expo (E3)** in 2012.

In 2012, Oculus VR launched a Kickstarter campaign to fund the development of the Rift. The campaign was a massive success, raising over $2.4 million, significantly surpassing its initial goal of $250,000. This demonstrated the potential and public interest in VR gaming and laid the groundwork for future developments in the industry.

Emergence of major players and acquisition: 2013–2014

In 2013, Valve Corporation, the company behind the popular gaming platform Steam, started working on its VR technology. Valve would eventually partner with HTC to create the HTC Vive, a VR headset that would compete with the Oculus Rift.

In 2014, Oculus VR was acquired by Facebook for $2 billion, a move that signaled the social media giant's interest in the potential of VR. This acquisition provided Oculus with the resources and support needed to further develop its technology and expand its reach.

Oculus VR released two development kits, the DK1 in 2013 and the DK2 in 2014, allowing developers to create content for the platform.

The launch of the first consumer VR headsets: 2015–2016

In 2016, the first consumer versions of the Oculus Rift and HTC Vive were released. These headsets provided users with immersive, high-quality VR experiences, and their launch marked a turning point in VR.

In 2014, Sony had announced Project Morpheus, which would later become **PlayStation VR (PSVR)**. Released in October 2016, PSVR was the first VR headset compatible with a gaming console—the PlayStation 4.

More information on VR headsets including current ones can be found in the *History of VR* and *VR Devices* subsections.

Evolution of VR games and experiences

Early VR games focused on simple mechanics and interactions, such as shooting gallery-style games. As VR technology advanced, games incorporated more complex mechanics and interactions, allowing players to perform a wider range of actions in-game.

The VR landscape has expanded to include a variety of genres, from horror to action, adventure, and puzzle games. Additionally, VR experiences have emerged, such as educational applications, simulations, and virtual tours.

The growth of VR gaming has been further fueled by the support of game developers. Major developers, such as Ubisoft, Bethesda, and Valve, have released VR titles, while independent developers have created innovative, VR-exclusive experiences.

Notable VR games and experiences (2016-Present) include the following:

- **Job Simulator (2016)**: Owlchemy Labs' *Job Simulator* is a humorous and highly interactive VR game that simulates various jobs, such as office worker, gourmet chef, and convenience store clerk. The game's unique concept and interactivity made it an early standout in the VR market.

- **Superhot VR (2016)**: *Superhot VR*, developed by Superhot Team, is a unique first-person shooter where time only moves when the player moves. The game's minimalist graphics and strategic gameplay made it a popular choice for VR enthusiasts.

- **The Elder Scrolls V: Skyrim VR (2017)**: Bethesda's popular open-world RPG, *The Elder Scrolls V: Skyrim*, was adapted for VR in 2017. Skyrim VR allowed players to fully immerse themselves in the game's expansive world, complete with motion-controlled combat and exploration.

- **Beat Saber (2018)**: Developed by Beat Games, *Beat Saber* is a rhythm-based game where players use lightsabers to slash incoming blocks in time with the music. Its addictive gameplay and immersive visuals quickly made it a standout title in the VR space.

- **Supernatural (2020)**: Developed by Within, *Supernatural* is a fitness app that became very popular quickly. In 2021, Meta bought Within mostly to be able to have the rights to Supernatural, but it wasn't until 2023 that they were assured ownership after the FTC lost its case filed against Meta, which it filed with concerns regarding competition.

- **Half-Life: Alyx (2020)**: Developed by Valve, *Half-Life: Alyx* is a first-person shooter set in the Half-Life universe. As a highly anticipated title, it pushed the boundaries of VR gaming with its detailed graphics, physics-based interactions, and engaging story.

Other popular titles include *The Walking Dead: Saints & Sinners*, *Pistol Whip*, and *Boneworks*, each showcasing different aspects of VR gaming's potential.

Location-based VR (LBVR)

LBVR refers to a type of virtual reality experience that takes place at a specific physical location, usually in a dedicated VR center, arcade, or another similar venue. These locations are typically equipped with specialized VR hardware, such as headsets, motion controllers, haptic feedback devices, and sometimes even large-scale props or environmental elements that can enhance the sense of immersion.

In LBVR, users can participate in individual or multiplayer experiences that are designed to take advantage of the unique features of the venue. These experiences often include interactive games, simulations, educational content, or even immersive movies. The main goal is to provide users with an enhanced and immersive VR experience that may not be possible to achieve with home-based VR systems due to limitations in space, hardware, or software.

The rise of LBVR was fueled by advancements in VR headsets, such as the Oculus Rift, HTC Vive, and Sony PlayStation VR, which were released in 2016. These devices provided higher resolution, lower latency, and more accurate tracking.

Wireless VR systems, such as the TPCAST for HTC Vive, enabled users to move freely without being tethered to a computer. This development was crucial for the growth of LBVR.

Here are some pioneers in LBVR:

- **The Void**: The Void was one of the first LBVR companies, launching in 2015. It combined physical sets, VR headsets, and haptic feedback to create unique, immersive experiences for users. In 2017, The Void partnered with Disney to launch the *Star Wars: Secrets of the Empire* experience, attracting mainstream attention and paving the way for further partnerships.

- **Zero Latency**: Zero Latency, an Australian company, was another early pioneer in LBVR. It opened its first location in 2015, offering free-roaming VR experiences for up to eight players.

- **Dreamscape Immersive**: Backed by major Hollywood studios, it opened its first location in 2018. It offered immersive, cinematic VR experiences that combined storytelling with advanced technology.

- **Sandbox VR**: Launched in 2016, Sandbox VR first focused its locations in Asia and had just started to try to expand in the U.S. when COVID-19 hit.

Up until COVID-19, LBVR companies were expanding globally—The Void and Zero Latency especially—and establishing locations in Europe, Asia, and the Americas.

The impact of COVID-19 (2020–2021)

The COVID-19 pandemic forced many LBVR venues to temporarily or permanently close, causing significant financial strain on the industry.

There were several adaptations and innovations that the LBVR industry experienced.

- **Hygiene measures**: LBVR venues implemented strict hygiene measures to ensure customer safety, such as sanitizing headsets and frequently touched surfaces.
- **Socially distanced experiences**: Companies designed socially distanced VR experiences, such as VR escape rooms and single-player games, to adapt to the new normal.
- **Remote VR experiences**: Some LBVR companies, such as Exit Reality, pivoted to remote VR experiences, allowing users to rent VR equipment for at-home use.

LBVR has experienced somewhat rapid growth since its inception in 2015, with advancements in technology and strategic partnerships propelling the sector forward. The Void shut down for other reasons besides COVID-19, and Zero Latency is still in action, but both Dreamscape Immersive and Sandbox VR suffered greatly due to the pandemic, with both rebounding to some degree. Sandbox VR received more funding to continue with its growth, and Dreamscape Immersive, though continuing with LBVR, is also branching out into other areas, such as education. Time will tell if LBVR will have a sustainable future.

Social VR and multiplayer experiences

Social VR and multiplayer experiences have gained significant momentum since 2015, fueled by advancements in VR technology and the growing popularity of online social interaction. These immersive environments allow users to interact with one another in real time using virtual avatars to represent themselves. Players can participate in shared activities, engage in conversations, and explore virtual worlds together, fostering a sense of community and connection among users.

In the early days of social VR, platforms such as AltspaceVR (which was recently closed down by Microsoft) and VRChat began to lay the groundwork for virtual social experiences. AltspaceVR, launched in 2015, allowed users to create and customize avatars, attend live events, and interact with others in various virtual spaces. VRChat, released in 2017, took the concept further, allowing users to create and share user-generated content, including custom avatars and worlds, fostering a creative and inclusive environment for a diverse range of experiences.

As VR technology evolved, new platforms and experiences emerged, each with its unique features and offerings. Facebook launched its social VR platform, Facebook Horizon Worlds (formerly Facebook Horizon), in 2020. It allowed users to create, explore, and socialize within a variety of user-generated spaces. Other popular platforms such as Rec Room and Bigscreen emphasized different aspects of social interaction, such as gaming, collaboration, and media consumption.

Simultaneously, multiplayer VR games gained traction, with titles such as Echo Arena, Onward, and Population: ONE delivering engaging and competitive experiences that encouraged teamwork and communication among players. These games span various genres, from shooters to sports, and leverage VR's immersive capabilities to create unique experiences.

However, despite the growth and excitement surrounding social VR and multiplayer experiences, challenges and ethical concerns have also arisen. Issues such as harassment, privacy, and addiction have prompted developers and platform owners to implement safety measures and community guidelines. Maintaining a balance between user freedom and safety remains an ongoing challenge in the world of social VR.

The future of social VR and multiplayer experiences is likely to be shaped by advancements in technology, increased adoption, and the emergence of new platforms and experiences. As VR hardware becomes more accessible and affordable, a greater number of users will be able to participate in these immersive experiences, driving innovation and growth within the industry.

In 2019, VR gaming entered the mainstream with the release of the Oculus Quest, a standalone VR headset that didn't require a PC or console. This development made VR more accessible, leading to a surge in interest and adoption. The Oculus Quest 2, released in 2020, further refined the standalone VR experience with improved performance, resolution, and ergonomics.

As VR entered the 2020s, it continued to push the boundaries of interactive entertainment. Developers experimented with new ways to blend storytelling, interactivity, and immersion, promising even more captivating experiences in the future.

Despite its growth and advancements, VR gaming and social VR still face challenges such as limited content, high hardware costs, and motion sickness for some users. However, the industry continues to evolve, with new devices such as the PlayStation VR 2 and HTC Vive Pro 2 offering higher fidelity experiences and addressing some of the technology's limitations.

While this overview only scratches the surface of the VR gaming and social VR landscape, it highlights the rapid advancements and exciting developments that have shaped this burgeoning industry. From its early beginnings to its current status as a growing entertainment medium, VR gaming and social VR have continued to evolve, promising a future filled with innovative and immersive experiences.

Our next section looks at headsets in terms of their features, specifications, and downsides.

VR and mixed reality (MR) devices

We have included MR devices here in the same section as VR devices rather than including them in the AR section. Our reason for doing so is that even though MR devices can do both VR and AR, the type of AR that MR headsets can do is called **passthrough AR**. Passthrough AR, also known as **passthrough mode** or **camera passthrough**, is a feature in some virtual reality headsets that allows users to view their real-world surroundings through the headset's built-in cameras. This mode essentially *passes* *through* the video feed from the cameras to the headset's display, enabling users to see and interact with their physical environment without needing to remove the headset.

Passthrough mode can be useful in various scenarios, such as checking on your surroundings, finding objects, or interacting with other people in the room while still wearing the VR headset. It can also enhance safety by helping users avoid potential hazards and obstacles in their environment during VR experiences. Now, this is the traditional use of passthrough mode; Apple, with their Vision Pro MR headset, is using passthrough as its core feature, with almost all of the apps shown in their announcement presentation using passthrough AR.

In this section, we first review the major features and desirable specifications of VR/MR headsets, followed by some negatives. Then, we provide the specifications of some notable headsets.

Here are some major features and desirable specifications to consider when comparing different VR and MR headsets:

- **Display resolution**: This refers to the number of pixels used to display the VR environment. The resolution of the display affects the image quality and clarity of the VR experience. The higher the resolution, the more detailed and realistic the virtual environment will appear.

- **Field of view (FOV)**: FOV refers to the extent of the virtual environment that is visible to the user. Horizontal FOV is a more useful measure than vertical FOV for VR. A wider horizontal FOV provides a more immersive experience by allowing the user to see more of their virtual surroundings and is less likely to cause motion sickness.

- **Refresh rate**: The refresh rate refers to the number of times per second the display refreshes. A higher refresh rate results in smoother and more fluid motion, reducing motion sickness and improving the overall VR experience.

- **Tracking**: Tracking refers to how accurately the VR headset tracks the movement of the user's head and body in real time. Higher-quality tracking allows for more natural and immersive interactions with the virtual environment.

- **Controllers**: These are the input devices used to interact with the virtual environment. Some controllers feature motion tracking, haptic feedback, and other advanced features.

- **Audio**: VR headsets may feature built-in headphones or require external audio devices. High-quality audio can enhance immersion and realism.

- **Connectivity**: A few VR headsets require a connection to a PC, while others are standalone and do not require a computer. Connectivity can affect portability and ease of use.

- **Comfort**: Comfort is an important consideration, as VR headsets are worn on the head for extended periods. The weight, padding, and adjustability of the headset can all affect comfort.

- **Content library**: The number and quality of available VR apps and games can also be an important factor when comparing VR headsets. A larger and more diverse content library allows for more varied and engaging experiences.

- **Price**: VR headsets can vary greatly in price, and the cost may be a significant factor for many users. The most expensive headsets tend to offer the highest quality and most advanced features, while more affordable options may provide a more basic VR experience.

Since we've covered some comparative points, it's time to review some negative points.

Some cautionary things about VR

VR technology has been touted as a game-changer for entertainment, education, and even healthcare. However, like any technology, it also has its downsides. Here are some of the negative aspects of VR:

- **Motion sickness**: One of the most common issues associated with VR is motion sickness. This is because VR devices simulate movements that are not happening in the real world, which can cause dizziness, nausea, and headaches. Some people are more susceptible to motion sickness than others, and this can limit their ability to use VR.

- **Eye strain and fatigue**: Using VR for extended periods of time can lead to eye strain and fatigue, as the eyes have to constantly adjust to the focus and distance of the virtual environment. This can cause discomfort, headaches, and even vision problems over time.

- **Social isolation**: VR can be a very immersive experience, but it can also be isolating. Users are often cut off from the real world while in VR, which can make it difficult to connect with others and can lead to feelings of loneliness and isolation.

- **Addiction**: Like many forms of technology, VR can be addictive. Users can become so immersed in the virtual world that they neglect other aspects of their lives, such as work, relationships, and physical health.

- **Cost**: VR devices and software can be expensive, which can be a barrier to entry for many people. Additionally, there are ongoing costs associated with maintaining and upgrading VR equipment, which can be prohibitive for some.

- **Privacy concerns**: VR technology collects a lot of data about users, including their movements, behaviors, and preferences. This data can be used for targeted advertising, and there are also concerns about how it could be used for more nefarious purposes, such as surveillance or identity theft.

- **Physical limitations**: VR requires users to be physically active, which can be a challenge for those with disabilities or mobility issues. Additionally, some VR experiences may be too intense or physically demanding for some users, which can limit their ability to fully engage with the technology.

- **Risk of injury**: Because VR involves simulating movements and environments, there is always a risk of injury if users are not careful. Users can trip, fall, or bump into objects in the real world while they are immersed in the virtual environment.

- **Unrealistic expectations**: VR can create very realistic simulations of the real world, but it can also create unrealistic expectations. For example, users may expect that they can do things in the virtual world that are not possible in the real world, which can lead to disappointment and frustration.

- **Addiction to violent content**: VR games and simulations often involve violence and aggression, which can desensitize users to these behaviors and make them more likely to engage in them in real life.

In summary, while VR technology has many benefits, it also has its downsides, including physical health risks, psychological effects, addiction, social isolation, cost, limited content, technical limitations, and ethical concerns. It is important to be aware of these potential drawbacks and to use the technology responsibly.

Notable VR/MR headsets

VR/MR headsets have emerged as cutting-edge technologies. In this section, we will explore some of the most notable VR/MR headsets available on the market, discussing their particular features and specifications.

Some of the major headsets are as follows:

Meta Quest 3

- Standalone MR headset
- Powered by the Qualcomm Snapdragon XR2 Gen 2 platform
- LCD display with "4K+ infinite Display"
- 2064 x 2208 pixels per eye resolution
- Refresh rate of 90Hz, 120Hz (experimental)
- 110 degree horizontal field of view
- Built-in audio and microphone
- Compatibility with Air Link and Oculus Link for PC VR gaming
- Hand tracking and voice commands, but no eye tracking
- Passthrough AR (full color)

PlayStation VR (PSVR)

- Tethered VR headset for PlayStation 4 and 5
- Powered by PlayStation console
- 5.7-inch OLED display
- 960 x 1080 pixels per eye resolution
- Refresh rate of 90Hz or 120Hz
- 96-degree horizontal field of view
- 3D audio support
- PlayStation Move and DualShock 4 controller compatibility
- PlayStation camera for tracking

PlayStation VR2 (PSVR2) released in 2023

- Tethered VR headset for PlayStation 5
- Powered by PlayStation console
- 6.4-inch OLED display
- 2000 x 2040 pixels per eye resolution
- Refresh rate of 90Hz or 120Hz
- 110-degree horizontal field of view
- 3D audio support
- Sense controllers with haptic feedback and adaptive triggers
- Inside-out tracking (four built-in cameras)
- Eye tracking
- Passthrough AR (grey scale)

Valve Index

- Tethered VR headset for PC
- 6.0-inch OLED display
- 1440 x 1600 pixels per eye resolution
- Refresh rate of 80Hz, 90Hz, 120Hz, or 144Hz
- 120-degree horizontal field of view

- Integrated off-ear headphones
- External SteamVR 2.0 base stations for tracking
- Index Controllers (formerly known as "Knuckles") for finger tracking

HP Reverb G2

- Tethered VR headset for PC
- Developed in collaboration with Valve and Microsoft
- 5.78-inch OLED display
- 2160 x 2160 pixels per eye resolution
- Refresh rate of 90Hz
- 98-degree horizontal field of view
- Integrated off-ear headphones designed by Valve
- Inside-out tracking with four camera sensors
- Windows Mixed Reality controllers

HTC Vive Pro 2

- Tethered VR headset for PC
- 5.76-inch OLED display
- 2448 x 2448 pixels per eye resolution
- Refresh rate of 90 Hz or 120 Hz
- 120-degree horizontal field of view
- Integrated on-ear headphones
- Can be used with the SteamVR platform

Pico Neo 3

- Standalone VR headset
- 5.5-inch LCD display
- 2880 x 1700 pixels per eye resolution
- Refresh rate of 90 Hz or 120 Hz
- 98-degree horizontal field of view

- Passthrough AR (grey scale)
- Pico Neo 3 and the updated version, Pico Neo 4, are not available in the U.S. (available in Europe, China, Japan, South Korea, and Malaysia)

Pico Neo 3 Pro

- Standalone VR headset
- 5.5-inch LCD display
- 3664 x 1920 pixels per eye resolution
- Refresh rate of 90 Hz or 120 Hz
- 98-degree horizontal field of view
- Passthrough AR (grey scale)
- Pico Neo 3 Pro is available in the U.S., but the updated version, Pico 4 Enterprise, is not yet available in the U.S. (available in Europe, China, Japan, South Korea, and Malaysia), and Both Pico Neo 4 and Pico 4 Enterprise have color passthrough AR.

Apple Vision Pro (available Spring 2024)

- Tethered external (battery pack) MR headset
- Two micro-OLED displays (about an inch in diameter) with over 4K resolution each
- 23 million pixels (per-eye resolution of around 3400×3400 for a square aspect ratio, or around 3680×3140 for the 7:6 aspect ratio)
- Refresh rate of 90Hz or 96Hz (when watching 24fps content, avoiding frame pacing judder)
- Passthrough AR (color)
- visionOS (spatial operating system)
- An Apple M2 processor and a specialized R1 processor for cameras and other visuals
- Face and eye tracking
- Personalized ray-traced spatial audio
- Six microphones
- Twelve cameras and five sensors (LiDAR and TrueDepth sensors) for monitoring hand gestures

More on the Apple Vision Pro

The Apple Vision Pro was extremely anticipated. It will be ready to ship in Spring 2024 after having been announced in June 2023.

These days, we are heading into **spatial computing**, where humans, robots, virtual beings, and virtual content make use of sensors, AI, and computer vision to interpret, move through, and/or inhabit spatial areas. This maps to the human mind much more closely.

With the Apple Vision Pro, it is the use of eye movement, hand movements (especially gestures), and voice in a spatial computer on your face that brings a much more human-centered way of computing. Apple made some key choices here to put us deeply into this new type of computing.

First, they didn't do hardware controllers like Meta does. Hardware controllers are liked by high twitch video gamers who are playing games such as Call of Duty that require very fast hand–eye coordination.

Apple went with no controllers, although it did show that gamers can hook one up as an option. It knew that most people wouldn't want to deal with using a controller just to get in and watch a movie or talk with their friends. It is the default that matters because now developers who build experiences for the Vision Pro know that they must develop for hands, eyes, and voice first.

Apple learned that VR devices are often used in homes with other people around who aren't in VR devices. The problem is that VR devices separate people from others in the room. Talking to someone with a VR headset on feels weird because you can't see their eyes. Apple fixed that by putting a screen that faces outward that shows digitally recreated eyes of the person wearing it.

It is important to Apple that its headset isn't seen as removing its wearers from others. This is an important signaler about what the Vision Pro is for—not isolating yourself from reality but integrating with it.

Avatars are another example of this. Apple's avatars are very realistic when compared to those on other systems. Apple spent a lot of time including a 3D sensor on the front of Vision Pro that is used during setup to make a 3D scan of your face that then produces the avatar and your eyes, which are shown to others on the external screens.

This will be very important in the future as developers release experiences that you can play with other family and friends and apps for working together with your business colleagues. A raft of experiences and apps is coming, from FaceTime to games, that will use these new Apple avatars.

The Vision Pro is a mixed reality headset, so why did Apple feature passthrough AR much more than its VR capabilities during its presentation announcement?

For several reasons. During the announcement presentation, Apple focused on showing off the practical uses of the headset. AR allows the user to see 2D and 3D visuals without taking them away from their present real environment. So, people would be able to use AR apps on the Vision Pro to work, study, explore, or reminisce while remaining accessible to others. These others could be family members, friends, or other workers of a company if the headset is used for remote work.

With AR, there could be the placement of several 2D screens or 3D objects with flexible sizing within the headset user's view. Hundreds of established apps, including those made by Apple, are already available (such as FaceTime) and utilize virtual ultra-realistic versions of people on the call called "personas." Screens from an existing Mac or iPad could be "lifted" to the user's space directly in front of them or wherever they'd like, creating virtual screens with text that is crisp and can readily be read.

It appears that Apple is more interested in focusing on practical human uses while allowing people to be able to be completely immersed if they want to be. Focusing on AR with the Vision Pro also fits their longer-range plans of producing AR glasses.

But why even have a product like the Vision Pro before Apple's AR glasses are available? A purported $40 billion was put towards research and development of the Vision Pro, with the project many times being put aside or discontinued only to be picked up again as a viable product. Since the plan was always to produce AR glasses, why not use a product that had so much money and effort already put into it to help prepare the market and herald a new element into their ecosystem?

When Apple talked about immersion during the presentation, it did not mean the immersion that is typically associated with VR, but rather just a background screen that portrayed, for example, a beautiful mountain scene that would fill the headset wearer's entire view.

There are places and instances where the Vision Pro will be inappropriate to use. There's a chance that the headset will not faithfully reproduce reality enough for such situations:

- **Dangerous jobs**: If you have your hands near a saw blade in a factory job, for instance, you will not want to wear a passthrough AR device, which is what Vision Pro is. It is clear that Apple has an AR-centric vision that is being launched with the Vision Pro, which will be continued with further iterations and carried to fruition with their AR glasses.

- **Driving a car**: Even being in a moving car will be resisted by Apple until it can bring devices to us that let us see the real world and avoid sickness or a lack of awareness of the environment. Autonomous cars are starting to come to consumers, but Apple will discourage this kind of use for a while.

- **Shooting a gun**: If you are a police officer or a military person, you will not want to use a passthrough device. It isn't precise enough to properly see what you are aiming at.

- **Moving fast**: Riding a mountain bike, skiing, and running won't be things that a Reality Pro will do well or safely.

- **Outside in bright sunlight**: The Vision Pro's sensors were designed for indoor use, not for outside. While some uses might work, the Vision Pro wasn't designed for you to walk around your neighborhood to play some new kind of augmented reality game.

- **Anything that would make the device wet**: Swimming, walking in the rain, wearing it in the shower, and other uses that could make the device wet won't be good for the device.

- **Jobs, like surgery or dentistry, that require seeing the real world in the finest detail possible**: Yes, you can use the Vision Pro for training for these jobs, but Apple will generally discourage these kinds of uses until it can bring devices out that let you "see through" the device into the real world.

The reason for this is because this device is a passthrough device that requires a ton of sensors to properly capture reality and, if the device fails, the wearer will be blind and in a black box until they take it off or reboot the device.

Competitors such as Magic Leap and HoloLens will be more appropriate for these kinds of uses, at least for now, but we do expect the Vision Pro to be a developer device to build services that can be used in future devices for these kinds of use cases.

Other than these jobs, the Apple device has a lot of invisible technology inside that makes it better for pretty much every other use case than its competitors can deliver.

In our next section, we cover what good VR can do in areas other than gaming. More on particular use cases in these areas can be found in *Part 3*, *Consumer and Enterprise Use Cases*.

Everything else but gaming

VR has seen significant advancements in recent years, leading to a wide range of innovative applications across various industries. The immersive nature of VR has made it a powerful tool for both small and large enterprises, offering unique solutions to business challenges and fostering growth. Here, we will review some major applications for VR outside of gaming, including training and education, design and prototyping, marketing and sales, remote collaboration, healthcare, and real estate.

Training and education

One of the most prominent enterprise uses of VR is in training and education. Companies across multiple sectors leverage VR to deliver realistic, immersive, and risk-free training experiences. Some examples include:

- **Safety training**: VR enables workers to experience hazardous situations without any real danger. This is particularly useful for industries such as construction, mining, and oil and gas, where on-the-job accidents can have severe consequences.

- **Soft Skills Training**: VR can also be used to teach and develop soft skills such as negotiation, conflict resolution, and customer service by simulating realistic scenarios that require employees to interact with virtual avatars.

- **Military and Law Enforcement**: VR is widely used to train soldiers and police officers in various combat and crisis situations, improving their decision-making and reaction times in high-pressure environments.

Design and prototyping

VR technology has revolutionized the way companies design and prototype products. With the ability to create realistic 3D models, designers can do the following:

- Visualize and manipulate designs in real-time, enabling rapid iteration and reducing time to market

- Collaborate with stakeholders and clients, receiving immediate feedback and avoiding costly redesigns later in the process

- Test product functionality and ergonomics, ensuring a better user experience and identifying potential design flaws

Marketing and sales

VR has opened up new possibilities for marketing and sales teams, who can now create immersive experiences to showcase products and services. Examples include:

- **Virtual showrooms**: Companies can create virtual showrooms that allow customers to explore and interact with products in a realistic environment, regardless of their physical location.

- **Product demos**: VR can be used to demonstrate complex products and technologies, enabling potential clients to gain a better understanding of their capabilities and benefits.

- **Immersive advertising**: Brands can create VR experiences that engage customers on a deeper level, fostering stronger emotional connections and increasing brand recall.

Remote collaboration

As remote work becomes increasingly popular, VR has emerged as a valuable tool for collaboration. Virtual meeting spaces can be created to do the following:

- Facilitate face-to-face communication between geographically dispersed teams, increasing engagement and fostering stronger working relationships

- Enable real-time collaboration on projects and designs, improving efficiency and decision-making

- Provide a more immersive and interactive experience compared to traditional video conferencing, reducing "Zoom fatigue" and improving overall satisfaction

Healthcare

The healthcare industry has embraced VR for various applications, including:

- **Medical training**: Medical students can practice surgeries and other procedures in a virtual environment, improving their skills before working with real patients.

- **Physical therapy and rehabilitation**: VR can be used to create engaging and motivating exercise programs for patients recovering from injuries or surgeries, speeding up the recovery process.

- **Pain management and mental health**: VR has been shown to reduce pain and anxiety for patients undergoing certain medical procedures, as well as providing effective treatment for conditions such as PTSD and phobias.

Real estate

In the real estate sector, VR is transforming the way properties are marketed and sold. Some key applications include:

- **Virtual tours**: Prospective buyers can explore properties in 3D without the need for physical site visits, saving time and resources.

- **Property Staging**: VR allows real estate agents to virtually stage properties with different furniture and décor options, helping potential buyers visualize their future homes.

- **Architectural Visualization**: Developers can use VR to showcase architectural designs and planned developments, giving investors and buyers a better understanding of the finished project.

Tourism and hospitality

The tourism and hospitality industry has also benefited from VR technology in the following ways:

- **Virtual travel experiences**: Travel agencies and tourism boards can create immersive VR experiences, allowing potential tourists to explore destinations and attractions before booking a trip.

- **Hotel Previews**: Hotels can use VR to showcase their rooms, facilities, and amenities, helping guests make informed decisions and increasing the likelihood of bookings.

- **Training and Onboarding**: VR can be used to train new staff members in various hospitality roles, ensuring a high level of customer service and reducing the learning curve for new employees.

Automotive industry

VR has made the following significant inroads into the automotive sector:

- **Vehicle design**: Automotive designers can use VR to visualize and iterate on vehicle designs more efficiently, reducing development time and costs.

- **Manufacturing**: VR can be employed to simulate and optimize manufacturing processes, leading to increased efficiency and productivity on the assembly line.

- **Showrooms and test drives**: Car dealerships can offer virtual showrooms and test drives, allowing customers to experience vehicles without leaving their homes.

Retail industry

The retail industry has also seen the benefits of incorporating VR into their operations:

- **Virtual stores**: Retailers can create virtual stores, allowing customers to browse and shop for products in a realistic and engaging environment.

- **Customer Analytics**: VR can provide valuable insights into customer behavior and preferences, enabling retailers to optimize their store layouts, product offerings, and marketing strategies.

- **Employee training**: VR can be used to train retail staff in various scenarios, such as customer interactions, sales techniques, and store management.

Entertainment and events

Finally, the entertainment and events industries have embraced VR for a variety of purposes:

- **Concerts and festivals**: VR can be used to live stream concerts and festivals, providing a more immersive experience for fans who cannot attend in person.

- **Film and television**: VR technology is increasingly being used in film and television production, both for visual effects and as a storytelling medium.

- **Virtual events**: Companies and organizations can host virtual conferences, trade shows, and other events using VR, reducing costs and increasing accessibility for attendees.

To summarize, VR technology has made significant strides in recent years, leading to a multitude of enterprise applications across various sectors. From training and education to design and prototyping, marketing and sales, remote collaboration, healthcare, real estate, tourism, automotive, retail, and entertainment, VR is revolutionizing the way businesses operate, resulting in increased efficiency, reduced costs, and improved customer experiences. As VR technology continues to evolve, we can expect to see even more innovative applications in the years to come.

Many of those innovations will go towards how VR will be used in the Metaverse, which we will discuss in our next section.

VR and the Metaverse

In the previous chapter, we addressed how AR will be used in the Metaverse, and we made it clear that there was a solid place there for AR. The reason why this point had to be made is because the Metaverse initially was thought of as a VR-only domain. This comes directly from Neal Stephenson's 1992 science fiction novel, *Snow Crash*, where people use VR to access the Metaverse, a term that Stephenson coined himself. So, when we talk about the Metaverse started bubbling up after the Oculus DK1 headset came about, it was assumed that VR was the unique way to access the Metaverse.

What is it about the VR that still clearly anchors it in the Metaverse? The idea is that the Metaverse is an alternate 3D reality where you can immerse yourself, taking you away from the real world. This alternate reality could be a completely fantastical one where the rules of physics do not apply, where there are creatures that have been created by humans, and where there are lands, skies, and suns that have no resemblance at all to the real world.

It could also be an alternate reality that mimics the real world, one where friends and family located in different parts of the world could virtually meet, share 2D and 3D photos and videos, play multiplayer games, spontaneously create visuals and games together using generative AI, and play in them. They could experience a virtual roller coaster together, talk and walk together in a park, go into a store to virtually try on clothes together, and talk about the kinds of clothes they like. You would be able to navigate the Metaverse using your voice to take you to whichever virtual store or location you would want to go to.

You would potentially be able to go to a virtual car dealership with a friend to buy a real car or any other kind of business with a product. This is the merging of the social and retail capabilities that the Metaverse can bring.

In terms of businesses that have products to sell, the Metaverse is a kind of nirvana. We had talked about how AR could achieve this, and together with VR, the Metaverse will give rise to new business models and economic opportunities. Virtual real estate, digital goods, and services can be bought, sold, and traded, creating a thriving virtual economy. This environment also enables creators and developers to monetize their work through various means.

People will also be meeting in the Metaverse to virtually work and learn there, taking the capabilities of a Zoom call much, much farther.

VR in the Metaverse can be used in several more ways, transforming many industries and aspects of everyday life. Here is a summary of some of the key areas to watch for in the VR landscape:

- **Immersive experiences**: The development of more immersive, realistic, and high-quality experiences will be critical for VR's growth. This will likely include advancements in haptic feedback, motion tracking, and 3D audio, which will all contribute to creating an increasingly lifelike virtual world.

- **Gaming**: The gaming industry will continue to be a major driving force for VR, with developers producing more advanced and sophisticated games that push the boundaries of what is possible within the medium. As VR gaming becomes more mainstream, we can expect to see a larger variety of genres and experiences available to players.

- **Social VR**: As VR technology becomes more accessible, the potential for shared experiences and social interactions in virtual environments will increase. This could lead to the creation of virtual meeting spaces, conferences, and entertainment venues, fostering a sense of presence and connection between users.

- **Education and training**: VR has the potential to revolutionize education and workforce training by offering immersive learning experiences that are more engaging and effective than traditional methods. As VR technology improves, we can expect to see wider adoption of VR-based educational tools and resources in schools, universities, and professional training programs.

- **Healthcare**: VR has already made significant inroads in the healthcare industry, and this trend is likely to continue. Applications for VR in healthcare include medical training and simulations, pain management, mental health treatment, and rehabilitation.

- **Architecture and design**: VR technology can provide architects and designers with a powerful tool for visualizing and interacting with their creations in a virtual environment. This can streamline the design processes, improve collaboration, and help clients better understand proposed projects.

- **Tourism and exploration**: VR can offer users the opportunity to explore far-off destinations or historical sites without ever leaving their homes. As VR experiences become more realistic, this form of virtual tourism is likely to become increasingly popular.

- **Film and entertainment**: The film and entertainment industries will continue to experiment with and adopt VR technology, offering new and innovative ways for audiences to experience storytelling. This could range from virtual movie theaters to immersive narrative experiences that place the viewer at the center of the action.

- **Enterprise applications**: VR has the potential to change how businesses operate, with applications ranging from remote collaboration and virtual conferences to product design and employee training.

- **Accessibility and affordability**: As VR technology becomes more affordable and accessible, it will open up new opportunities for people who may have been previously excluded from certain experiences due to physical, financial, or geographical limitations.

- **Ethics and regulations**: As VR continues to permeate society, ethical considerations and regulations will become increasingly important. This will involve addressing issues such as privacy, data security, and the potential for addiction, as well as establishing standards and guidelines to protect users and ensure responsible development.

Particular use cases that illustrate VR's place in the Metaverse in more detail can be found in *Part 3, Consumer and Enterprise Use Cases.*

Summing this up, the future of VR in the Metaverse is incredibly promising, with the potential to transform industries, enrich daily life, and create new opportunities for connection, exploration, and learning. As technology continues to advance and VR becomes more widely adopted, we can expect to see a wealth of new experiences and applications that further blur the lines between the virtual and the real.

Summary

This chapter threw light on the history of VR and how gaming spurred its development. We showed you which VR/MR headsets were notable, what the general features and desirable specifications are, and what some of the cautionary things to be aware of are. VR and the Metaverse were then discussed.

VR and the Metaverse are two of the most significant technological advancements of our time. VR allows us to experience and interact with digital worlds in a more immersive way than ever before, while the Metaverse promises to create a shared virtual space where people can work, play, learn, and socialize together.

The importance of VR lies in its ability to simulate real-world experiences, explore new environments, and interact with virtual objects in a more engaging and interactive way.

The Metaverse, on the other hand, has the potential to change the way we live, work, and play by creating a virtual space where people from all over the world can come together and interact. It could revolutionize the way we think about socializing, learning, and working, opening up new opportunities for collaboration and creativity.

Overall, VR and the Metaverse represent a significant shift in the way we interact with technology and each other.

In our next chapter, *Chapter 4, The Value of Using 3D Visuals to Interact*, we explain how using 3D visuals to interact will greatly enhance our ways of doing things, in addition to introducing new ways of doing them.

4

The Value of Using 3D Visuals to Interact

Augmented reality (AR) and **virtual reality (VR)** technologies used in the Metaverse enable significant new ways of seeing and doing. Existing and professionally made 3D images and video, as well as created ones that are now easily made by consumers with **generative AI (genAI)**, will be available to you for display and use in the Metaverse. All the 3D visuals that can be found there will provide you with a gamut of information and entertainment, and they can also be searched using AI for objects appearing there, as well as for particular language used in videos. Interactivity made possible by apps and digital assistants will greatly enhance the way you do things, in addition to introducing new ways of doing them.

In this chapter, we're going to cover the following main topics:

- What is meant by 3D **synthetic reality (SR)** and how relevant genAI is

- How the Metaverse is bringing about a massive amount of easily available 3D visual information and why it is so valuable

- Why machine-to-human interaction greatly increases with the Metaverse and how it benefits human beings, businesses, and economies

3D SR

3D SR refers to a computer-generated, three-dimensional environment that replicates aspects of the real world or creates entirely new ones. It employs VR or AR technologies that aim to create immersive, interactive experiences for users. The primary objective of 3D SR is to produce a lifelike and captivating experience that blends virtual and physical realms.

Machine-to-human interaction greatly increases with the use of 3D technology to create and use 3D SR in the Metaverse for several reasons. The Metaverse, as an immersive, interconnected digital universe, relies on 3D technology to create realistic and engaging experiences for users. These experiences foster more meaningful and intuitive machine-to-human interactions, as detailed next:

- **Immersive environments**: 3D technology is essential for creating lifelike virtual worlds within the Metaverse. The spatial depth and realism that 3D provides enhance users' sense of presence and make it easier for them to engage with the virtual environment, leading to increased machine-to-human interactions.

- **Realistic avatars**: 3D technology enables the creation of realistic avatars that users can control and customize. These avatars serve as digital representations of users, allowing them to interact with others and the environment in a more natural and intuitive way.

- **Spatial computing**: 3D technology facilitates spatial computing, which allows users to interact with virtual objects and environments in a way that mimics real-world interactions. This makes machine-to-human interactions more seamless and user-friendly as it feels closer to how humans naturally interact with their surroundings.

- **Enhanced simulations**: The use of 3D technology in the Metaverse enables more accurate and engaging simulations, which can be used for training, education, and entertainment purposes. These simulations offer users more opportunities to interact with machines and learn from them.

- **Haptic feedback and gesture recognition**: 3D technology in the Metaverse can be combined with haptic feedback and gesture recognition systems to make machine-to-human interactions more tangible and realistic. This allows users to feel the sensation of touch and use gestures to control their virtual environment, making interactions feel more natural and intuitive.

- **Multi-sensory experiences**: The integration of 3D technology with other sensory inputs, such as audio and smell, can create multi-sensory experiences within the Metaverse. These experiences further enhance machine-to-human interactions by making them more immersive and engaging.

- **AI-powered interactions**: 3D technology can be combined with AI to create smarter, more personalized interactions between machines and users. This can include AI-powered avatars, virtual assistants, or adaptive learning environments, providing users with more opportunities to interact with machines in meaningful ways.

3D technology is a crucial component of the Metaverse as it enables the creation and use of 3D SRs. With Apple having announced the Vision Pro MR headset, the high probability of massive buy-in from the public for Apple MR headsets and glasses almost assures an intense need for the consumption of 3D visuals. Advancements in computer graphics, AI, and real-time rendering have contributed to the development of increasingly realistic and complex 3D SRs. These environments often incorporate elements such as physics, lighting, and sound to create a more convincing experience for the user. 3D SR gets even better with the use of a particular kind of AI, genAI, to create unique, dynamic worlds or environments on the fly.

Here are some key aspects of 3D SR:

- **Immersion**: 3D SR aims to create an immersive experience for users by generating a realistic sense of depth, scale, and spatial awareness. This allows users to feel as though they are truly present within the virtual environment.

- **Interactivity**: Users can interact with objects and elements within the 3D SR environment, making it more engaging and dynamic than traditional 2D visual experiences.

- **Realism**: Advanced rendering techniques and graphical capabilities are used to create realistic textures, lighting, and shadows, enhancing the visual quality of the 3D SR environment.

- **Real-time rendering**: 3D SR often relies on real-time rendering, which means that the virtual environment is continuously updated and rendered as the user interacts with it. This creates a fluid and responsive experience.

GenAI brings the potential for a radically different way of computing that is something closer to *Star Trek*'s Holodeck than Microsoft Windows. Such a Holodeck will be fed by potentially dozens of **genAIs**. What are genAIs? They are AIs that use **unsupervised learning** (UL) algorithms with human inputting of text prompts to create new virtual photos, videos, text, code, or audio.

UL can identify previously hidden or unclear patterns in data. As an example of this, genAIs can currently create 2D and 3D still images and transform an existing 2D video into a new 3D one. Runway Gen-2, genAI software, can currently generate short-form 2D video using text, images, or video clips. Soon, genAI will be able to create long-form consumer-ready 2D and 3D video using the same media. And there is work currently being done to make genAI video capable of streaming.

If you tear apart how you would build a Holodeck, it really can be broken down into component parts that genAIs could serve. You have a 3D environment around you that can be instantly changed. "Hey, OpenAI, take us to Tokyo," you might tell the Holodeck of the future, and the system will instantly take you there.

Just such a request might invoke dozens of different genAIs – one generating the scene, another generating your clothes, another generating the script of the story you are about to experience, and yet another generating all the furniture and interactive objects. Then, you think about what to do there.

"Hey, OpenAI, can we join a whodunit mystery in Tokyo?" Now, you need ways to get your friends into the Holodeck and meet up with something akin to a new kind of play – one with generated costumes and clothes, furniture, props, music, non-human characters and robots, and more.

Break this all down, and you will see each being worked on by the industry that's building the genAI systems that are all the rage now.

Consumers of the future who have such a Holodeck will be different from consumers today, who mostly are watching TV and movies or playing games in their homes.

This new consumer won't be satisfied with merely lying back and watching a movie or a TV show, but instead will be one who expects to be "inside" the movie and, even, able to create or interact with key pieces of it. We already are seeing the beginnings of this new "creator consumer" and the AI culture that it is bringing.

Just look at the AI art that is being shared on social media, and you'll see a community of creative people who are, today, creating videos and photos with genAI systems such as Midjourney, DALL-E, or Stable Diffusion.

The possibilities of what we can do with a virtual environment such as this are endless. Imagine being able to step into your favorite movie or TV show and interact with the characters and the world around you, or having a virtual amusement park where you can ride the latest and greatest roller coasters without ever leaving your living room (after you design your own, of course, which you are riding on). Or being able to attend concerts or live events in a virtual environment, where you can be front row and center without ever having to worry about ticket prices or crowds.

To better grasp the concept of genAI and what it could do for 3D SR used in the Metaverse, it is essential to first comprehend the nature of genAI in more detail.

GenAI is taught to interpret and process natural content in machine language by employing **machine learning** (**ML**) models. The training of AI programs, chatbots, or virtual assistants in genAI involves the utilization of several ML models. A description of some of these models and their outputs follows.

GenAI versus discriminative models

To train AI using the discriminative model, a human supervisor is involved to guide the AI in distinguishing between different objects in a given input sample. For example, when the AI is presented with 10 pictures of 10 different animals, the discriminative model helps it correctly differentiate between all the animals.

On the other hand, the generative model allows the AI to create objects using sample data with minimal or no supervision. A generative ML model aids the AI in understanding the input data and retains this knowledge in its **neural network** (**NN**) memory, enabling it to recall the experience if faced with a similar task in the future.

Generative adversarial networks (GANs)

This AI training method combines generative and discriminative models, where the generative model produces samples based on input vectors such as keywords and inquiries, and then the discriminative model verifies the authenticity of the generated sample. If the sample is found to be fabricated, the generative model generates a new output for evaluation by the discriminative model. This process is repeated iteratively until the generative model can create convincing samples that cannot be distinguished from the original input by the discriminative model.

Transformer-based models

AI models that utilize the transformer architecture employ **deep NNs (DNNs)** to analyze input vectors and generate possible outputs, including predicting words that can construct meaningful sentences before or after the input, even if they are unrelated.

The transformer architecture is made up of an encoder that captures all the input sequence's features and transforms them into input vectors. The decoder then scrutinizes these input vectors, extracting context from the data to produce an output sequence.

Numerous AI models that incorporate the transformer architecture have exhibited exceptional performance.

ChatGPT, also known as a **Generative Pre-trained Transformer model 3**, is a language model designed for dialogue applications. Similarly, **LaMDA** is another language model built on Google Transformer. With the help of these models and other advanced technologies, developers have successfully created various functional genAI programs that can produce impressive outputs from simple inputs such as texts, images, audio, and so on.

For instance, these AI programs can generate images of non-existent human beings by referencing inputs from magazines, websites, Google image search, and so on. They can also convert sketches into real images, transfer artistic styles from one art to another, and even synthesize a CT scan from an MRI as input.

AI programs such as OpenAI's DALL-E 3, which is integrated into ChatGPT, can create remarkable images that display accurate lettering from simple texts, while DeepMind and Amazon Polly can generate human-like speeches from texts. Another AI technology that was created by AI Music, now owned by Apple, can transform public domain music into soundtracks that could be used for 3D SR experiences created for the Metaverse.

GenAI changes the game because it can, on the fly, create literally infinite stories, visuals, voices, and music. A new "War of the Worlds" will soon be possible where genAI shows you something like an alien attack on your neighborhood. That just isn't possible using current Hollywood technology. So, it's easy to see that having genAI able to spontaneously create 3D images and video in the future would be spectacular.

NeRF

Neural Radiance Fields (**NeRF**) is a groundbreaking technique introduced in 2020 by researchers from UC Berkeley and Google Research. It involves using **deep learning** (**DL**), which is a form of AI, to synthesize photorealistic, novel views of 3D scenes from a set of input 2D images. NeRF combines principles from both **computer vision** (**CV**) and computer graphics to generate high-quality, realistic renderings of 3D scenes. NeRF is already being used to create realistic 3D environments for VR and AR immersive experiences and is looking to be a very relevant technology for the Metaverse.

Here are some relevant technical features about NeRF generation:

- **Scene representation**: NeRF represents a 3D scene as a continuous volumetric function where each point in the scene (x, y, z) is associated with a color (RGB) and density (sigma) value. This continuous function is called the radiance field, and it encodes the color and opacity of the scene at every point in 3D space. Unlike traditional 3D representations, such as meshes or point clouds, NeRF's continuous representation allows for more accurate modeling of complex geometry, intricate textures, and realistic lighting effects.

- **Network architecture**: NeRF utilizes a fully connected NN (also known as a **multilayer perceptron** or **MLP**) to learn the continuous volumetric function that represents the scene. The network takes the 3D coordinates of a point in the scene and a viewing direction as input, and it outputs the color and density values for that point. The network is designed to be translationally invariant, which means that it can model complex scenes without overfitting to specific viewpoints.

- **Training**: To train the NeRF network, a dataset of 2D images captured from different viewpoints around the scene is required. During training, the goal is to optimize the network parameters such that the rendered images closely match the input images. This optimization is achieved through a process called **backpropagation**, which minimizes the differences between the rendered images and the input images using a loss function. The loss function typically used in NeRF is a combination of the color difference and the density difference between the rendered and input images. The optimization process is iterative, and it continues until the network converges to a solution that best models the given scene:

Figure 4.1 – Different aspects of a NeRF made at Casa de Fruta in Hollister, California
Source: Robert Scoble

- **Rendering**: Once the NeRF network is trained, it can be used to render novel views of the scene. Rendering with NeRF involves casting rays from a virtual camera into the scene and sampling points along these rays. For each sampled point, the network predicts the color and density values, which are then combined using volume rendering techniques to produce the final image. NeRF's rendering process can handle complex lighting effects, such as shadows, reflections, and refractions, which contributes to the high quality and photorealism of the generated images.

NeRF has demonstrated its profound impact on the field of 3D rendering through its innovative technical features. By leveraging NNs, sparse input data, and efficient volumetric representation, NeRF has revolutionized the way we generate and interact with intricate digital scenes.

The greatness of NeRF

NeRF has garnered positive attention in the world of computer graphics and AI due to its remarkable ability to recreate intricate 3D scenes with stunning detail. As a cutting-edge DL technique, NeRF offers a unique blend of realism and environments that has quickly attracted both users and investors.

Here are some of the major positive features of NeRF:

- **High-quality 3D rendering**: NeRF stands out for its ability to generate photorealistic 3D scenes using only a sparse set of input images. By leveraging DL, NeRF can capture complex geometry, materials, and lighting conditions, producing results that rival traditional computer graphics techniques. The resulting scenes are so detailed that they often defy human perception, making it difficult to distinguish between real and generated content.

- **Data efficiency**: NeRF's brilliance lies in its ability to generate 3D models from a minimal number of 2D images. Traditional methods require extensive data inputs or time-consuming manual modeling. In contrast, NeRF can create high-quality 3D representations using just a handful of photos, making it an efficient and practical solution for various applications, from gaming to VR.

- **Novel view synthesis**: One of the most impressive aspects of NeRF is its capacity for novel view synthesis. This allows the model to generate new images of a scene from viewpoints that were not included in the initial input set. This capability has significant implications for industries such as entertainment, where it can be used to create immersive experiences in VR and AR, as well as in professional domains such as architecture and product design.

- **Scalability and adaptability**: NeRF's architecture is both scalable and adaptable, making it suitable for a wide range of applications. Researchers have already begun to explore extensions of the original NeRF model to address specific challenges, such as handling dynamic scenes, improving rendering speed, and refining details. This adaptability ensures that NeRF will continue to evolve and find new applications across diverse domains.

As we witness rapid advancements in technology, it is essential to recognize and embrace the transformative potential of NeRF. Its robustness, scalability, and ability to integrate with other technologies are just a few of the many positive features that make NeRF a game changer in the realm of 3D reconstruction and rendering. With ongoing research and development, we can expect NeRF to further define and revolutionize the way we interact and perceive the digital world.

Current limitations of NeRF

Despite its many advantages, NeRF also has some limitations that are currently being addressed by companies and academic researchers (another technology called 3D Gaussian Splatting, detailed in *Chapter 4, The Value of Using 3D Visuals to Interact*, may overcome NeRF's limitations, but it is still emerging):

- **Computational efficiency**: The training and rendering processes in NeRF can be computationally expensive, particularly for high-resolution images and large scenes. Researchers are actively working on improving the efficiency of NeRF through techniques such as hierarchical sampling, adaptive sampling, and leveraging hardware acceleration.

- **Dynamic scenes**: NeRF is primarily designed for static scenes, and it struggles to handle scenes with moving objects or changing lighting conditions. Ongoing research aims to extend NeRF to handle dynamic scenes and incorporate temporal consistency in the generated views.

- **Generalization**: While NeRF can model a specific scene quite well, it does not inherently generalize to other scenes or objects. There is ongoing research on developing techniques that can transfer the knowledge learned from one NeRF model to another, enabling the creation of more general 3D scene understanding systems.

- **Integration with traditional 3D representations**: NeRF's continuous volumetric representation is distinct from traditional 3D representations such as meshes, point clouds, or voxel grids. Integrating NeRF with these representations remains an active area of research, with the goal of combining the strengths of both approaches for even more accurate and efficient 3D modeling and rendering.

In short, NeRF is a powerful technique that leverages DL to generate photorealistic novel views of 3D scenes from a set of 2D images. Its continuous volumetric representation, robust network architecture, and efficient training and rendering processes make it particularly well-suited for handling complex scenes with intricate geometry, textures, and lighting effects.

3D Gaussian Splatting – an emerging alternative

In the ever-evolving landscape of 3D modeling and rendering, an innovative and promising alternative emerges – 3D Gaussian Splatting. This groundbreaking approach introduces a range of substantial improvements and pioneering innovations, providing a compelling solution to the limitations experienced with NeRF. Departing significantly from NeRF's established reliance on continuous volumetric representations, 3D Gaussian Splatting follows a unique path, with the following advantages:

- **Enhanced computational efficiency**: The first notable advantage of 3D Gaussian Splatting is its superior computational efficiency, effectively addressing a significant challenge faced by NeRF. NeRF's training and rendering processes can be computationally intensive, especially for high-resolution images and expansive scenes. In contrast, 3D Gaussian Splatting introduces a more streamlined approach. It accomplishes this by projecting intricate 3D points onto 2D images using Gaussian functions, substantially reducing the computational load. This increased efficiency is particularly vital for real-time and resource-intensive applications.

- **Versatile handling of dynamic scenes**: NeRF is primarily designed for static scenes and may struggle to adapt to dynamic environments characterized by moving objects or evolving lighting conditions. In contrast, 3D Gaussian Splatting demonstrates remarkable proficiency in managing dynamic scenes. It achieves this by adaptively projecting 3D data onto 2D images, making it exceptionally well suited for scenarios where objects are in motion or lighting conditions change dynamically.

- **Improved generalization and adaptability**: While NeRF excels in modeling specific scenes, it often faces challenges when attempting to generalize its knowledge to other scenes or objects. On the other hand, 3D Gaussian Splatting exhibits a higher degree of generalizability. Its unique approach allows for a more versatile representation of scenes, simplifying the application of knowledge learned from one model to another. This paves the way for developing more universally applicable 3D scene understanding systems.

- **Alternative perspective on scene representation**: Another significant advantage of 3D Gaussian Splatting is its capacity to offer an alternative perspective on scene representation. It achieves this by projecting 3D data onto 2D images, providing a unique viewpoint that proves particularly advantageous in scenarios characterized by moving objects and evolving lighting conditions. This alternative perspective significantly enhances the quality and realism of the generated scenes.

In summary, 3D Gaussian Splatting represents a remarkable leap forward in the field of 3D modeling and rendering. Its numerous advantages, including enhanced computational efficiency, adept handling of dynamic scenes, improved generalization, an alternative scene representation, and seamless integration with traditional 3D representations, make it a compelling alternative to NeRF. As technology continues to advance, the coexistence of NeRF and 3D Gaussian Splatting, alongside other emerging techniques, promises to be pivotal in the ongoing evolution of the digital landscape, ensuring an innovative, dynamic, and cross-disciplinary future.

Current limitations of 3D Gaussian Splatting

Within the realm of 3D modeling and rendering, the innovative approach of 3D Gaussian Splatting has gained attention for its numerous benefits. However, it is important to acknowledge that, as with any technology, 3D Gaussian Splatting has its limitations that should be considered:

- **Limited application scope**: 3D Gaussian Splatting, while offering numerous advantages, may not be universally applicable. Its proficiency is more pronounced in specific areas, and it may not be the most optimal choice for highly complex scenes or applications where different techniques are better suited.

- **Data sensitivity**: The effectiveness of 3D Gaussian Splatting can be influenced by the quality and quantity of input data. In scenarios with sparse or noisy data, the technique may struggle to produce accurate results. Therefore, ensuring high-quality data acquisition is crucial for optimal performance.

- **Computational demands**: Despite its enhanced computational efficiency when compared to some methods, 3D Gaussian Splatting may still require significant computational resources for specific applications, particularly when dealing with large or highly detailed scenes. Efficient computation remains an important consideration.

- **Integration challenges**: Although 3D Gaussian Splatting integrates with traditional 3D representations more seamlessly than some other techniques, challenges may arise when harmonizing it with specific existing systems or workflows. Ensuring compatibility and integration may necessitate additional efforts.

- **Dynamic scene handling**: While 3D Gaussian Splatting excels at handling dynamic scenes, it is not immune to challenges in scenarios with rapid, highly complex, or extreme variations in motion and lighting. The method may not consistently deliver the desired results in such dynamic and challenging environments.

- **Quality control and realism**: Achieving photorealistic results with 3D Gaussian Splatting may require meticulous fine-tuning and calibration. Ensuring the highest level of realism may involve careful adjustment and optimization, which can be time-consuming.

- **Ongoing research and development**: It is important to note that 3D Gaussian Splatting, as with any technology, is subject to ongoing research and development. As it gains wider adoption, refinements and advancements are expected, necessitating a continuous commitment to keeping up with the latest developments in the field to ensure optimal results.

In conclusion, while 3D Gaussian Splatting offers significant benefits in the field of 3D modeling and rendering, understanding its limitations and addressing them through further research and development is a crucial step in maximizing the potential of this innovative technique.

With regard to business, the potential productivity gains and cost savings due to the use of genAI, including NeRF and 3D Gaussian Splatting technologies, are very great. Luma AI is a company that has software that could generate both NeRF and 3D Gaussian Splatting, and we expect that there will be many more that will do so. A reduction in time needed to create and/or augment is the largest benefit. This reduction in time spent allows more work to be focused on other areas that are needed. And, magnified to several, several tens, hundreds, or thousands of employees depending on the company (not to mention outsourced work that could be delivered so much quicker), this can represent an incredible increase in productivity that previously would have seemed like fiction.

The benefits of 3D visual information

3D SR, with or without the creative use of genAI, has the potential to greatly benefit both consumers and businesses due to the visual information that is provided by it.

The advent of 3D visual information and the Metaverse has heralded a veritable paradigm shift in the digital landscape, revolutionizing the way we interact with one another and the environment around us. From immersive VR experiences to seamless online interactions, this technological innovation has transformed the domain of human communication, fostering the creation of an interconnected ecosystem with far-reaching implications.

The marriage of 3D visual information and the Metaverse has far-reaching benefits, spanning across various sectors and industries, and has the potential to redefine our understanding of communication, collaboration, and connectivity. By breaching the barriers of physical limitations, we are now able to dive into a world of endless possibilities, where the only constraint on our experience is the limit of our imagination.

Consider the potential impact of 3D visual information on the education sector. Gone are the days when students were confined to the drabness of 2D textbooks; today, they can explore historical sites, delve into the intricacies of biological systems, and unravel the mysteries of the cosmos, all in vivid 3D. Envision students walking through the ancient streets of Rome, witnessing firsthand the architectural grandeur of the Colosseum, or participating in an interactive lesson on the intricacies of the human body. Not only does this enhance their educational experience by bringing it to life, but it also fosters an environment of boundless curiosity and creativity.

Moreover, the Metaverse has provided a fertile ground for the evolution of the arts and entertainment industry. With the lines between the physical and digital worlds increasingly blurred, artists, musicians, and filmmakers have found themselves with an array of new tools to create immersive experiences that transcend the limitations of traditional media. Picture the magnificent spectacle of a virtual art gallery where visitors can stroll through halls adorned with masterpieces, pausing to scrutinize the delicate brushstrokes of a Van Gogh painting or the luminescent grace of a Monet landscape. Similarly, musicians can perform live concerts to fans scattered across the globe, all gathered in a virtual space that echoes the vibrant melodies of their art.

Furthermore, the Metaverse has facilitated the emergence of new business models and economic opportunities. In this digital expanse, brands can create experiential showrooms, allowing consumers to virtually explore products and services before making a purchase. Imagine stepping into a virtual automobile showroom and taking a test drive in the latest electric vehicle, all from the comfort of your living room. Such immersive experiences can prove instrumental in fostering customer engagement and driving brand loyalty.

The Metaverse has the potential to bring people together, transcending geographical boundaries and fostering a sense of global community. The social aspect of this virtual realm offers opportunities for individuals to connect, collaborate, and share experiences in ways that were previously unimaginable. Visualize a virtual conference where attendees from every corner of the Earth can gather in a shared space, exchanging ideas and insights and forging connections that may have been otherwise unattainable.

3D visual information offers numerous benefits over traditional 2D representations, especially in fields such as education, entertainment, design, and scientific research.

Here are some of the general key advantages:

- **Enhanced spatial understanding**: 3D visualization provides a more accurate representation of objects and environments, enabling users to gain a deeper understanding of spatial relationships, dimensions, and proportions. This can be particularly useful in fields such as architecture, engineering, and medicine, where accurate visualization of complex structures is crucial.

- **Improved engagement and immersion**: 3D visuals can create more engaging and immersive experiences, making them ideal for entertainment, such as movies, video games, and VR and AR.

- **Improved decision-making**: In industries such as urban planning, construction, and product design, 3D visualizations can help stakeholders make more informed decisions by providing a better understanding of the project's or product's appearance and functionality. This can lead to better designs and reduced costs due to fewer errors and changes during the development process.

- **Enhanced communication and collaboration**: 3D visualizations can effectively convey complex information to a diverse audience, making it easier to share ideas and collaborate. This can be particularly beneficial for remote teams, where 3D models and simulations can provide a shared understanding of a project or product, streamlining communication and reducing misunderstandings.

- **Customization and personalization**: 3D visualization enables users to customize and personalize objects and environments according to their preferences, leading to unique and tailored experiences. This can be especially valuable in industries such as fashion, interior design, and automotive, where clients often seek products that reflect their individual tastes and preferences.

- **Interactive learning**: In educational settings, 3D visual information can make learning more engaging and interactive. By simulating real-world environments and scenarios, learners can develop a better understanding of concepts and improve their problem-solving skills. This is particularly relevant in fields such as biology, chemistry, and physics, where 3D models can help students visualize complex structures and processes.

- **Improved accuracy in scientific research**: 3D models can help researchers visualize complex structures, such as molecules and proteins, leading to a better understanding of their properties and interactions.

- **Medical applications**: 3D visual information is invaluable in medical imaging, enabling physicians to examine internal structures and diagnose conditions more accurately.

- **Real estate and urban planning**: 3D visualization can help stakeholders envision proposed developments or changes to existing structures, allowing for better decision-making and public engagement.

- **Remote exploration and inspection**: 3D imaging techniques, such as LiDAR and photogrammetry, enable remote exploration and inspection of hard-to-reach areas, such as archaeological sites, hazardous environments, or large-scale infrastructure.

- **Accessibility**: 3D visual information can make information more accessible to people with disabilities or different learning styles. For instance, 3D models and simulations can provide tactile and auditory cues, enabling users who are visually impaired to better understand and interact with the content.

- **Preservation and restoration**: In cultural heritage and archaeology, 3D visualization can help preserve and restore historical sites, artifacts, and monuments. By creating accurate 3D models, researchers and conservationists can document, analyze, and share valuable information, ensuring the preservation of our shared cultural heritage.

3D visual information offers a wide range of benefits across various industries and applications. From improved understanding and decision-making to enhanced communication and interactive learning, 3D visualization is revolutionizing the way we perceive and interact with the world around us.

Business benefits

3D visual information and technology used in the Metaverse offer a wide range of benefits across various industries and applications. They can help businesses in various ways by offering new opportunities for growth, customer engagement, and innovation.

Here are some key areas where businesses can benefit:

- **Virtual storefronts and showrooms**: Businesses can create immersive 3D virtual stores or showrooms, allowing customers to browse and purchase products in an engaging environment. This can help businesses reduce overhead costs associated with physical locations and expand their customer reach globally.

- **Marketing and advertising**: The Metaverse offers a new platform for businesses to showcase their products and services through immersive 3D advertisements, product placements, or sponsored events. These can be highly engaging and offer an interactive way to connect with customers.

- **Collaboration and remote work**: 3D virtual environments can provide businesses with a platform for remote collaboration, allowing employees to work together in a shared virtual space. This can help reduce travel expenses and improve communication across teams.

- **Employee training and development**: VR and 3D environments can be used to simulate real-world scenarios for training purposes. This can help businesses provide more effective, hands-on training for employees and reduce costs associated with traditional training methods.

- **Product design and prototyping**: 3D modeling and VR tools can help businesses visualize and test product designs in a virtual environment before investing in physical prototypes. This can save time and resources in the product development process.

- **Customer support**: Businesses can leverage 3D avatars and virtual assistants to provide interactive and personalized customer support experiences, improving customer satisfaction and retention.

- **Networking and events**: Businesses can host virtual conferences, trade shows, or networking events in the Metaverse, attracting a global audience and creating new opportunities for collaboration and partnership.

- **Data visualization and analytics**: 3D data visualization tools can help businesses better understand and analyze complex datasets, leading to more informed decision-making.

- **Entertainment and experiences**: Businesses can create unique and engaging virtual experiences or games for customers, generating new revenue streams and enhancing brand loyalty.

- **Intellectual property (IP) and digital assets**: Businesses can capitalize on the growing market for virtual goods and services by creating and selling digital assets such as virtual real estate, in-game items, or **non-fungible tokens (NFTs)**.

By integrating 3D visual information and technology and the Metaverse into their operations, businesses can adapt to the changing digital landscape and unlock new opportunities for growth, innovation, and customer engagement.

Retail and customer service examples

In terms of retail, many companies have been creating 3D visualizations of their products using various techniques and technologies starting over 10 years ago.

Here are some examples:

- **Amazon** uses a mix of 3D modeling software and high-quality photography to create 3D visualizations of its products. The company has been investing heavily in this technology, with plans to use it across a range of products, including furniture, appliances, and clothing.

- **Walmart** has also been working on 3D visualizations. In 2018, the company acquired a VR start-up called Spatialand, which has been working on developing tools to create immersive shopping experiences. Walmart is continuing to actively explore ways to use VR and AR to create more engaging shopping experiences for its customers.

- **IKEA** has also been using 3D visualizations to showcase its products. The company was one of the first to employ 3D visualization in the consumer retail community, using photorealistic 3D renderings of its products on its website and in its catalog. IKEA is known for promoting its AR technology to customers. IKEA developed a 3D visualization tool called IKEA Place that allows customers to virtually place furniture items in their homes before purchasing. By using AR, the app offers a more accurate representation of how the products will look and fit in their living spaces, reducing the likelihood of returns and dissatisfaction. This innovative customer support approach has also helped IKEA build a stronger connection with its customers by streamlining the decision-making process.

- **Shopify** has incorporated 3D visualization and AR into its platform, allowing merchants to showcase their products in 3D and enable customers to visualize items in their homes. This innovative approach has not only enhanced the shopping experience but has also led to increased sales and customer satisfaction, resulting in financial gains for Shopify and its merchants.

- **Apple** has embraced 3D visualizations to improve customer support for its products. Apple's support website offers detailed 3D models of its devices, accompanied by interactive guides that walk users through common troubleshooting steps. These visual aids make it easier for customers to identify and resolve issues, reducing the need for phone or in-person support.

- **Tesla** has an online configurator that offers customers a detailed 3D model of their desired vehicle, allowing them to visualize various color and feature combinations before committing to a purchase. This interactive tool not only simplifies the car-buying process but also provides a more personalized experience, ensuring that customers feel confident in their decisions.

- **GE HealthCare**, a leading provider of medical equipment and services, has also harnessed the power of 3D visualizations to improve customer support. The company utilizes interactive 3D models of its medical devices to train technicians on maintenance and repair procedures. By offering a more engaging and intuitive learning experience, GE Healthcare ensures that its customers receive the best possible support, ultimately resulting in better patient outcomes.

These companies are using 3D visualizations to create more engaging and immersive shopping and customer service experiences for their customers. By using a mix of 3D modeling, AI, photography, and VR and AR technologies, they are able to provide customers with a more realistic representation of their products, which can help increase sales and reduce returns.

Employee training and development examples

3D visualizations are increasingly being used in employee training and development because they offer several advantages over traditional training methods. These visualizations provide a realistic and interactive environment that allows trainees to practice skills and apply knowledge in a safe and controlled environment. Additionally, they provide a level of immersion that helps trainees retain information better and have a more engaging experience.

Here are two examples:

- **Walmart** has used VR simulations to train its employees in various areas, including customer service, compliance, and safety. For example, it used a simulation called **Black Friday** to prepare its employees for the rush of customers and potential safety hazards that occur during the busiest shopping day of the year. This simulation allowed employees to practice their skills in a realistic environment, making them better prepared to handle the actual event.

- **Boeing** uses VR and AR simulations to train pilots, mechanics, and other employees. For example, it uses a simulation called **Maintenance Training Device** to teach mechanics how to repair and maintain aircraft engines. This simulation provides a realistic environment that allows mechanics to practice their skills and learn new ones in a safe and controlled environment.

Overall, 3D visualizations offer a powerful tool for employee training and development. They provide a realistic and engaging environment that allows trainees to practice skills and apply knowledge in a safe and controlled environment. With advances in technology, more and more companies are likely to adopt these visualizations to improve their training programs and enhance employee performance.

Product design and prototyping examples

3D visualizations for product design and prototyping are widely used across various industries because they offer numerous benefits that streamline and enhance the design process. These visualizations allow designers to create highly realistic representations of their products, enabling them to identify potential design flaws, optimize functionality, and improve aesthetics before physical prototypes are made.

Here are some examples:

- **Apple** is a leading company known for its innovative designs and cutting-edge technology. By employing 3D visualizations, Apple's design team can create highly detailed models of devices such as the iPhone, iPad, and MacBook. This enables them to make informed decisions about component placement, material usage, and overall form factor. Moreover, they can simulate the user experience, ensuring that their products are not only visually appealing but also user-friendly and functional.

- **IKEA** uses 3D visualizations in its product design and development process. This approach allows the company to create photorealistic images of its furniture and home accessories, which can then be used for marketing purposes and in its digital catalog. Furthermore, these visualizations enable IKEA to test various design configurations and material options, ensuring that the final products are both visually appealing and cost-effective.

- **Tesla** and **Ford** rely on 3D visualizations for designing their vehicles. Using advanced software, they can create highly accurate models of their cars, enabling them to optimize aerodynamics, structural integrity, and overall performance. These models can be used to conduct virtual crash tests and simulate different driving conditions, ensuring that their vehicles meet stringent safety standards while also being aesthetically pleasing.

3D visualizations for product design and prototyping are an essential tool in modern industry. They allow companies such as Apple, IKEA, Tesla, and Ford to create highly realistic representations of their products, enabling them to optimize designs, improve functionality, and enhance aesthetics. By leveraging these visualizations, companies can save time and resources, streamline their design processes, and ultimately create better products for their customers.

Data visualization and analytics examples

3D data visualization and analytics is the process of visualizing and analyzing data using 3D models and graphics. This approach enables users to represent complex data in a more intuitive and visually appealing way, making it easier to identify patterns, relationships, and trends that may not be apparent in traditional 2D charts and graphs.

To create 3D data visualizations, data is first processed and analyzed using advanced analytics techniques such as ML and data mining. The resulting insights are then mapped onto three-dimensional models, allowing users to visualize and interact with the data in a virtual environment.

3D data visualization and analytics can be used in a variety of fields, including engineering, architecture, healthcare, and finance, to name a few. For example, it can be used to analyze and optimize the design of complex products or buildings, to explore medical data and identify patterns in patient health, or to analyze financial data and detect anomalies or trends.

Here are some examples:

- **Airbus** uses 3D data visualization and analytics to analyze and optimize the design and manufacturing process of its planes. By creating 3D models of its aircraft components and running simulations, it can identify areas where it can reduce weight and improve fuel efficiency.

- **Autodesk** has developed a 3D data visualization tool called BIM 360 that enables architects, engineers, and construction professionals to collaborate on building projects in real time. The tool allows users to visualize 3D models of buildings and track project progress in real time, making it easier to identify and resolve issues before they become costly problems.

- **Siemens** has developed a 3D data visualization and analytics platform called **Simcenter** that is used in a variety of industries, including aerospace, automotive, and energy. The platform enables engineers and designers to simulate and optimize the performance of their products using 3D models and data analytics, helping to reduce development costs and **time-to-market** (TTM).

- **GE Aerospace** has developed a 3D data visualization and analytics tool that is used to optimize the maintenance and repair of aircraft engines. By creating 3D models of engine components and analyzing data from sensors, the tool can identify potential issues before they cause a breakdown, reducing maintenance costs and downtime.

- **NVIDIA** has developed a 3D data visualization and analytics platform called Omniverse that is designed for collaborative virtual design and engineering. The platform allows multiple users to interact with 3D models in real time, making it easier to collaborate on complex design projects and identify potential issues before they occur.

Overall, 3D data visualization and analytics is a powerful tool for exploring complex datasets and gaining new insights, and it has the potential to revolutionize the way we understand and interact with data in the future.

Helpful use cases in new ways of social interaction, virtual and on-site work, 3D and 2D content forms and creation, and retail can be found in *Part 3, Consumer and Enterprise Use Cases.*

Economic benefits

The Metaverse, which combines elements of 3D visualization, VR, and AR, has witnessed rapid growth in recent years. While it is challenging to provide exact figures for the entire Metaverse market, some segments have shown promising potential.

For instance, a study by PwC estimates that the global market for virtual goods, which is a key aspect of the Metaverse economy, could reach $67.2 billion by 2025.

The Metaverse, with 3D visualization and technologies, can contribute to the growth and development of national economies in various ways.

Here are some key areas where these technologies can have a positive impact:

- **Job creation**: As the demand for 3D- and Metaverse-related technologies increases, there will be a need for skilled professionals, such as developers, designers, content creators, and managers. This can lead to job creation and new career opportunities.

- **Boosting innovation**: The development of 3D and Metaverse technologies can foster innovation, leading to new products and services, which in turn can contribute to a country's economic growth.

- **Attracting investments**: The growth of the Metaverse and 3D technologies can attract both domestic and foreign investments, contributing to the overall economic development of a nation.

- **Expanding markets**: The Metaverse enables businesses to reach a global audience, leading to increased trade and export opportunities for local products and services.

- **Enhancing productivity**: 3D technologies and virtual environments can improve productivity by enabling more efficient remote work and collaboration, reducing costs, and streamlining workflows.

- **Education and workforce development**: The use of 3D and Metaverse technologies in education can help develop a more skilled workforce, better equipped to handle the challenges of the digital age. This can, in turn, contribute to a nation's economic competitiveness.

- **Infrastructure development**: The growth of the Metaverse will require improvements in network infrastructure and data centers to support the increased demand for data and processing power. This can lead to investments in digital infrastructure, boosting the economy.

- **Tourism and cultural promotion**: The Metaverse can be used to promote national heritage sites and cultural attractions, potentially attracting more tourists and increasing revenue from tourism-related activities.

- **Environmental sustainability**: 3D and Metaverse technologies can contribute to a more sustainable economy by reducing the need for physical transportation, lowering carbon emissions, and promoting the use of virtual goods and services that require fewer natural resources.

- **Tax revenue generation**: As businesses and individuals generate income from 3D- and Metaverse-related activities, governments can collect taxes, which can be used to fund public services and contribute to overall economic growth.

The Metaverse, with 3D visualization and technologies, can provide significant opportunities for national economies by fostering innovation, creating new jobs, attracting investments, and expanding markets. By embracing these technologies, countries can strengthen their economic competitiveness and improve the overall quality of life for their citizens.

Hyper interactivity

Hyper interactivity refers to the level of engagement and interaction between a user and a technological system. It is associated with the use of advanced technologies that allow for real-time feedback and responses, such as VR, AR, and AI.

From a technical perspective, hyper interactivity involves creating user interfaces that enable a high degree of responsiveness and personalization, with features such as real-time messaging, live streaming, and dynamic content that updates in real time. This requires sophisticated algorithms and backend systems that can process vast amounts of data and respond quickly to user input.

In essence, hyper interactivity is about creating a seamless, immersive experience that blurs the lines between the user and the technology, allowing for a more natural and intuitive interaction. It has become increasingly important in modern technology, as users expect more personalized and engaging experiences from the products and services they use.

The benefits of hyper interactivity

The Metaverse is an expansive, interconnected virtual space that brings together various digital worlds and experiences, enabling users to interact with each other in real time. As it emerges as a new frontier for human interaction and digital innovation, one of its most compelling features is the potential for hyper interactivity with 3D visuals. This heightened level of engagement has the power to revolutionize the way we experience and navigate virtual environments, ushering in a new era of immersive experiences.

Here are some key benefits of having hyper interactivity with 3D visuals in the Metaverse:

- **Enhanced user experience**: The combination of hyper interactivity and 3D visuals allows for a more immersive and engaging user experience. Users can seamlessly navigate and explore virtual environments, interact with digital objects, and participate in virtual events. This level of interactivity creates a sense of presence and agency that is crucial for a successful Metaverse experience.

- **Improved collaboration and communication**: In a 3D environment, users can communicate more effectively using gestures, body language, and facial expressions that are not possible in traditional 2D platforms. This enhances collaboration and social interaction, making it easier for users to work together on projects, share ideas, and build relationships.

- **Real-time responsiveness**: The Metaverse thrives on real-time interaction and responsiveness. Hyper interactivity enables 3D visuals to react to user input instantaneously, creating a fluid and seamless experience.

- **Greater accessibility**: Incorporating hyper interactivity with 3D visuals can make the Metaverse more accessible to people with different needs and abilities. For example, users with limited mobility can interact with the environment using assistive devices or customized avatars, while those with visual or auditory impairments can benefit from tailored sensory experiences.

- **Increased creativity and innovation**: The combination of hyper interactivity and 3D visuals fosters a more dynamic and creative environment. Users can experiment with new ideas, create unique content, and explore novel ways of solving problems. This can lead to innovative applications across various industries, including entertainment, education, healthcare, and more.

- **Personalization**: Hyper interactivity allows users to customize their avatars, environments, and experiences in the Metaverse. This personalization increases the value of using 3D visuals, as users can create unique, tailored experiences that reflect their preferences and needs.

- **Education and training**: The use of 3D visuals and hyper interactivity in the Metaverse can lead to innovative educational and training opportunities. Users can practice and learn new skills in a safe, interactive environment, accelerating their learning and enhancing retention.

- **Economic growth and new opportunities**: As the Metaverse evolves, the demand for hyper-interactive 3D experiences will create new opportunities for developers, artists, and entrepreneurs. This can lead to the growth of new industries and markets, driving economic development and job creation.

Tying hyper interactivity to using 3D visuals in the Metaverse leads to a more immersive, engaging, and accessible digital world. By fostering creativity, innovation, and collaboration, this combination has the potential to unlock new opportunities and drive economic growth in the Metaverse ecosystem.

10 examples of beneficial hyper interactivity

3D visualization and technologies, along with the development of the Metaverse, are greatly increasing hyper interactivity by creating immersive and interconnected experiences that connect people in unprecedented ways.

Here are 10 distinct examples that showcase this phenomenon:

- **The gaming industry** has embraced the power of 3D visualization and VR, leading to groundbreaking titles such as *Half-Life: Alyx*. In this game, players can interact with objects and the environment in realistic ways, creating a more immersive and engaging experience than traditional 2D gaming.

- **The real estate industry** is leveraging 3D technologies to provide virtual tours of properties. Prospective buyers can explore homes or commercial spaces from anywhere in the world, experiencing the properties as if they were there in person, leading to more informed decision-making and increased interactivity.

- **The film and entertainment industry** is also taking advantage of 3D visualization to create immersive experiences for audiences. For example, in VR cinemas, viewers can be transported into the movie's world, feeling as if they are part of the action and interacting with the environment and characters.

- **In the field of architecture and urban planning**, 3D visualization tools such as Autodesk's Revit and SketchUp are used to create detailed virtual models of buildings and cityscapes. This enables architects and planners to collaborate more effectively and make more informed decisions, as they can virtually walk through and interact with their designs before they are built.

- **Museums and art galleries** are utilizing 3D visualization, VR, and AR to offer virtual tours, enabling people from around the world to explore and interact with exhibits and artworks without physically visiting the location. This has expanded access to culture and education while providing a highly interactive experience for visitors.

- **In the sports and fitness realm**, 3D visualization and the Metaverse are enabling new forms of interactivity through platforms such as Peloton and Supernatural. Users can engage in immersive workouts, receive real-time feedback, and connect with others in a virtual environment, fostering a sense of community and motivation.

- **The healthcare sector** is also benefiting from increased interactivity through 3D visualization technologies. For instance, doctors can use tools such as Microsoft HoloLens to overlay digital information on a patient's body during surgery, providing valuable guidance and improving the precision of their procedures.

- **In the world of fashion and design**, 3D visualization and technologies such as AR are used to create virtual fashion shows. Designers can showcase their collections in immersive environments, allowing viewers to interact with the clothing and see how it moves and fits on virtual models.

- **The automotive industry** is utilizing 3D visualization to create virtual showrooms, enabling customers to explore and customize vehicles in a fully interactive digital environment. This allows users to make more informed decisions and provides a more engaging experience than traditional sales methods.

- **The travel and tourism industry** is incorporating 3D visualization and VR to offer virtual tours of popular destinations, providing a way for users to explore new places and interact with the environment before they commit to a trip. This not only increases interactivity but also promotes more sustainable tourism practices by reducing the environmental impact of travel.

These 10 examples illustrate how 3D visualization and technologies, along with the development of the Metaverse, are transforming various industries and aspects of our lives by increasing hyper interactivity and fostering more immersive and engaging experiences.

Potential drawbacks of hyper interactivity

Using 3D visuals to interact can offer an immersive and engaging experience, but there are also some potential drawbacks to consider:

- **Hardware and software requirements**: 3D visuals often require powerful hardware and up-to-date software to run smoothly. Users with lower-end devices or outdated software may experience performance issues or be unable to access the content altogether.

- **Complexity and learning curve**: 3D environments can be more complex and harder to navigate compared to 2D interfaces. Users may need time to adapt and learn how to interact effectively in these spaces, which could lead to initial frustration or disorientation.

- **Motion sickness**: Some users may experience motion sickness, discomfort, or disorientation when interacting with 3D visuals, especially in VR or AR environments. This can limit the accessibility of 3D experiences for certain individuals.

- **Development and production costs**: Creating high-quality 3D visuals can be more time-consuming and expensive compared to 2D assets. This might be a barrier for smaller developers or companies with limited budgets, but as mentioned previously in this chapter, genAI technologies could very well remove this barrier.

- **Accessibility**: 3D environments can be less accessible for people with disabilities, such as visual impairments or motor control issues. Designers must be mindful of accessibility concerns to ensure a broad range of users can engage with the content.

- **Distraction and information overload**: In some cases, 3D visuals can be overly stimulating or distracting, making it harder for users to focus on the task at hand. The added complexity may hinder the user experience if not carefully managed.

- **Social isolation**: While 3D visuals can enable new forms of communication and interaction, excessive use may contribute to social isolation, as users may prioritize virtual environments over face-to-face interactions.

It's essential to weigh these potential drawbacks against the benefits of using 3D visuals to interact, considering the specific context and target audience.

Summary

We started this chapter discussing the aspects of 3D SR, going into more detail on genAI, including how it differs from discriminative models, what GANs, transformer-based models, NeRF, and 3D Gaussian Splatting are, and some current limitations of NeRF and 3D Gaussian Splatting. Next, we reviewed what the benefits of 3D visual information are, including business benefits, some particular examples, and economic benefits. Finally, we addressed how hyper interactivity, 3D visualization, and the Metaverse are intrinsically connected, along with 10 examples displaying this connection and the potential drawbacks of it, as well.

In conclusion, this chapter aimed to provide a comprehensive understanding of 3D SR, its underlying technologies, and the potential benefits and challenges it brings to the world. By exploring the intersection of hyper interactivity, 3D visualization, and the Metaverse, we hope we have shed light on the future possibilities and the impact these advancements may have on various industries and our everyday lives.

In *Chapter 5, Understanding Perception Technologies*, we cover AI, CV, and tracking and capture technologies that allow users to perceive and interact with the virtual or augmented environment.

Part 2:
Key Technologies That Power the Metaverse

In *Part 2*, we explore perception technologies, the cornerstones of AR, VR, and the evolving Metaverse. These technologies encompass AI and computer vision, driving intelligent interactions and object recognition.

The Metaverse's growth demands substantial computational power, including the cloud, the edge, distributed, and decentralized computing, which are all covered in this part.

Additionally, reconciling older and incompatible software with newer software within the Metaverse, as well as newer software interfacing, necessitates **application programming interfaces** (**APIs**). Software tools, including 3D modeling and AI-driven content generation, enrich the Metaverse's 3D environment. We will explore the evolution of avatar creation, from basic to realistic representations.

Lastly, we will analyze the distinctions in **User Experience** (**UX**) and **User Interface** (**UI**) design between AR and VR and their relevance within the Metaverse.

This part has the following chapters:

- *Chapter 5, Understanding Perception Technologies*
- *Chapter 6, The Different Types of Computing Technologies*
- *Chapter 7, Where Are APIs Needed*
- *Chapter 8, Making and Using 3D Models and Integrating 2D Content*
- *Chapter 9, Understanding User Experience Design and User Interface*

5
Understanding Perception Technologies

Perception technologies are essential to enable AR, VR, and the Metaverse. We will present here what the main technologies are and how they work, in order for you to understand what the Metaverse is capable of and how businesses can benefit. Areas covered in this chapter include artificial intelligence, computer vision, and tracking and capture technologies.

In this chapter, we will cover the following main topics:

- What the main perception technologies are for the Metaverse
- How and why these perception technologies are essential to the Metaverse
- How businesses can directly benefit from the use of these technologies and to what degree

The role of artificial intelligence

Artificial intelligence (**AI**) is fundamentally transforming the development and potential of the Metaverse. As an advanced digital ecosystem that integrates VR, AR, and a range of other interactive technologies, the Metaverse represents an evolution of digital interaction that goes beyond anything we've previously seen. AI plays a critical role in this ecosystem, enhancing the capabilities of the Metaverse and enabling increasingly sophisticated, immersive, and personalized experiences for users.

AI is the engine that drives the Metaverse's many features and functions, including real-time interactions, user-generated content, digital economies, and multi-experience interfaces. From enhancing the realism of virtual environments with generative AI models, and facilitating advanced user interactions with **natural language processing** (**NLP**), to powering smart digital assistants, AI technologies are fundamental to the vision and operation of the Metaverse.

With the power of AI, the Metaverse can evolve from being a simple virtual space into a highly interactive, intelligent, and adaptive environment. This environment can understand and respond to individual users, offer personalized experiences, and even learn and adapt over time. Whether it's for gaming, socializing, commerce, or work, AI makes the Metaverse more immersive, engaging, and useful.

AI also drives the creation of new opportunities and business models within the Metaverse. By leveraging AI technologies, businesses can gain deeper insights into user behavior, develop innovative products and services, and create more engaging and effective marketing campaigns.

In essence, AI is the enabling technology that allows the Metaverse to become an intelligent, dynamic, and responsive environment, thereby shaping the future of our digital experiences.

The Metaverse relies on various types of AI to function effectively. Some of the AI technologies and applications commonly used in the Metaverse include the following:

- **NLP**: To mimic human interaction between users and AI-driven entities, such as chatbots or virtual assistants, NLP is utilized. Interpreting user input involves the application of various NLP methods, such as sentiment analysis, part-of-speech tagging, tokenization, and named entity recognition.

- **CV**: Computer vision techniques help AI systems identify, analyze, and interpret visual information from the digital environment. Applications include object recognition, facial recognition, and gesture recognition. **Convolutional neural networks** (**CNNs**) and other deep learning architectures are commonly employed. Models such as YOLO, Mask R-CNN, and EfficientNet are popular for object detection and recognition.

- **Spatial AI**: Spatial AI combines computer vision, sensor fusion, and machine learning techniques to create realistic simulations of physical spaces and objects. This technology enables precise navigation, mapping, and interaction with a virtual environment.

- **Generative AI**: Focused on creating unique and unprecedented data points, and based on the regularities and patterned information found in existing data, generative AI is a specific branch of AI techniques and systems. By grasping the underlying data distribution from their training sets, these models are adept at producing a variety of outcomes, including images, written text, music, or verbal language. Significant generative AI methods encompass LLMs (ChatGPT, GPT-4, Bard, and others mentioned earlier), **Generative Adversarial Networks** (**GANs**), **Variational Autoencoders** (**VAEs**), Transformer-based architectures, and other autoregressive models. (Much more detail on generative AI besides the summaries here can be found in *Chapter 4, The Value of Using 3D Visuals to Interact*.)

- **GANs**: GANs are used to generate content, including virtual environments, character models, and 3D objects, for the Metaverse. GANs are instrumental in the creation of new digital assets such as avatars, apparel, and diverse environments. Comprising a pair of neural networks known as a generator and a discriminator, GANs function in a competitive dynamic, with the aim to create high-quality content. GAN model examples are StyleGAN, CycleGAN, and BigGAN.

- **VAEs**: In the realm of machine learning, VAEs are a distinct category of generative models. By synthesizing deep neural networks with statistical techniques, they can decipher complex data structures. The two primary elements that make up VAEs are an encoder, responsible for transforming input data into a latent, or hidden, representation, and a decoder, tasked with recreating the original data using this latent representation. The notable ability of VAEs to construct a consistent and well-ordered latent space enables the effective generation of novel, realistic samples. The mechanism of VAEs is dependent on maximizing a lower bound on the data likelihood, which strikes a balance between faithful data reproduction and preserving regularity in the latent space. These features render VAEs suitable for a variety of applications, including unsupervised learning, image generation, and outlier detection.

- **Machine Learning (ML)**: There are three main types of ML – supervised learning, where algorithms are trained using labeled data; unsupervised learning, where learning is derived from unlabeled data; and reinforcement learning, which involves an agent learning from interactions with an environment, improving its decision-making skills through a system of rewards or penalties. ML, a particular area of study in the wider field of AI, consists of algorithms and models that provide computers the ability to "learn" and infer from data. The effectiveness of ML algorithms is amplified across iterations through a process known as training. This procedure involves adjusting internal parameters based on the examples of inputs and their corresponding outputs. Through this, a machine can learn to better understand patterns and make accurate predictions.

- **Deep Learning (DL)**: Utilizing artificial neural networks to model complex patterns in data, DL, a subset of ML, plays a crucial role in the Metaverse. It powers various applications such as content generation, image and speech recognition, and language translation.

- **Neural networks**: The operational mechanisms and structure of biological neural networks in the human brain serve as the inspiration behind the design of computational neural networks. These networks are formulated as assemblies of interconnected nodes or neurons, arranged across multiple layers, and through weighted connections, they are proficient in processing and conveying information. These neural networks are utilized to carry out complex tasks, such as decision-making by learning from examples, pattern recognition, and NLP, leveraging the learning derived from provided examples. During a training process, connection weights are altered using specific algorithms, such as backpropagation, empowering neural networks to adjust themselves and incrementally boost their performance.

- **Graph Neural Networks (GNNs)**: GNNs are a type of neural network that can work with graph-structured data. They are used in the Metaverse to model social networks, understand user preferences, and make personalized recommendations. Models such as GCN, GAT, and GraphSAGE are popular in this field.

- **Neural rendering**: This technology uses DL techniques to enhance computer graphics and create photorealistic environments. Neural rendering can improve the visual quality of the Metaverse, reduce latency, and optimize rendering performance.

- **Reinforcement Learning (RL)**: AI agents are empowered to learn from their environmental interactions and independently make decisions through RL. In the Metaverse, RL has widespread use in managing **Non-Player Characters (NPCs)**, establishing adaptive game mechanics, and refining simulations of virtual worlds. Various techniques such as **Proximal Policy Optimization (PPO)**, Q-learning, and **Deep Q-Networks (DQNs)** are commonly utilized for this purpose.

- **Speech recognition and synthesis**: AI-powered speech recognition systems are used to transcribe and understand spoken language, enabling voice-based interactions in the Metaverse. Conversely, speech synthesis techniques allow AI agents to generate human-like speech for more natural communication.

- **Emotion and sentiment analysis**: AI technologies can detect and analyze users' emotions and sentiments based on their text, voice, facial expressions, and biometric data, helping create more personalized experiences and allowing developers to design emotionally responsive environments.

- **AI-driven procedural generation**: Procedural generation algorithms are used to create environments, objects, and events dynamically in the Metaverse, allowing for unique and personalized experiences. Techniques such as Perlin noise, L-systems, and cellular automata are used for terrain generation, object placement, and procedural storytelling.

- **Multi-agent systems**: Multi-agent systems involve multiple AI agents interacting with each other to achieve a common goal or competing. In the Metaverse, multi-agent systems enable complex simulations, NPC interactions, and collaborative or competitive gaming experiences.

- **AI-based animation and simulation**: Generating realistic animations and interactions among virtual characters and objects in the Metaverse is accomplished through AI-driven algorithms, such as inverse kinematics, physics-based simulations, and motion capture. This substantially contributes to overall immersion and believability. Complex movement patterns can be learned and generated by ML models – for instance, **Long Short-Term Memory (LSTM)** networks and VAEs.

- **Distributed AI systems**: Given the massive scale of the Metaverse, distributed AI systems are employed to handle the computational load and ensure seamless performance. Techniques such as federated learning and edge computing are used to train and deploy AI models across multiple devices and servers.

- **AI-driven content moderation**: AI is used for content moderation in the Metaverse, ensuring a safe and inclusive environment for users. Techniques such as sentiment analysis, text classification, and computer vision models are employed to detect and filter inappropriate content or behavior.

- **AI-based content curation**: AI systems can be used to personalize user experiences in the Metaverse by recommending relevant content, events, or virtual spaces, based on user behavior and preferences. Collaborative filtering and content-based filtering are common techniques used for these tasks.

The Metaverse stands on the precipice of a major transformation, largely driven by AI technology. Such advancements hold great promise in enriching our engagement with the digital realm. AI has the potential to bring an unprecedented level of realism to virtual environments and avatars, as well as enhance the quality of our social interactions within this digital expanse.

It is evident that, as AI technologies progress, the scope for the Metaverse's evolution is virtually boundless. This encompasses everything from journeying into unexplored virtual territories to forging connections with friends and professional acquaintances or immersing ourselves in pioneering forms of digital amusement.

Undeniably, the pivotal role of AI in crafting the future of the Metaverse cannot be understated. It's clear that, as we traverse this expansive digital landscape, AI will be instrumental in shaping our experiences.

For the rest of this section on AI, we will expound more on some of the AI technologies and applications mentioned here.

NLP

NLP is geared toward equipping computers with the capacity to comprehend and produce human language. The objective encompasses enabling human-like interaction between computers and humans, extracting valuable data from written content, and facilitating deep text analysis.

Here's what NLP includes:

- **Syntax analysis**: This is about understanding the structure of sentences. It includes the following:

  - **Tokenization**: Breaking up text into smaller parts such as words or phrases

  - **Part-of-Speech (POS) tagging**: Determining whether a word is a noun, verb, adjective, and so on

  - **Parsing**: Studying the grammatical structure of a sentence

- **Semantic analysis**: This is about understanding the meaning of sentences. It includes the following:

  - **Named Entity Recognition (NER)**: Finding and categorizing certain things in text, such as people's names or places

  - **Word Sense Disambiguation (WSD)**: Figuring out the correct meaning of a word, based on how it's used

  - **Semantic Role Learning (SRL)**: Understanding the relationships between words in a sentence

- **Pragmatics**: This is about understanding language in context. It includes the following:

  - **Coherence Resolution**: Linking together mentions of the same thing in a text

  - **Discourse Analysis**: Studying the structure and coherence of a text

- **Text generation and summarization**: This is about creating and condensing text. It includes the following:

  - **Machine translation**: Translating text from one language to another

  - **Text summarization**: Creating a short summary of a longer text

  - **Dialogue systems**: Having a conversation in natural language with humans

To do all this, NLP uses different methods and algorithms, such as the following:

- **Rule-based approaches**: These are systems that use rules (created by humans) to process and analyze text – for example, regular expressions or grammar.

- **Statistical methods**: These use ML and a lot of data to build models that can analyze and generate text – for example, hidden Markov models or N-grams.

- **Neural network-based approaches**: These use advanced techniques (such as DL) to model and process language – for example, **Recurrent Neural Networks** (**RNNs**) or models such as GPT-4 and BERT.

Often, to get the best results, NLP applications need to use a mix of these methods. The field is always advancing, thanks to progress in ML, DL, and the availability of lots of language data.

GANs

To explain GANs, let's start with a simple analogy. Imagine two characters – a forger and a detective. The forger's goal is to create counterfeit art pieces so flawless that they can pass off as originals. Meanwhile, the detective's task is to discern between genuine and counterfeit artwork. As the forger gets better at creating convincing fakes, the detective also needs to improve their discernment skills to distinguish them from the real ones. Over time, both of them improve, thanks to this adversarial relationship.

Here are the main components of a GAN:

- **Generator**: In the context of working with images, such as faces, the role of the generator is to generate new data, starting from an arbitrary point often referred to as "random noise." The generator could be likened to a counterfeiter; however, its craft is not forging art but, rather, creating counterfeit data. The aim is to produce data that closely resembles the type we are focused on, such as generating new images that convincingly appear as real faces.

- **Discriminator**: Conversely, the discriminator acts much like a detective. It's responsible for scrutinizing both the original, real data and the counterfeit data manufactured by the generator. Its task is to differentiate between real and fake data, making a judgment on the authenticity of each data piece it encounters. The discriminator's performance is gauged by its accuracy in correctly identifying real versus fake data.

The generator and the discriminator are in a constant loop of competition. The generator is continuously getting better at producing realistic data, and the discriminator is continuously getting better at telling whether the data is real or fake. This competitive process drives both of them to improve over time.

During each round of training, the following occurs:

- The generator creates a batch of data, which the discriminator then evaluates along with a batch of real data. Then, based on the discriminator's assessments, both the generator and discriminator are updated.

- This is achieved by using a kind of feedback called backpropagation, which is used across many types of ML algorithms. Essentially, backpropagation involves adjusting the internal parameters of the generator and the discriminator (called weights) to make them better at their jobs.

- In the case of the discriminator, we want to adjust its weights so that it is more likely to classify real data as real and fake data as fake. For the generator, we want to adjust its weights so that it is more likely to create data that the discriminator classifies as real.

- Interestingly, while we're training the generator, we keep the discriminator's weights fixed, and vice versa. This is to ensure that each part of the GAN can learn from a stable version of its adversary.

Challenges with GANs

While GANs are undoubtedly powerful, they also have their fair share of challenges:

- **Mode collapse**: This is a situation where the generator begins to produce the same output (or very similar outputs) over and over again, regardless of the input, because it has found a kind of *loophole* that tricks the discriminator. This limits the diversity of the generated data.

- **Training instability**: GANs are notoriously difficult to train. Since the generator and discriminator continuously learn from each other, it's possible for the system to become unstable, resulting in poor results.

- **Lack of a clear objective evaluation metric**: Evaluating the performance of GANs can pose a challenge, due to the absence of a concrete success measure, unlike traditional ML models, which provide explicit evaluation metrics. Therefore, identifying the most efficient GAN model can often be complex, stemming from the fact that there's no universally acknowledged objective evaluation metric.

GANs have found their niche in a plethora of applications. They can generate convincing and intricate data forms, such as images, music, text, and speech, displaying their versatile capability. A compelling example of their application is the creation of realistic human face visuals. These GAN-generated faces bear striking resemblance to actual human faces, despite not being representative of any real individuals.

ML

Utilizing statistical analysis, algorithms are developed in ML, which predict an output by processing input data. As new data is introduced, these outputs are continually updated. This is the key idea behind ML, a specific form of AI. It allows computers to acquire knowledge from data autonomously, without needing any specific programming for this purpose. The automated generation of analytical models through data analysis is an essential aspect of this process.

There are several types of ML, including the following:

- **Supervised learning**: Primarily acknowledged as the most common method, supervised learning is a technique where an algorithm evolves through interaction with a training dataset. It's akin to a learner acquiring knowledge under guidance, thus justifying the terminology *supervised*. Employing this strategy, models harness the capability to predict results corresponding to particular input data.

- **Unsupervised learning**: Unlike supervised learning, no labels are provided, and the model learns the structure of the input data on its own. It can be used for clustering (grouping similar items together) or dimensionality reduction (simplifying data while keeping it meaningful).

- **Reinforcement learning**: Through interaction with its environment, an algorithm gradually acquires knowledge. This process involves taking specific actions and meticulously evaluating the resulting outcomes, which may be favorable or unfavorable. As it accrues experience, the algorithm refines its decision-making strategy to favor actions that yield the highest benefit.

The crux of ML is to equip computers with the ability to learn in a manner similar to human beings. This involves a continuous, autonomous refinement of their learning capabilities. This process heavily relies on the analysis of practical experiences and empirical data. Consequently, these systems grow to make more informed decisions over time, echoing human decision-making processes.

DL

DL, a branch of ML, utilizes neural networks as its primary instrument. These networks, characterized by multi-layered arrangements, mimic the operational mechanism of the human brain in a simplified manner. Abundant layers within these networks give rise to the term *deep* in DL.

Artificial Neural Networks (**ANNs**), CNNs, and RNNs are prevalent types of these intricate networks. Interconnected nodes, often referred to as *neurons*, manage input data and transmit it to the subsequent layer. This iterative process continues until a decision or prediction is generated.

The main objective of DL is to develop algorithms capable of recognizing patterns in data. By comprehending data representations instead of relying solely on task-specific algorithms, DL offers a comprehensive approach to ML, enabling informed judgments.

One distinguishing aspect of DL is its ability to automatically extract features from raw input data. In contrast, traditional ML typically requires manual feature extraction. This capability to learn from extensive datasets enhances the efficiency of DL algorithms.

DL plays a vital role in the field of AI and serves as the foundation for various applications and services. These include image and voice recognition, autonomous vehicles, and machine translation. The extensive potential of DL across diverse fields is exemplified by its capacity to accomplish analytical and physical tasks without human intervention. These advancements in automation are a testament to the progress facilitated by DL technologies.

Neural networks

The interlinked networks of neurons, or *neural networks*, ensure a steady information flow, providing an effective platform for data analysis and decision-making. Similar to a well-coordinated production line, each neuron functions as a dedicated worker within its layer, contributing to the overall processing of information.

Here's how this sophisticated system operates – just as tasks are passed down along a production line, each neuron receives inputs from its preceding layer. It then applies various mental computations and operations to these inputs, closely scrutinizing and processing the data at hand. Once a neuron completes its assigned tasks, it forwards the results to the next set of workers, or neurons, in the succeeding layer.

This methodical progression echoes the sequential flow of a production line. As the information traverses through the network, the system progressively discerns increasingly complex patterns within the data. This iterative process can be likened to gradually piecing together a jigsaw puzzle, with each step revealing a more detailed and comprehensive image.

Through the extraction of these sophisticated patterns, the neural network broadens its knowledge base, becoming more insightful over time. This is similar to the deep understanding acquired from accumulating and scrutinizing substantial amounts of information. Eventually, the profound understanding it gains enables the network to make well-informed decisions, based on the intricate data representation it has developed.

In summary, neural networks facilitate a smooth exchange of information through their interconnected layers of neurons, enabling the extraction of intricate patterns from data. This leads to an improved decision-making ability within the network, as it gradually develops a deeper understanding of the data it processes.

Here's a breakdown of the key components:

- **Neurons**: The neurons, the cornerstone of a neural network, are vested with the responsibility of accepting, processing, and forwarding input data. The relative importance of each input is expressed via an assigned weight. Once these weighted inputs are combined with a bias, the neuron employs an activation function. The latter then becomes the determining factor in whether the signal gets advanced, and to what extent. The fundamental components of a neural network are its neurons. They undertake the process of taking in input, executing the necessary computations, and subsequently transmitting the results to their fellow neurons. Each input is tagged with a weight to highlight its relevance. Following the aggregation of all these weighted inputs and a bias, the neuron administers an activation function. This function is essentially the deciding factor of whether, and to what degree, the signal will progress.

- **Layers**: Constructed of different levels, a neural network functions by receiving information through its initial stage, known as the input layer. Following this, computations and processing take place predominantly in the intermediary stages, generally referred to as hidden layers. The culmination of this process yields the results, which are made accessible via the final layer, also known as the output layer.

- **Connections and weights**: The significance of a neuron's input is denoted by a weight that is attached to each individual connection within the network. The network's composition is made up of these links, each serving to transfer the output from a specific neuron into the input of another.

- **Activation function**: A neuron's output is determined by a specific mathematical function that operates on its input or series of inputs. This function ultimately interprets the input to generate a corresponding output.

- **Training**: Guided by the input data and expected outcome, the biases and weights of a neural network are progressively fine-tuned. This is typically achieved through backpropagation, in conjunction with stochastic gradient descent or an equivalent optimization strategy.

- **Loss function**: During the training phase, a neural network's predictions are assessed by a loss function, which determines a *penalty* for incorrect predictions. This penalty, crucial in the backpropagation process, helps in fine-tuning the weights and biases.

The versatility of neural networks extends to areas such as speech and image recognition, suggestion systems, and NLP. A key strength that enhances their practicality is their ability to self-correct by learning from errors.

RL

RL falls under the larger category of ML. It is a process where an agent refines its actions by interacting with an environment and making decisions, all for the purpose of achieving a certain goal. This agent's learning is directed by the feedback it gets in the shape of rewards or penalties.

The environment, states, actions, and the agent are the four main elements in RL. The environment is where the actions of the agent, which is the decision-maker, take place. The situations in which the agent may find itself in the environment are called states, and the potential steps the agent may choose to perform are known as actions.

In order to maximize the overall reward over time, the main goal of the agent is to develop a policy. A policy is a tool used by the agent to determine which actions to take in different states. To achieve this, the agent must find a balance between exploration, where it tries out new actions to learn the results, and exploitation, in which the agent selects actions believed to provide the most rewards.

The learning process generally involves these steps:

1. The agent initiates the process by taking an action according to its existing knowledge and policy.
2. The environment responds to the action, leading to a transition to a new state.
3. The agent then observes the new state of the environment, gathering information about the changes that occurred.
4. Finally, the agent uses the observed outcome to update its knowledge and policy, making adjustments based on the feedback received from the environment.

There are several algorithms in RL, each with its unique approach to learning optimal policies. Some of the prominent ones include the following:

- **Q-learning**: A model-free, off-policy algorithm that learns the optimal action-value function by iteratively updating Q-values, using the Bellman equation
- **DQNs**: An extension of Q-learning that uses deep neural networks as function approximators to estimate action-value functions, making it suitable for high-dimensional state spaces
- **Policy gradients**: A family of model-free, on-policy algorithms that directly optimize the policy by computing gradients with respect to policy parameters
- **Actor-critic**: A hybrid approach that combines value function estimation and policy optimization, using two separate networks (the "actor" for policy and the "critic" for value estimation) to learn the optimal policy

RL continues to make remarkable strides, despite encountering various challenges. These challenges encompass the need to enhance the sample efficiency of learning algorithms, formulate advanced exploration strategies, and fortify learned policies to withstand changing environments or adversaries. Nonetheless, RL remains highly promising, as evidenced by its successful applications across a wide range of domains, including game playing, robotics, NLP, and recommendation systems.

Speech recognition and synthesis

Let's break down the two major topics here: **speech recognition** and **speech synthesis**. Both use AI extensively, especially ML techniques such as DL.

Speech recognition

AI plays a significant role in the field of speech recognition, also referred to as **Automatic Speech Recognition (ASR)**. Its primary function revolves around the conversion of spoken language into written text:

- **Data collection**: Large amounts of spoken language data (audio files) and their text transcriptions are collected.

- **Feature extraction**: The audio data is then converted into a numerical format, often using a technique called **Fourier Transform**, to extract **features** (characteristic patterns or components) from the sound waves. These features might include pitch, duration, and intensity.

- **Model training**: Using the collected data and their transcriptions, an ML model is trained. The most commonly used models are DL, RNNs, LSTM networks, and more recently, Transformer models. These models learn to predict the text corresponding to the given audio features.

- **Decoding**: The trained model can now take new spoken language input, extract features, and predict the corresponding text. Decoding strategies can vary, from choosing the most likely word at each timestep to optimizing over a whole sequence.

- **Improvement**: Techniques such as transfer learning and active learning are often used to continuously improve a model's performance, by leveraging knowledge from related tasks and iteratively focusing on hard-to-learn examples, respectively.

Speech synthesis

AI has emerged as a crucial factor in the realm of speech synthesis, commonly referred to as **Text-to-Speech (TTS)**. Its fundamental purpose lies in the transformation of written text into spoken language:

- **Data collection**: Similar to speech recognition, large amounts of spoken language data (audio files) and their text transcriptions are collected.

- **Text analysis**: The input text is analyzed and broken down into smaller components, often **phonemes** (distinct units of sound). Additional details, such as punctuation and word emphasis, are also noted to assist in producing more natural-sounding speech.

- **Model training**: DL, a sophisticated subset of ML, predominantly employs a variety of models. These models, including but not limited to CNNs, RNNs, LSTM, and Transformer models, are renowned for their effectiveness. In a distinct context, ML models are geared to effectively derive audio characteristics from the interplay of text. This essentially means that such models are taught to make predictions concerning audio elements, based solely on textual input.

- **Waveform generation**: The predicted audio features are then used to generate the sound wave of the spoken language. This step can be achieved using a variety of methods, including vocoders or directly generating the waveform, using models such as WaveNet.

- **Improvement**: In order to facilitate the model's evolution over time, transfer learning focuses on and analyzes tasks of a similar nature. Meanwhile, the strategy of active learning concentrates on intricate situations, leading to a steady enhancement of the model's performance. As a whole, by implementing the methods of transfer learning and active learning, the model's effectiveness is gradually optimized, either by learning from tasks bearing a resemblance or by tackling complex examples, respectively.

Worth mentioning is the fact that the *standard* methodologies I've elaborated cannot be uniformly applied across every instance. Certain situations might demand added stages or modifications tailored to the unique requirements of the task at hand. Simultaneously, the keystone role of AI should not be ignored – it aids significantly in deciphering the complex, non-linear correspondence between verbal sounds and their written counterparts.

Emotion and sentiment analysis

The process of examining sentiments and emotions leverages the prowess of tools such as computational linguistics, biometrics, text mining, and NLP, enabling an in-depth exploration and study of human affective states and subjective perceptions. Various resources, including customer reviews, feedback surveys, online and social media content, and even health-related documents, often undergo this analysis. The technique's adaptability finds it a place in numerous sectors, such as customer care, promotional activities, and the healthcare field.

AI, particularly ML and DL, plays a significant role in emotion and sentiment analysis. Here's how:

- **NLP**: AI algorithms can be used to understand sentiments in text format, such as social media posts, customer reviews, or any text data. NLP techniques can extract key phrases and words, classify text as positive, negative, or neutral, and even understand context and sarcasm.

- **ML**: An exemplar application of supervised learning models can be seen in sentiment prediction tasks. After being trained on a dataset with texts labeled for sentiments, these models can effectively predict sentiments in unlabeled text. They work by considering features in the text, which could be as small as words or as large as sentences.

- In addition to this, a few renowned models are extensively used for such sentiment analysis tasks. These include the likes of support vector machines, Naïve Bayes, and decision trees. These models have proven their efficiency in understanding and predicting sentiments through their widespread usage and success.

- It's worth noting that all these techniques are encapsulated under the umbrella of AI training. The ability to recognize and understand emotions and sentiments in text data is a powerful application of AI.

- **DL**: More recently, DL techniques such as CNNs and RNNs, and especially Transformer models such as BERT and GPT, have been used for sentiment analysis. These models can understand the context and sequence of words, making them more accurate at sentiment analysis.

- **Transfer learning**: AI models can be pre-trained on large datasets, learning general language understanding, and then fine-tuned for sentiment analysis on a smaller specific dataset. This allows models to leverage knowledge learned from large amounts of data, even when the specific sentiment analysis dataset is relatively small.

- **Emotion AI**: This is a specific subfield of AI dedicated to recognizing and interpreting human emotional signals. It uses a variety of modalities, including text (as in sentiment analysis), voice tone, facial expressions, and even physiological signals.

Based on news sentiment, AI technologies have demonstrated their value in predicting shifts in the stock market. Further uses for these frameworks include scrutinizing customer feedback, monitoring social media mood, and overseeing brand reputation. Their adaptability lends them to a broad spectrum of applications.

How businesses can benefit

With its potential to revolutionize industries and reshape the way we conduct business, the Metaverse has become a fertile ground for the integration of AI. Here, we will focus on areas such as data analysis, customer engagement, virtual workforce management, and the creation of new revenue streams.

The power of AI-driven data analysis in the Metaverse

By harnessing the power of AI in their Metaverse plans, companies are able to draw from a wealth of data, from both users and devices, to predict what lies ahead with impressive accuracy. This dataset, rich with consumer preferences, behaviors, and interactions, grants companies the ability to shape their strategies based on deep insights. The foresight provided by AI algorithms gives businesses a unique leg up in forecasting market trends and consumer demands, thus strengthening their competitive edge and capacity for quick, effective responses.

Enhancing customer engagement through AI-powered personalization

Customer engagement is critical for business success, and AI technologies have the potential to revolutionize how businesses interact with their customers in the metaverse. Personalization is a key aspect of this transformation, as AI-powered algorithms can analyze user data to create tailored experiences that cater to individual preferences and needs. This could manifest in the form of personalized product recommendations, targeted promotions, or even immersive experiences that leverage VR and AR technologies. By offering a more engaging and customized experience, businesses can foster stronger relationships with their customers, driving brand loyalty and boosting revenue.

Managing virtual workforces with AI-enabled tools

As businesses increasingly operate within the metaverse, they will need to manage virtual workforces that include both human and AI-powered entities. The use of AI in workforce management can streamline processes, enhance collaboration, and improve overall productivity. For example, AI-driven tools can assist with task allocation and prioritization, ensuring that resources are deployed effectively and that objectives are met in a timely manner. Furthermore, AI can facilitate communication and collaboration between remote teams, breaking down barriers and fostering innovation. By adopting AI in virtual workforce management, businesses can unlock new levels of efficiency and effectiveness, allowing them to thrive in the competitive landscape of the metaverse.

Unlocking new revenue streams through AI-generated content and services

The Metaverse presents an abundance of opportunities for businesses to generate new revenue streams, through the development and provision of AI-generated content and services. Virtual goods, digital art, and immersive experiences are just a few examples of the types of products that can be created and monetized using AI. Additionally, businesses can offer AI-driven services, such as virtual personal assistants, automated customer support, and personalized recommendations. By harnessing the power of AI, companies can not only create new sources of income but also diversify their offerings, attracting new customer segments.

CV

The critical part played by computer vision – an integrated matrix of systems, methodologies, and algorithms enabling computers to interpret, evaluate, and understand visual or multidimensional data – in the establishment and functioning of the Metaverse is unassailable. Computer vision serves as the essential tool that equips these digital worlds to sense and engage with users instantly, thus leading to the creation of lively and fully realized virtual realities. It serves as the bedrock for features such as facial and gesture recognition, reconstruction of objects and scenes, AR overlays, and real-time avatar animation – all crucial elements in crafting an authentic and captivating Metaverse experience.

Here are some key aspects of computer vision used in the Metaverse:

- **Image and video processing**: Computer vision relies on advanced image and video processing techniques to analyze, understand, and process visual information from the virtual world. Techniques such as filtering, edge detection, segmentation, and feature extraction are used to interpret the environment and enable interaction between users and the Metaverse. Examples include Gaussian blurring, edge detection, and image pyramids.

- **Feature extraction**: Feature extraction techniques, such **as Scale-Invariant Feature Transform (SIFT)**, **Speeded-Up Robust Features (SURF)**, and **Oriented FAST and Rotated BRIEF (ORB)**, are widely employed to discern and meticulously describe essential characteristics in images and video footage. These traits play a vital role in identifying objects, tracking them subsequently, and understanding the overall context of a scene. This process is of utmost significance when it comes to object identification and grasping the setting of the scene. Various procedures, including but not limited to SIFT, SURF, and ORB, are commonly utilized to extract these features.

- **Object detection and recognition**: To ensure a seamless experience in the Metaverse, the identification and classification of diverse objects, characters, and elements within a virtual environment are paramount. Several techniques, including CNNs, **Region-Based CNNs (R-CNNs)**, **Single Shot MultiBox Detector (SSD)**, and **You Only Look Once (YOLO)**, are utilized extensively to accomplish this task effectively. These techniques play a vital role in identifying and categorizing various elements present in a virtual environment, ensuring a smooth and immersive experience in the Metaverse.

- **DL and CNNs**: Tasks such as semantic segmentation and GANs commonly employ CNNs. These networks, including vital layers such as pooling, convolutional, and fully connected layers, have brought about a transformative shift in computer vision. Their deployment has led to remarkable advancements in accurate and efficient image recognition, segmentation, and generation, revolutionizing the field of DL.

- **3D reconstruction and scene understanding**: Creating realistic virtual spaces heavily relies on techniques that transform 2D images into 3D models of objects and environments. Some prominent examples encompass **Multi-View Stereo (MVS)**, **Structure from Motion (SfM)**, and the utilization of depth sensing. These approaches enable the generation of a comprehensive 3D representation from both images and videos.

- **AR and VR integration**: Marker-based and markerless tracking, feature-based tracking, and sensor data fusion are integral to achieving a seamless user experience in computer vision applications that merge digital content with the physical world, such as AR and VR. **Visual Inertial Odometry (VIO)** and **Simultaneous Localization and Mapping (SLAM)** are two noteworthy techniques employed to track and map in these contexts. These methods ensure the synchronization of digital elements with real-world environments, enabling captivating and immersive virtual experiences.

- **Optical flow and motion estimation**: Estimating motion between frames in a video sequence enables tracking, stabilization, and prediction of object movements. Techniques include the Lucas-Kanade method, the Horn-Schunck algorithm, and DL-based approaches.

- **Scene understanding and semantic segmentation**: To provide context-aware experiences, computer vision algorithms are used to understand and interpret the environment. Semantic segmentation techniques, such as **Fully Convolutional Networks (FCNs)**, U-Net, and DeepLab, are employed to classify each pixel in an image, enabling rich scene understanding and object interaction.

- **Facial recognition and emotion detection**: For personalized and interactive experiences in the Metaverse, computer vision algorithms are used to identify users, detect facial expressions, and analyze emotions. Techniques such as **Active Appearance Models (AAM)**, Eigenfaces, Fisherfaces, **3D Morphable Models (3DMMs)**, and DeepFace, along with emotion recognition algorithms such as FER2013, are employed.

- **Gesture recognition and motion tracking**: To enable natural user interactions in the Metaverse, computer vision algorithms are used to recognize gestures and track motion. Techniques such as optical flow, **Dynamic Time Warping (DTW)**, and LSTM networks are commonly used for these tasks.

- **Gaze estimation**: Understanding where users are looking and predicting the direction of their gaze in a virtual environment helps improve user experiences. Gaze estimation techniques such as 3D gaze estimation and **Pupil Center Corneal Reflection (PCCR)** are employed for this purpose. Gaze estimation is featured in the upcoming Apple Vision Pro, although the exact technology has not been disclosed.

- **Pose estimation**: Estimating the pose of 3D objects or characters (i.e., position and orientation) within a virtual environment is essential for natural interactions. Techniques such as Procrustes analysis, **Perspective-n-Point (PnP)** algorithms, and the **iterative closest point (ICP)** methods are used for this purpose. Human pose estimation and tracking methods, such as OpenPose, DensePose, and AlphaPose, utilize DL and part-based models to estimate 2D and 3D human body keypoints, allowing for natural user movement and interaction within a virtual environment.

- **ML and data-driven approaches**: Supervised, unsupervised, and reinforcement learning techniques are used to train and optimize computer vision models for various tasks. **Support Vector Machines (SVMs)**, random forests, and DQNs are examples of different learning methods.

- **Optimization and acceleration**: As the Metaverse requires real-time processing, techniques to optimize and accelerate computer vision algorithms are essential. Hardware accelerators, such as GPUs, FPGAs, and ASICs, as well as software optimization techniques, such as pruning, quantization, and distillation, are employed to enhance performance.

- **Multi-agent systems and collaborative AI**: In the Metaverse, multiple AI agents may need to collaborate and interact with each other. Computer vision is utilized to enable these agents to perceive and understand their surroundings, communicate, and coordinate their actions effectively.

Computer vision technologies play a vital role in creating the Metaverse experience. Techniques such as DL, 3D reconstruction, object recognition, pose estimation, facial analysis, and scene understanding are combined to create a seamless, interactive, and immersive virtual environment for users to explore and interact with.

How businesses can benefit

Incorporating computer vision within the Metaverse can provide several benefits for businesses, such as the following:

- **Product visualization**: Businesses can use computer vision to create 3D models of their products for customers to interact with in the Metaverse. This not only makes for a more engaging shopping experience but also enables customers to get a much better understanding of a product before they make a purchase.

- **Customer analytics**: Just as businesses use computer vision in physical stores to track customer movements and interactions, the same can be done in the Metaverse. Computer vision can help businesses understand how customers interact with their virtual stores, products, or experiences, which can provide valuable insights for optimization and personalization.

- **Virtual try-ons**: For businesses in the fashion and accessory industries, computer vision can enable virtual try-ons. Customers can see how clothes, glasses, jewelry, or even makeup will look on them virtually before making a purchase.

- **Real-time customization**: With computer vision, businesses can offer real-time customization options. For example, customers could change the color or features of a product and see those changes reflected immediately.

- **Improved accessibility**: Computer vision can be used to improve accessibility in the Metaverse. For instance, it can help translate sign language into text or speech, provide descriptive audio for visually impaired users, or enable easier navigation for users with mobility issues.

- **Interactive marketing**: Businesses can create interactive advertisements that respond to a user's actions or even their emotions, as determined through computer vision.

- **Workforce training**: Businesses can use the Metaverse to train employees, with computer vision enabling virtual simulations that replicate real-world scenarios. This can lead to more effective and efficient training programs.

- **Security**: Just like in the physical world, security will be crucial in the Metaverse. Computer vision can be used for identity verification, monitoring virtual spaces for any suspicious activity, and ensuring compliance with community standards or rules.

By leveraging computer vision in the Metaverse, businesses can offer more immersive, personalized, and interactive experiences to their customers, which can lead to increased engagement, satisfaction, and ultimately, sales. Furthermore, the data gathered through computer vision can provide valuable insights that can drive innovation and improvement in products and services.

Tracking and capture technologies

The proliferation of the Metaverse – a sprawling, interconnected virtual realm where digital and physical realities converge – owes much of its dynamism to tracking and capture technologies. These technologies, which include techniques such as motion capture, eye tracking, and object tracking, are critical to enhancing user immersion and enabling complex interactions within this digital frontier.

The Metaverse seeks to provide an experience that extends beyond the realm of traditional 2D interfaces, creating a space where users can engage and interact in ways that feel as real and vibrant as their physical-world experiences. To achieve this level of immersion and interactivity, the Metaverse relies heavily on tracking and capture technologies.

These technologies serve as a conduit, translating the movements and actions of users in the physical world into their digital counterparts in the Metaverse. Motion capture, for instance, can record and digitally reproduce a user's physical movements to control an avatar or interact with virtual objects. Similarly, eye-tracking technologies can provide intuitive control mechanisms and enhance the realism of avatars, by replicating a user's gaze and blinks.

The bridging of the physical and digital realms within the Metaverse heavily relies on tracking and capturing technologies that mirror human behavior and interaction. These technologies serve as a foundational layer, enabling users to immerse themselves in the Metaverse and creating a more natural, intuitive, and lifelike experience. By seamlessly integrating these technologies, the Metaverse becomes a realm where human presence is authentically projected and interactions feel genuinely immersive.

Without these technologies, the Metaverse would lose much of its potential, transforming from a vibrant, immersive digital realm into a static, two-dimensional interface. In essence, tracking and capture technologies form the backbone of the Metaverse's immersive potential, turning a static virtual world into an engaging and interactive universe.

Required tracking and capture technologies

Some of the key tracking and capture technologies needed for the Metaverse are covered in the following subsections.

Motion capture (MoCap)

In the Metaverse, the utilization of motion capture is paramount in capturing and recording the intricate movements of individuals or objects. This process involves employing various sensors such as optical, inertial, or magnetic sensors to collect data. Subsequently, this data is harnessed to animate digital characters or manipulate virtual objects within the Metaverse. **Inertial measurement units (IMUs)**, optical marker-based systems, and time-of-flight cameras are among the prevalent technologies utilized in the realm of MoCap:

- **IMUs**: These devices utilize a blend of magnetometers, accelerometers, and gyroscopes to deduce a subject's position and motion. By employing advanced sensor fusion methodologies such as complementary filters and Kalman filters, these devices handle data assimilation from diverse sensors and minimize errors. This approach is fundamental to consolidating and aligning data collected from multiple sensors, thus delivering precise and dependable evaluations of pose and movement.

- **Optical marker-based systems**: Using high-speed cameras to track retro-reflective markers placed on a subject's body, 3D coordinates are calculated using triangulation. **Direct Linear Transformation (DLT)** and bundle adjustment algorithms are employed for accurate pose estimation.

- **Time-of-flight cameras**: These cameras capture depth information using the time it takes for emitted light to reflect off an object and return to a camera. The captured depth maps can be used to reconstruct 3D models and estimate poses.

Hand and gesture tracking

Tracking the movement and gestures of users' hands is crucial for natural interactions in the Metaverse. Technologies such as Ultraleap and HaptX use a combination of infrared cameras, computer vision algorithms, and ML to recognize and track hand movements and gestures:

- **CV algorithms**: CNNs and RNNs are used to detect and recognize hand gestures from 2D images or depth maps.

- **Point cloud processing**: Depth sensors generate point clouds that are then analyzed and matched, using methods such as **Normal Distribution Transform (NDT)**, **Iterative Closest Point (ICP)**, and **Random Sample Consensus (RANSAC)**.

Eye tracking

Eye-tracking technology measures the point of gaze, allowing for gaze estimation and interaction with the virtual environment using eye movements. This technology enhances the immersion and enables foveated rendering, which focuses computational resources on the area where the user is looking. Companies such as Tobii and Pupil Labs offer eye-tracking solutions, using infrared cameras and computer vision algorithms:

- **Pupil Center Corneal Reflection (PCCR)**: Infrared cameras and light sources are used to track the position of the pupil center and corneal reflection. The vector between the two points is used to estimate the gaze direction.

- **3D gaze estimation**: Using multiple cameras, 3D gaze estimation calculates the intersection of the gaze vectors from both eyes to estimate the point of gaze in 3D space.

Facial capture

Facial capture technology records facial expressions and emotions, which can be used to animate digital avatars or drive virtual characters' expressions in the Metaverse. Techniques such as marker-based facial capture, depth cameras, and photogrammetry are employed for this purpose. Apple's ARKit and FaceID technology are examples of facial capture systems.

- **Constrained Local Models (CLMs) and Active Appearance Models (AAMs)**: These statistical models represent the shape and appearance of a face and are used to track facial features and expressions in 2D images.

- **Blendshape models**: Using a set of predefined facial expressions, blendshape models linearly interpolate between these expressions to create a wide range of facial animations.

Haptic feedback

Haptic feedback technology provides users with tactile sensations to simulate touch, texture, and force in a virtual environment. Devices such as haptic gloves, suits, and vests use actuators, sensors, and vibration motors to recreate the sense of touch in the Metaverse:

- **Piezoelectric actuators**: These actuators generate force or vibration in response to an applied voltage, providing precise haptic feedback.

- **Electroactive Polymers (EAPs)**: EAPs change their shape or size when stimulated by an electric field, enabling them to simulate various textures and sensations.

Positional tracking

Users' movements within a digital space are meticulously recorded through the crucial aspect of positional tracking, which serves to elevate their sense of being and immersion. Numerous technologies are employed to keep track of and map the users' location within this virtual realm, including GPS-based tracking, **Visual Inertial Odometry (VIO)**, and **Simultaneous Localization and Mapping (SLAM)**:

- **SLAM**: Algorithms such as FastSLAM, LSD-SLAM, and ORB-SLAM, under the umbrella of SLAM, utilize data from sensors to concurrently generate an environmental map and gauge a user's location within it.

- **Visual Inertial Odometry (VIO)**: VIO combines visual data from cameras and inertial data from IMUs to estimate a user's position and orientation. Algorithms such as **Multi-State Constraint Kalman Filter (MSCKF)** and **Visual-Inertial Bundle Adjustment (VIBA)** are employed.

Brain-Computer Interfaces (BCIs)

BCIs facilitate an intimate, two-way dialogue between a user's neural activity and the digitally constructed universe, thus crafting deeper, more intuitive interactions. The intricate task of gathering the brain's signal transmissions involves deploying technologies such as functional **Near-Infrared Spectroscopy (fNIRS)**, **Electroencephalography (EEG)**, and **Magnetoencephalography (MEG)**, which then interpret these signals as comprehensible commands for the Metaverse.

- **EEG**: By placing electrodes onto the scalp, EEG captures the brain's electrical dynamism. Crucial data characteristics are then extracted through signal processing methodologies such as **Fast Fourier Transform (FFT)**, **Independent Component Analysis (ICA)**, and wavelet transforms.

- **ML algorithms**: ML, supervised and unsupervised, uses models such as ANNs, SVMs, and DL to differentiate brain signals into diverse cognitive states or directives for the Metaverse.

Spatial audio

Spatial audio technology creates a realistic audio environment in the Metaverse by simulating the way sound behaves in the real world. Technologies such as ambisonics, binaural audio, and object-based audio rendering are employed to provide users with a more immersive audio experience:

- **Ambisonics**: This is a full-sphere surround sound technique that encodes audio in a spherical harmonic domain, allowing for flexible speaker configurations and realistic sound field reproduction. **Higher-order Ambisonics (HOA)** increases the spatial resolution for an enhanced audio experience.

- **Binaural audio**: Binaural audio simulates the way humans perceive sound by using **Head-Related Transfer Functions (HRTFs)** to model the filtering effects of the head, ears, and torso. This technique creates a realistic sense of directionality and spatial awareness when using headphones.

- **Object-based audio rendering**: Audio objects are assigned metadata describing their position, size, and other properties. Rendering engines such as Dolby Atmos and MPEG-H 3D Audio use this metadata to adaptively render audio for different playback systems and environments.

Sensor fusion

Combining data from multiple sensors, such as accelerometers, gyroscopes, and magnetometers, sensor fusion algorithms provide a more accurate and robust tracking solution for the Metaverse. Techniques such as Kalman filters and particle filters are used to integrate data from different sources, improving the overall tracking performance:

- **Kalman filters**: These recursive algorithms estimate the state of a dynamic system by fusing noisy sensor data with a predictive model. They are widely used in tracking and navigation systems due to their computational efficiency and robustness.

- **Particle filters**: These non-parametric algorithms use a set of particles to represent the probability distribution of a system's state. They are particularly useful for non-linear and non-Gaussian estimation problems, where the system's dynamics cannot be accurately modeled by a Gaussian distribution.

Networking and synchronization

To ensure a seamless and interactive experience in the Metaverse, efficient networking and synchronization technologies are required. These technologies enable real-time communication and collaboration among users, maintaining consistency across different devices and platforms:

- **Network protocols**: Protocols such as WebRTC, QUIC, and SpatialOS enable low-latency, real-time communication and synchronization between users in the Metaverse, ensuring a smooth and consistent experience across different devices and platforms.

- **Distributed systems**: Consistency algorithms such as the Paxos and Raft consensus protocols are used to maintain a consistent state across multiple servers or nodes, facilitating seamless interactions and collaboration in the Metaverse.

In summary, tracking and capture technologies for the Metaverse rely on advanced techniques and algorithms to provide highly immersive and interactive experiences. These technologies enable users to interact with the virtual world using their physical movements, gestures, and expressions, creating a strong sense of presence and facilitating seamless interactions between users and a digital environment.

How businesses can benefit

Tracking and capture technologies can offer a wealth of benefits to businesses operating in the Metaverse, thanks to their ability to record and analyze user movements, interactions, and behaviors. Here are some ways these technologies can benefit businesses:

- **User behavior analytics**: By tracking and capturing user movements and interactions, businesses can gain a better understanding of user behavior, preferences, and habits. This data can be used to improve product designs, enhance user experiences, and create more effective marketing strategies.

- **Personalized experiences**: Tracking technologies can help businesses offer more personalized experiences to users. For example, a virtual store might change its layout or the products it displays based on the preferences and past behaviors of the user currently visiting.

- **Performance optimization**: Businesses can use tracking and capture technologies to identify any issues or bottlenecks in their virtual environments. By understanding how users navigate and interact with these environments, businesses can make improvements to enhance performance and usability.

- **Security and compliance**: Tracking and capture technologies can also help ensure security and compliance in the Metaverse. Businesses can monitor user behavior to detect any fraudulent activities or violations of terms and conditions.

- **Immersive marketing and advertising**: Businesses can use tracking technologies to create immersive and interactive advertisements. For instance, an ad could respond to a user's movements or interactions, making for a more engaging experience.

- **User testing**: Businesses can use tracking and capture technologies to conduct user testing in the Metaverse. By observing how users interact with a new product, service, or feature, businesses can gather valuable feedback and make necessary improvements before a full launch.

- **AR and VR applications**: These technologies are essential to create immersive AR and VR experiences in the Metaverse. For instance, businesses can offer virtual tours, training sessions, or conferences, where users can interact with an environment and each other in a realistic way.

- **Real-time collaboration**: Tracking and capture technologies can facilitate real-time collaboration in the Metaverse, which can be particularly beneficial for businesses with remote teams. Users can interact with shared virtual objects, and their movements and changes can be seen by all participants in real time.

By leveraging tracking and capture technologies, businesses can offer more engaging, personalized, and effective products and services in the Metaverse, leading to increased user satisfaction and business success.

Summary

Perception technologies – spanning a diverse spectrum of systems such as computer vision, object recognition, auditory processing, and haptic feedback – serve as pivotal cornerstones for the Metaverse. Their importance is twofold, augmenting both the interaction capabilities of users within this expansive digital universe and the sophistication of the virtual environment itself.

Perception technologies enable the Metaverse to accurately interpret user behaviors, actions, and intentions. By capturing and translating real-world movements, gaze, speech, and other human intricacies in the digital space, these technologies create a conduit for intuitive, natural interaction within the Metaverse. This leads to a heightened level of immersion, as users can navigate and interact within the Metaverse in a manner akin to their experiences in the physical world.

Simultaneously, perception technologies allow the Metaverse to dynamically adapt and respond to user input, enhancing the richness and realism of a virtual environment. Be it through real-time object detection and spatial mapping for AR overlays, or auditory processing for spatial sound effects, these technologies allow the Metaverse to mimic the complexity of the real world, generating a holistic, engaging sensory experience.

Without perception technologies, the Metaverse's potential would be significantly curtailed, relegating it to a two-dimensional, limited form of digital interaction. However, with these technologies, the Metaverse is brought to life, providing an expansive, interactive, and immersive digital universe – a new frontier where the limitations of the physical world give way to the limitless possibilities of imagination and innovation.

In conclusion, perception technologies serve as the senses of the Metaverse. They are the vital components that convert a static, lifeless digital space into a vibrant, engaging, and immersive universe that transcends the boundaries of reality, enabling a new form of human-computer interaction that redefines our conception of what the digital realm can be.

In the following chapter, *Chapter 6, The Different Types of Computing Technologies*, we will review the kinds of high computational resources that enable realistic visuals, physics simulations, and AI algorithms, enhancing user engagement in the Metaverse.

6

The Different Types of Computing Technologies

The most comprehensive form of the Metaverse will demand more computing power than anything we've seen so far. In this analysis, we'll be examining the various types of computing power that will play significant roles. We'll kick things off with cloud computing, detailing how it serves as the Metaverse's key structure for storage and processing capacities. Then, we'll discuss the importance of edge computing in reducing delays, enhancing user experiences, and facilitating real-time engagement. Next, we'll review the role of distributed computing in effectively spreading workloads across numerous systems for maximum performance and scalability. Finally, we'll delve into the value of decentralized computing, emphasizing its essential part in promoting a secure, user-driven, and democratic Metaverse.

In this chapter, we're going to cover the following main topics:

- Learning what the main different types of computing technologies are for the Metaverse
- Learning about the advantages of using one type of computing technology over the other from business and tech perspectives
- Understanding the ramifications of particular computing technologies in terms of their power capability, management, and data privacy

Cloud computing

The foundation of the Metaverse is constructed upon cloud computing, which provides dynamic scalability, universal accessibility, and immense processing power. This technology enables the creation of virtual realms that are vivid, interactive, and accessible from any device, at any location and time. By managing the extensive computational demands of the Metaverse, cloud computing has paved the way for incredibly realistic digital environments, AI-driven experiences, and worldwide synchronous interactions.

Cloud computing refers to a computing model where internet-based resources such as servers, storage, databases, networking, software, analytics, and intelligence are utilized to deliver flexible resources, foster innovation, and achieve economies of scale. Users can store and process data on remote servers, accessing applications and services remotely.

Cloud computing is revolutionizing various sectors, including gaming, social media, eCommerce, and more, within the Metaverse. It is transforming our digital existence and shaping how we live, work, and engage.

Cloud computing and the Metaverse

In this section, we'll provide a detailed exploration of the connection between cloud computing and Metaverse.

Infrastructure and scalability

The Metaverse necessitates an immense amount of computational capacity and storage due to its inherent characteristics. It is an ever-present and ever-changing universe where numerous users, potentially numbering in the millions or billions, engage in real-time interactions. The scale of this demand surpasses the capabilities of conventional computing infrastructure. Cloud computing, offering virtually boundless scalability and worldwide accessibility, furnishes the essential foundation for sustaining the Metaverse. It enables swift resource allocation to meet increasing demands and the flexibility to scale down during periods of reduced demand.

Latency and connectivity

To ensure a smooth and captivating experience in the Metaverse, it is essential to have fast and responsive connections with minimal delays. The utilization of cloud computing infrastructure, along with innovations such as edge computing and 5G technologies, plays a vital role in reducing latency and facilitating rapid data transmission. These advancements are especially important for enabling real-time interactions and immersive encounters within the Metaverse.

Real-time data processing and analytics

The Metaverse produces an immense volume of data. Each interaction, movement, and conversation generates data that must be promptly processed and evaluated to offer users a smooth and captivating experience. Cloud computing plays a pivotal role in enabling this by leveraging cutting-edge analytics and machine learning. It enables instantaneous data processing and facilitates the extraction of valuable insights, which can be employed to enrich the user's journey.

Artificial intelligence and machine learning

The Metaverse deeply utilizes **artificial intelligence (AI)** and **machine learning (ML)** to formulate intelligent landscapes and evolve AI entities, personalize user activities, and analyze user trends. These tech solutions generally leverage cloud computing, given the vast processing strength needed for building and introducing AI models.

Decentralization and blockchain

Many cloud providers now offer **Blockchain as a Service (BaaS)**, which provides the necessary infrastructure for developing and running blockchain applications. This significant advancement can greatly contribute to the creation of a decentralized Metaverse, empowering users with significantly greater control over their data and virtual assets. There is a growing trend toward the decentralization of the Metaverse, with blockchain technology often seen as a key tool for achieving this.

Accessibility and ubiquity

Cloud computing plays an essential role in ensuring the Metaverse's persistence, enabling it to evolve and exist continuously, regardless of whether a user is actively logged in or not. Moreover, with the support of cloud computing, the Metaverse is readily available on any internet-connected device, providing consistent access and usability to users, irrespective of their geographical location.

Security and privacy

Cloud computing offers robust security measures that guard data, applications, and the underlying infrastructure against possible dangers. This is especially vital in the Metaverse, where the challenge is to safeguard the interactions and shared information of potentially billions of users. In such a context, the importance of security and privacy cannot be overstated.

Development and innovation

To create an immersive Metaverse experience, high-quality graphics are essential, and they necessitate substantial computational power. This is where cloud computing comes into play, serving as a crucial catalyst for the innovation and development necessary for the Metaverse's realization. Moreover, cloud platforms equip developers with the necessary tools and services, enabling the rapid and efficient creation and deployment of applications. These platforms also provide the vast computational resources required to manage complex simulations.

The concept of a fully immersive, persistent, and interactive Metaverse wouldn't be possible without the foundation laid by cloud computing. Its scalability, accessibility, data processing abilities, security measures, development tools, and other features are integral for the realization of this seemingly science-fiction dream. Indeed, the Metaverse isn't just linked to cloud computing – it's fundamentally empowered by it.

Cloud computing technology overview

Cloud computing involves delivering on-demand computing services over the internet, including servers, storage, databases, networking, software, analytics, and intelligence. Instead of owning their own computing infrastructure or data centers, companies can rent access to anything, from applications to storage, from a cloud service provider.

Let's look at a detailed, technical explanation of how cloud computing works.

Components of cloud computing

Cloud computing encompasses many different components, and understanding these helps us appreciate how the system works:

- **Clients**: Clients are the end user devices that interact with cloud data storage. Clients can be thin clients or thick clients. Thin clients run a web browser and do not have any installed applications, while thick clients run standalone applications.

- **Applications**: Cloud applications can be diverse, from photo editing software to business applications such as CRM and ERP.

- **Platform**: The platform provides a runtime environment where a cloud application can run. This could include an operating system, programming language execution environment, database, web server, and more.

- **Infrastructure**: The infrastructure in the cloud is the collection of hardware and software that runs the entire system. This includes servers, storage, networks, and the virtualization software that makes cloud computing possible.

- **Service provider**: The service provider makes all these services available to the user. The service provider ensures the services are up and running and takes care of maintenance and updates.

Cloud computing services

Cloud computing offers various services that are mainly divided into three categories:

- **Infrastructure as a Service (IaaS)**: This is the base layer of the cloud stack. It provides access to fundamental resources such as physical machines, **virtual machines** (**VMs**), virtual storage, and so forth. Cloud service providers manage the infrastructure while clients are responsible for managing their data, applications, runtime, middleware, and operating systems.

  Examples: **Amazon Web Services** (**AWS**), **Google Cloud Platform** (**GCP**), and Microsoft Azure.

- **Platform as a Service (PaaS)**: Built on top of IaaS, PaaS provides an environment for developing, testing, and managing applications. It abstracts much of the management of underlying infrastructure, including operating systems, server hardware, and networking resources. The cloud service provider manages everything up to the middleware, runtime, operating system, virtualization, servers, storage, and networking. The client just manages the applications and data.

Examples: Google App Engine, AWS Elastic Beanstalk, Microsoft Azure App Services.

- **Software as a Service (SaaS)**: This is the most comprehensive layer and is where a complete product/application is offered to the user as a service on demand. The service provider hosts both the application and the data, and the end user is free to use the service from anywhere. The service provider manages all aspects of the cloud service, including data, middleware, applications, runtime, operating systems, servers, storage, and networking.

  Examples: Google Apps, Dropbox, and Salesforce.

Deployment models

Cloud deployment models classify and characterize a given cloud environment, dictating how and where it's located based on elements such as ownership, size, and access methods. By establishing these factors, they ultimately shape the cloud's function and nature, providing detailed insight into its accessibility and intended use.

The following are three major cloud computing deployment models:

- **Public cloud**: These are owned and operated by third-party cloud service providers, who deliver their computing resources over the internet. Microsoft Azure and Amazon AWS are examples of public clouds.

- **Private cloud**: This refers to cloud computing resources used exclusively by a single business or organization. Private clouds can be physically located at the organization's on-site data center or hosted by third-party service providers.

- **Hybrid cloud**: As the name suggests, hybrid clouds combine public and private clouds, allowing data and applications to be shared between them. Businesses get the flexibility and benefits of both worlds, enabling them to move workloads around as costs, needs, and technologies change.

In terms of the infrastructure setup, here is how cloud computing works:

- **Data center facilities**

  The backbone of cloud computing is a vast network of physical servers housed in data centers around the world. The servers are often racked and stacked with the latest technologies to support a vast amount of data storage and computing tasks.

  These facilities include redundant power supplies, cooling systems, security devices, and multiple network connections.

- **Virtualization**

  One of the essential techniques that's used in cloud computing is virtualization. Cloud computing utilizes virtualization technology to maximize the usage of physical resources.

  This technology allows you to create virtual instances of physical components such as a server, a desktop, a storage device, an operating system, or network resources.

  Using a software layer called a hypervisor, virtualization separates compute environments from the physical hardware, enabling multiple VMs to run on the same physical machine, each with its own operating system and applications.

- **Resource pooling and multi-tenancy**

  When a cloud service provider receives a service request, that request is directed to a particular data center. The data center's servers run a task-distributing program that fulfills the request using the available resources. If a single server isn't sufficient to process the job, the task can be distributed across multiple servers.

  Resources are pooled together in the data center and can be provisioned as needed to multiple customers – a model known as multi-tenancy.

  Each customer's data and applications remain isolated and secure from one another, even though they're using shared resources.

- **Self-service provisioning**

  End users can spin up compute resources for nearly any type of workload on demand. This self-service nature of cloud computing allows for flexibility and rapid elasticity.

- **Network connectivity**

  The servers in data centers are interlinked via a network (usually by high-speed, internal connections), and these servers connect to the end users over the internet.

  Data encryption is applied for secure transmission over the public network.

- **Service orchestration**

  Cloud service providers use orchestration systems to automate the management, coordination, and organization of complex services and microservices.

  This orchestration is often realized through the use of **application programming interfaces (APIs)** that interact with software-defined networks, servers, and storage.

- **Scalability and elasticity**

  Cloud computing allows for easy scalability and elasticity, which means services can be adjusted according to demand. Users can scale services up or down based on demand and pay only for the resources they use. If an application requires more resources due to a spike in user traffic or computational activities, a cloud application can automatically access more resources.

This can be achieved in two ways: vertical scaling (adding more power to an existing machine) or horizontal scaling (adding more machines to the network).

- **Load balancing**

  Cloud services distribute network or application traffic across multiple servers using a load balancer, enhancing both the availability and reliability of applications.

  Load balancing methods can be based on various algorithms, such as round-robin, IP hashing, or least connections.

- **Fault tolerance and disaster recovery**

  Fault tolerance is an attribute that enables a system to continue operating properly in the event of a failure surrounding some of its components.

  Disaster recovery is a set of policies and procedures to enable the recovery or continuation of vital technology infrastructure and systems following a natural or human-induced disaster.

  In cloud computing, fault tolerance and disaster recovery are often achieved through redundant systems and data replication across multiple locations.

- **Billing and metering of services**

  Cloud service providers monitor, control, and report resource usage in real time to provide transparency and billing information.

  Cloud providers often use a pay-as-you-go model, which can lead to unexpected operating expenses if administrators do not adapt to the cloud pricing model.

Edge computing

Edge computing is a distributed computational model that plays a crucial role in the realization of the Metaverse by enabling real-time, immersive experiences. The key function of edge computing is to perform processing at the "edge" of the network, which could encompass a local computer, an IoT device, or an edge server located closer to the origin of the data. The primary benefit of this methodology is the reduction in the need to shuttle huge volumes of data between the user's device and the cloud as it processes data nearer to its source.

In the context of the Metaverse, this is of substantial importance as it involves significant data processing to create immersive, real-time interactions. Such interactions traverse numerous data-intensive applications, including AI-driven virtual interactions, real-time gaming, VR, AR, 3D video, and more.

One of the significant concerns in this scenario is the potential latency issues that can arise when processing data in a centralized cloud. Such issues could disrupt the real-time and immersive environment of the Metaverse. However, edge computing can counteract this by boosting performance and lessening the load on the network infrastructure through local data processing.

Edge computing enhancements

As the Metaverse evolves, the incorporation of edge computing becomes essential for its optimal real-time functioning, thereby augmenting the foundational support offered by the cloud computing infrastructure. Compared to traditional cloud computing, edge computing enhances how you experience the Metaverse in these specific ways:

- **Latency reduction**:

  - Edge computing facilitates data processing close to its origin, substantially reducing latency.

  - The importance of this feature is notably amplified within the Metaverse, where instantaneous engagement is anticipated across gaming, social communication, and VR scenarios.

  - Reduced latency guarantees seamless, delay-free user experiences, which is a key requirement in use cases such as instantaneous gaming, live streaming, or virtual performances within the Metaverse.

- **Data localization and privacy**:

  - Through edge computing, data is processed in the vicinity or exactly where it originates, thereby achieving data localization.

  - The ability to process data locally, a unique characteristic of edge computing, reduces data transmission over networks, thereby potentially strengthening privacy and security measures – an important consideration given the personal and sensitive nature of data associated with Metaverse activities.

  - Adhering to data protection laws also becomes more straightforward as the data typically stays within the same jurisdiction.

- **Bandwidth efficiency**:

  - Edge computing enables the handling of substantial data quantities close to the origin, mitigating the necessity of transmitting data to and from the cloud.

  - Such an approach considerably minimizes the consumption of bandwidth and associated expenses, which is particularly crucial for high-bandwidth applications such as VR or 3D settings within the Metaverse.

  - Consequently, it enhances the comprehensive performance of Metaverse platforms, providing users with superior visual quality and interactive experiences.

- **Scalability and flexibility:**

  - The scalability of edge computing is enhanced as new nodes can be effortlessly integrated into the network to cater to more users or computational requirements.

- Such scalability simplifies the handling of the ever-expanding user base and the increasing demands for interactivity within the Metaverse.

- By allocating specific computational tasks to edge nodes, edge computing enhances flexibility, which subsequently frees up central resources for other operations.

- **Resiliency and reliability:**

    - By generating a decentralized network without a singular point of vulnerability, edge computing amplifies the robustness of Metaverse platforms.

    - If a single node collapses, backup nodes are available to sustain uninterrupted user experiences.

    - As edge computing brings data processing near to the users, it reduces susceptibility to major outages that may occur in centralized cloud systems.

- **Real-time analytics and AI:**

    - The Metaverse will heavily utilize real-time analytics, AI, and ML for aspects such as predicting user behavior, customizing content, and instant decision-making.

    - The application of edge computing enables these processes to happen near the origin, which offers swifter insights and responses, a critical requirement for real-time applications within the Metaverse.

Edge computing technology overview

Edge computing is a distributed computing paradigm that aims to bring data storage and computation closer to the devices or data sources where it's needed, improving response times and saving bandwidth. It decentralizes data processing and analysis, reducing the need for long-distance communication with central data centers or clouds. This approach is beneficial for situations where low latency is critical, or where data privacy and security are of concern, as it reduces the opportunity for data interception during transmission.

Devices involved

- Edge computing can involve various devices, such as sensors, IoT devices, mobile phones, laptops, edge servers, and more.

- A new device that uses edge computing is the Apple Vision Pro.

- These devices generate data that may need to be processed quickly, without the latency that could be involved in sending the data to a distant server or data center.

- In edge computing, a significant amount of processing happens on the edge devices themselves or on local edge servers.

Data generation and preprocessing

- Data is produced by various sources, such as sensors, users on their devices, or during the operation of automated systems.

- For many applications, it is impractical to send all raw data back to a central server due to the volume of data and the cost or time needed for transmission.

- Therefore, edge devices often preprocess data, filtering and summarizing it before sending it to the next step in the data processing chain.

Data processing at the edge

- Edge computing often involves real-time or near-real-time processing, with data being processed at the edge of the network instead of being sent back to a centralized server or cloud.

- Edge computing nodes (whether they're edge devices themselves or edge servers) perform more in-depth data processing.

- This near-source processing reduces the need for bandwidth and reduces the time required to obtain results (latency), which is essential for time-critical applications.

- The specifics of this processing vary by application but may include tasks such as trend detection, event correlation, anomaly detection, ML, and even predictive analytics.

- The results of this data processing can then be used to make decisions, inform users, or control other devices.

Infrastructure

- Edge infrastructure is typically composed of edge nodes that contain both computational and storage capabilities, similar to a traditional data center but on a smaller scale.

- These edge nodes can be standalone edge devices, edge servers, or micro data centers that service a specific geographic area or application.

- Each edge node typically operates autonomously but can connect with other edge nodes or a centralized server for coordination, backup, or additional processing power.

Communication and networking

- Communication protocols in edge computing are designed to provide robust, low-latency communication between edge devices, edge nodes, and central servers.

- The specific protocols used depend on the application but could include industrial communication protocols (for example, Modbus and DNP3), IoT protocols (for example, MQTT and CoAP), or more general network protocols (for example, TCP/IP).

- Edge networks can also include features such as network slicing (creating virtual networks for specific applications), multi-access edge computing (integrating wireless and wired networks), and software-defined networking (for flexible network management).

Distributed computing

The implementation of distributed computing, a concept that involves utilizing several networked computers' combined processing power, plays an integral role in the successful operation and management of the Metaverse. This technique offers a more efficient means of leveraging extensive processing capabilities than a single computer, especially for large-scale tasks.

The reason behind the Metaverse's immense computational demand lies in its complexity, which encompasses elements such as high-resolution visuals, real-time user interactions, and constantly evolving virtual landscapes. Distributed computing takes these intricate tasks, splits them among multiple computers, and then brings the results together, enhancing the efficiency and performance of the Metaverse.

Without distributed computing's capability to handle the Metaverse's complex, constantly changing, and real-time nature, its smooth operation would be virtually unattainable. As such, the Metaverse can offer an engaging and seamless user experience thanks to the combined processing might of multiple machines. Let's explore the basics of distributed computing, how it's used within the Metaverse, and its crucial role in ensuring the Metaverse's effective functioning.

Major reasons why distributed computing is crucial:

- **Scalability**: By dividing a task among numerous systems, distributed computing enhances the capacity to accommodate a significant number of users and manage hefty computation workloads. This computing approach could potentially support the Metaverse, a shared space anticipated to serve billions of people at the same time.

- **Real-time interactions**: Supporting real-time interactions among users and between users and the environment would be a requirement for the Metaverse. Distributed computing allows you to share processing load across multiple systems, thereby reducing latency and enabling real-time responses.

- **Fault tolerance and redundancy**: In the event of individual component failures, the Metaverse remains accessible and functional due to the enhanced reliability and availability provided by distributed systems. This is because, when one server or system breaks down, others are capable of assuming control.

- **Data management**: Distributed computing facilitates efficient management, storage, and processing of the vast amounts of data generated by the Metaverse, including user activities, AI interactions, and environmental changes.

- **Virtual reality rendering**: VR is a key component of the Metaverse, and rendering VR environments requires substantial computational power. Distributed computing can handle the computational load, allowing for more complex and detailed environments.

- **Global accessibility**: Users will access the Metaverse from different geographical locations as it is a global concept. To enhance user experience and reduce latency, a distributed computing architecture can ensure that data and processing power are situated near the users.

- **Security and privacy**: By spreading data and processing it over numerous systems, distributed computing makes it more challenging for malicious actors to breach a single system and obtain a substantial quantity of data. Consequently, distributed computing can potentially enhance privacy and security.

Distributed computing technology overview

A distributed system is a collection of autonomous computers that are connected through a network and distribution middleware, which enables the computers to coordinate their activities and share resources. This system is designed to tackle problems that can't be solved by a single computer due to their size, computational needs, or the requirement for concurrent processing and high availability. These systems work on the principle that many tasks can be done faster and more efficiently by splitting them up into smaller tasks that are then executed concurrently on different machines.

Principles of distributed systems

- **Transparency**: This means that the system shouldn't hide the fact that it's a distributed system from the user. It includes aspects such as access, location, migration, relocation, replication, concurrency, and failure.

- **Openness**: The system should be able to be flexibly configured and scaled, according to different protocols and interfaces.

- **Scalability**: A system is scalable if it can handle an increasing number of tasks, or if the system's performance improves after adding hardware resources such as memory, storage, or processor.

Components of distributed systems

- **Nodes**: The individual computers within the system are known as nodes. They can be personal computers, data center servers, or supercomputers.

- **Networks**: The computers are interconnected through a network. This could be a **local area network (LAN)** or a **wide area network (WAN)** such as the internet.

- **Middleware**: This is the software layer, and it provides a programming abstraction. It also masks the heterogeneity of the underlying networks, hardware, operating systems, and programming languages.

Distributed computing models

- **Client-server**: The client sends a request, and the server responds to the request. This is the most common model.

- **Peer-to-peer (P2P)**: Each node has equivalent responsibilities and communication capabilities. The most common example is blockchain technologies.

- **Master-slave**: One node (the master) distributes tasks among other nodes (slaves).

Process communication

- Nodes in a distributed system have to communicate with each other to coordinate their actions.

- They communicate by exchanging messages over the network. Messages can be a direct system-to-system communication or through a shared common area (message queue).

- There are two main communication methods: synchronous (blocking), where the process waits for the call to complete and returns control, and asynchronous (non-blocking), where the process doesn't wait for the call to complete and continues with other tasks.

Fault tolerance

- Fault tolerance is the capability of a system to continue functioning correctly even in the event of hardware or software failures.

- It can be achieved through redundancy, where each critical component has a backup. The system switches to the backup if the main component fails.

- Other strategies include error recovery (the system detects an error, corrects it, and continues the operation) and failure masking (the system hides the effects of a failure from the user.

Distributed algorithms

- Distributed algorithms are designed to run on multiple processors, without tight centralized control.

- They need to be very efficient in their communication to prevent overloading the network, and they need to handle the failures of individual nodes.

- Examples include distributed sorting algorithms, graph algorithms, and consensus algorithms.

Distributed databases and distributed filesystems

- Distributed databases are databases where the storage devices are not attached to a common processor but are dispersed over a network.

- Similarly, a distributed filesystem allows files to be stored on several nodes across a network but appear to users as if they are located on a local disk.

- The main advantages of distributed databases and filesystems are reliability (through redundancy), availability, and performance.

Security in distributed systems

- Distributed systems need to ensure confidentiality, integrity, and availability.
- Security protocols such as SSL/TLS are often used to encrypt the communication between nodes.
- Authentication mechanisms such as digital signatures, **public key infrastructure** (**PKI**), and Kerberos are used to verify the identity of the nodes in the system.

Challenges in distributed computing

- **Concurrency**: Multiple nodes must be able to work on the same task without conflict.
- **Latency**: There's always a delay when sending information across a network. The more the network is loaded, the larger the delay.
- **Data consistency**: When data is replicated across nodes, it's challenging to keep all the copies up to date.
- **Hardware or software heterogeneity**: Different nodes may have different hardware and software, making the development and maintenance of distributed systems challenging.
- **Scalability**: As a system grows, resources may become overused, leading to bottlenecks and performance degradation.

Distributed computing techniques and methods

- **Load balancing**: This is the process of distributing network traffic across several servers to ensure that no single server bears too much demand. This leads to enhanced responsiveness and availability of applications.
- **Caching**: This involves storing copies of data temporarily in high-speed media (such as RAM) closer to the application to reduce the data access times. Distributed caching can be used to store data across multiple nodes, to increase fault tolerance and availability.
- **Sharding**: This is a technique where data is partitioned into smaller, faster, more easily managed parts called shards. Sharding helps in the horizontal partitioning of data in a database or search engine. Each shard is held on a separate database server instance, which spreads the load, improving the performance of large systems.

Parallel processing and concurrency

- **Parallel processing**: This involves executing multiple parts of a program, on two or more processors, to execute it more quickly. Certain types of tasks, such as mathematical problems and AI training processes, can be divided into multiple smaller tasks and processed simultaneously.

- **Concurrency**: Concurrency is a property of systems in which multiple processes are executing at the same time, and potentially interacting with each other. Concurrency control is crucial in distributed systems to prevent conflicts and ensure data consistency.

Distributed hash tables (DHTs)

- DHT is a class of decentralized distributed systems that provides a lookup service similar to a hash table. Any participating node can retrieve the value associated with a given key.

- DHTs characteristically emphasize their potential for scalability and their ability to handle churn (continuous changes in the set of participants).

Distributed data structures

The data structure concepts of lists, queues, arrays, and more can be extended to distributed data structures, where data is stored across the distributed system. These provide a consistent and uniform view and operation, irrespective of the actual physical storage and distribution.

Virtualization in distributed systems

- Virtualization is the technique of creating virtual (rather than actual) versions of something, such as an operating system, a server, a storage device, or network resources.

- In distributed systems, it allows multiple applications to run concurrently on the same physical host. It can make the system more efficient and agile and reduce costs by consolidating resources.

Consensus algorithms in distributed systems

- A consensus algorithm is a process in computer science that's used to achieve agreement on a single data value among distributed systems. The most famous examples are the Paxos and Raft algorithms.

- Consensus algorithms are crucial in many distributed systems, such as blockchains and databases, to agree on the values and to ensure that the data is consistent across all nodes.

Security in distributed systems

- Distributed systems need to ensure confidentiality, integrity, and availability.

- Security protocols such as SSL/TLS are often used for encrypting the communication between nodes.

- Authentication mechanisms such as digital signatures, PKI, and Kerberos are used to verify the identity of the nodes in the system.

Synchronization in distributed systems

- Synchronization ensures that concurrent processes operate in the correct sequence to work correctly. The most common synchronization operations are locks, semaphores, and monitors.

- In a distributed environment, clock synchronization is also a significant issue as it helps to avoid issues related to the timing of events.

Distributed operating systems

- Distributed operating systems are an extension of the network operating system and support higher levels of communication and integration of the machines on the network.

- Examples include Amoeba, Plan9, and LOCUS. These systems have the flexibility of sharing resources and capabilities to provide users with a single and integrated coherent network.

Microservices architecture in distributed systems

- Microservices is an architectural style that structures an application as a collection of services that are highly maintainable and testable, loosely coupled, independently deployable, and organized around business capabilities.

- The microservices approach was the first realization of **service-oriented architecture** (**SOA**) and followed the introduction of DevOps. It is becoming more popular because of the advent of cloud computing.

Serverless computing

- Serverless computing is a cloud computing execution model where the cloud provider dynamically manages the allocation and provisioning of servers.

- In a distributed system, serverless computing can simplify the process of deploying code into production. Scaling, capacity planning, and maintenance operations may be hidden from the developer or operator. AWS Lambda is an example of serverless computing.

Future of distributed systems

- With the rise of IoT, 5G, edge computing, and AI, the number of devices connected to the internet and the amount of data generated are growing rapidly, increasing the need for efficient distributed systems.

- The future will see more complex distributed systems with higher levels of automation, improved fault tolerance mechanisms, and advanced data analytics capabilities.

Decentralized computing

Distributed control across multiple nodes is a characteristic of decentralized computing, a distinguishing factor that sets it apart from centralized systems, which are reliant on a single control point. This form of computing not only boosts the system's productivity by eradicating potential single points of failure but also promotes resilience, transparency, and heightened performance.

Within the context of the Metaverse, decentralized computing is instrumental in managing core operations. These include verifying asset ownership, enabling user interaction, and overseeing transactions. Through the introduction of blockchain and similar decentralized technologies, these functions have been fortified, pushing the Metaverse toward a future that leans more toward autonomy and a user-oriented approach.

In this section, we'll delve deeper into how decentralized computing operates within the Metaverse. Our discussion focuses on the importance of this technology, its various applications, and its capacity to influence the trajectory of the virtual world's future. The role it plays in the Metaverse's development and operation cannot be understated, serving as a crucial technological element that allows distributed networks to function seamlessly within the Metaverse systems.

Major advantages of decentralized computing

- **Increased resilience**: Decentralized computing reduces the dependence on a single central authority or server, distributing data and processing power across multiple nodes. This makes the system more robust and resistant to failures or attacks as the failure of one node does not result in the complete collapse of the system.

- **Enhanced security**: Centralized systems are more vulnerable to cyberattacks and data breaches because a single point of failure can compromise the entire network. Decentralized computing reduces this risk by dispersing data and processing it across multiple nodes, making it more difficult for malicious actors to target and infiltrate the system.

- **Improved scalability**: With decentralized computing, it is easier to scale the system as it can leverage the resources of multiple nodes. New nodes can be added to the network to increase processing power and storage capacity, allowing the system to handle growing demands without significant disruptions or bottlenecks.

- **Increased privacy**: Decentralization can enhance privacy by reducing the need for trusted intermediaries or central authorities that may have access to sensitive data. Distributed ledger technologies, such as blockchain, provide a decentralized approach to data management that ensures transparency and immutability while protecting user privacy.

- **Empowered individuals**: Decentralized computing can empower individuals by giving them greater control over their data and digital identity. Users can have direct ownership and control of their personal information, reducing reliance on centralized platforms that collect and monetize user data without their explicit consent.

- **Geographical independence**: Decentralized computing enables users to access and contribute to the network from anywhere in the world. This eliminates geographical restrictions and allows for global collaboration and participation, fostering innovation and inclusivity.

- **Lower costs**: Decentralized computing can potentially reduce infrastructure and operational costs by leveraging existing resources and avoiding the need for expensive centralized systems. It enables resource sharing, which can result in more efficient utilization of computing power, storage, and bandwidth.

- **Openness and transparency**: Decentralized systems are often built on open protocols and standards, fostering transparency and interoperability. This openness allows for peer-to-peer interactions and facilitates the development of **decentralized applications** (**DApps**) that can be built on top of existing decentralized networks.

Decentralized computing technology overview

In contrast to centralized systems where all resources, control, and decision-making processes are concentrated at a single point, decentralized systems distribute these elements across multiple nodes. This allows for greater resilience, potential redundancy, and more local decision-making power.

P2P systems are typically decentralized, with no central server. Each node (or "peer") in the network can act as both a server and a client, providing and consuming resources. This contrasts with traditional client-server models where servers provide resources consumed by clients.

The physical or logical structure of a network, commonly referred to as its topology, is one of the fundamental considerations in decentralized computing. Topologies such as Mesh, Ring, or Star, each with their unique characteristics, can enable decentralization to varying degrees.

A deeper technical dive

- **Decentralized data storage**: In a decentralized system, data is typically stored across multiple nodes. Each node might hold all the data (full replication), a portion of the data (partitioning), or some combination thereof (sharding). Data redundancy can provide fault tolerance and improve access speed.

- **Consistency models**: Ensuring data consistency in decentralized systems is challenging. Various models such as eventual consistency, strong consistency, or causal consistency are used, each with trade-offs between consistency, availability, and partition tolerance, as described by the CAP theorem.

- **Synchronization and coordination**: Protocols for ensuring all nodes in a decentralized system work in harmony are critical. Some commonly used protocols include Paxos, Raft, and Zab, which help achieve consensus in a network, a fundamental problem in distributed systems.

- **Communication protocols**: Decentralized systems use different types of protocols for peer-to-peer communication, such as TCP/IP, UDP, or protocols built on top of these such as BitTorrent for file sharing or blockchain protocols for decentralized ledger updates.

Security and trust

- **Decentralized identity systems**: These provide ways to verify the identity of users or nodes in a decentralized way, often using cryptographic techniques. **Decentralized public key infrastructure (DPKI)** and blockchain-based systems are examples.

- **Trust and reputation systems**: In a decentralized environment, systems need to assess the trustworthiness of other nodes. Reputation systems can use ratings, feedback, or even economic incentives to create a trust network.

- **Security and cryptography**: Given their open and distributed nature, decentralized systems often leverage cryptography for securing communication, ensuring data integrity, and verifying identities. Asymmetric encryption, hashing algorithms, and digital signatures are common.

Applications and implementations

- **Blockchain and cryptocurrencies**: Perhaps the most famous implementation of decentralized computing, blockchain provides a decentralized ledger that is secure, transparent, and resistant to modification. Cryptocurrencies such as Bitcoin use this technology.

- **Decentralized filesystems**: Systems such as **InterPlanetary File System (IPFS)** or Storj provide decentralized storage, allowing files to be stored and retrieved from multiple locations simultaneously.

- **Decentralized computing platforms**: Platforms such as Ethereum, which extends the blockchain concept with smart contracts, allow dApps to be built and run on a decentralized network.

- **Decentralized AI and ML**: AI models can be trained and run on decentralized networks, enhancing privacy and potentially leveraging untapped resources across the network.

Challenges and ongoing research

- **Scalability**: As the number of nodes in a decentralized system increases, the overhead for maintaining synchronization and consensus can also grow, leading to scalability challenges. Various solutions, such as sharding in blockchain, are being researched and implemented.

- **Privacy**: Despite its potential for enhancing privacy, decentralized computing also presents challenges. For example, in a public blockchain, all transactions are visible to all participants. However, privacy solutions based on **zero-knowledge proofs** (**ZKPs**) are now more often being included in blockchain offerings. Enabling confidential transactions, ZKPs allow users to execute transactions without disclosing underlying information such as the transaction amount and sender/receiver addresses.

- **Governance**: In a decentralized system, deciding on changes or upgrades to the system can be difficult, leading to debates about decentralized governance and decision-making mechanisms that are commonly used by DAOs or decentralized autonomous organizations.

Guidelines

In the context of a Metaverse application, the process of choosing the most appropriate computing technology hinges on various factors. These factors encompass the intended use case, the scale at which the application will be deployed, the geographic distribution of users, and the anticipated types of interactions that will take place within the Metaverse. Additionally, critical aspects such as security, privacy, and data sovereignty must not be overlooked.

Now, let's delve into an analysis of each computing technology mentioned earlier. We'll explore their strengths and weaknesses when they're applied to the dynamic realm of the Metaverse.

Cloud computing

- **Advantages**:

  - **Scalability**: Demand-driven adjustments to cloud services, either scaling up or down, can be executed swiftly.

  - **Cost-efficiency**: Businesses can economize by only incurring costs for the resources utilized thanks to pay-as-you-go models.

  - **Access to advanced services**: Numerous cloud vendors provide AI/ML toolsets, databases, and data analysis resources.

- **Disadvantages:**

  - **Latency:** High latency may arise in the Metaverse due to data's need to reach central servers, potentially impacting real-time interactions.

  - **Data privacy:** Potential privacy issues emerge as data resides on servers managed by cloud service providers.

  - **Internet dependence:** High-speed, steady internet is crucial for cloud computing.

Edge computing

- **Advantages:**

  - **Low latency:** Edge computing enables quick response times in the Metaverse, which is essential for providing a smooth, uninterrupted user experience, by processing data near its source.

  - **Improved privacy:** Data can undergo local processing and storage to minimize the potential for data breaches while in transit.

- **Disadvantages:**

  - **Limited resources:** Edge devices often lack the same level of computational power and storage capacity found in centralized cloud servers.

  - **Management challenges:** The relative increase in the number of devices requiring management and maintenance amplifies the complexity of the system.

Distributed computing

- **Advantages:**

  - **Resilience:** Tasks partitioned across numerous nodes in a distributed system enhance its resilience to failure.

  - **Scalability:** Enhancing computational capacity is achievable by adding more nodes.

- **Disadvantages:**

  - **Complexity:** The administration of decentralized networks presents notable challenges, predominantly in the alignment of responsibilities and facilitating interaction among units.

  - **Inconsistent performance:** In distributed systems, the overall performance can be significantly affected by the presence of even a single slow node within the network.

  - **Decentralized computing:** A subset of distributed computing is decentralized computing, where no single node is authoritative, thereby eliminating single points of failure.

Decentralized computing

- **Advantages**:

  - **Robustness**: The absence of a central authority enhances the system's resilience against attacks and failures.

  - **Data sovereignty**: Data control remains with users since there's no centralized authority.

- **Disadvantages**:

  - **Network latency**: When nodes are geographically distributed, network latency, similar to that encountered in distributed computing, may become a concern.

  - **Complex governance**: With no central authority, establishing rules and protocols for data exchange and interactions can be challenging.

Considering the characteristics at hand, a hybrid computing model could be the most appropriate solution for the Metaverse, exploiting the benefits offered by each component. Scalable resources and advanced services could be supplied through cloud computing, whereas edge computing may manage real-time interactions and ensure privacy for users. Concurrently, distributed and decentralized computing models can assure resilience and data sovereignty.

The ultimate decision, however, is contingent on the distinct requirements and limitations of your specific Metaverse application. For instance, in scenarios where real-time engagement and minimal latency are indispensable, the prominence of edge computing might be heightened. Conversely, if the priorities lie in the scalability and availability of sophisticated AI services, cloud computing may be the mainstay.

Concerning data privacy, an area of vital importance in the Metaverse, multiple layers of approach can be utilized. At the infrastructure stratum, employing edge computing and decentralized computing can bestow users with heightened control over their data. At the application stratum, the implementation of robust encryption and access controls is mandatory, and privacy should be a fundamental consideration throughout the design and operation phases (that is, privacy by design). Additionally, transparent communication to users about how their data will be utilized, along with offering them control over their data wherever feasible (that is, consent-based data sharing), is crucial.

Be aware that the development of new technologies and standards could influence these considerations. Network technologies such as 5G successors can diminish latency for cloud services, while sophisticated AI algorithms can assist in managing the complexity of distributed systems. Moreover, the progress in blockchain technologies can offer innovative methods to uphold data privacy and oversee decentralized systems.

To sum it up, the creation of a Metaverse is a multifaceted task that requires meticulous contemplation of various factors. The optimal strategy is likely to be a hybrid model that combines the advantages of diverse computing technologies and utilizes the latest progressions in networking, AI, and blockchain.

Summary

The Metaverse relies on a range of computing technologies that play significant roles in its operation. One such technology is cloud computing, which serves as the foundation of the Metaverse by offering scalable and on-demand virtual environments and services. By utilizing cloud resources, the Metaverse ensures it can easily adapt to the increasing demands of its users.

Another important technology in the Metaverse is edge computing, which brings computation closer to users. This proximity reduces latency and enables real-time interactions, greatly enhancing responsiveness and improving the overall user experience. Integrating edge computing into the Metaverse is crucial to minimize delays in interactions.

Distributed computing also plays a key role in enhancing performance and resilience within the Metaverse. By distributing processing tasks across multiple nodes, the system achieves greater efficiency and fault tolerance. This approach allows the Metaverse to handle large amounts of data and user interactions with ease.

Decentralized computing emerges as a critical component of the Metaverse in terms of trust and security. By removing central authorities, decentralized computing establishes a system where trust is distributed among participants. This approach ensures the security and autonomy of users while safeguarding sensitive data.

To effectively navigate the complexities of the Metaverse, certain guidelines are essential. First and foremost, leveraging cloud resources is crucial to accommodate the ever-expanding scale of the Metaverse. Additionally, optimizing edge computing is vital for minimizing latency and enabling seamless real-time interactions. Moreover, utilizing distributed computing helps balance the load and improve overall performance and fault tolerance. Lastly, implementing decentralized protocols is indispensable for upholding security, user autonomy, and trust within the Metaverse.

In our next chapter, *Chapter 7, Where Are APIs Needed*, we'll review what APIs are and where they are needed for the Metaverse to function.

7

Where Are APIs Needed

Applications built using older versioned software, as well as currently incompatible software, need to have **application programming interfaces (APIs)** built that allow for those applications to interface with new applications. Additionally, new applications in the Metaverse, for which there are and will be many, need APIs so that they can interconnect to function. Building APIs can be expensive in that many might be needed, so it is important to know where they are necessary. Here, we discuss the areas where APIs need to be built.

In this chapter, we're going to cover the following main topics:

- Why APIs are needed for the Metaverse
- Which APIs fix which issues
- The cost versus reward of using particular types of APIs

What are APIs?

APIs are the heart of modern software development. Think of them as essential diplomats who create communication channels between different software applications. They set rules for data exchange, making sure everything goes smoothly and securely. By interconnecting different software applications, APIs form a complex digital ecosystem that improves functionality and enhances user experiences.

Their widespread application in various areas of software development is what makes APIs particularly intriguing. They serve as bridges, linking diverse software platforms from complex desktop applications to convenient mobile apps, and from dynamic web applications to well-structured databases. In essence, APIs allow different software entities, such as operating systems and services, to work together more effectively, thereby boosting productivity.

Let's take a more detailed look at different types of APIs based on their functionality:

- **Web APIs:** These are the pathways that enable different web services to communicate. By leveraging the HTTP protocol, web APIs are essential for web and mobile application development.

- **Operating System APIs:** These provide a set of rules for software applications to use operating system resources. For instance, the Windows API includes functions for GUI, file management, and networking in the Windows environment.

- **Database APIs:** These are connectors between programs and databases. They offer functions such as data querying, updating, and transaction execution. A good example is the Java JDBC API, which interfaces with relational databases.

- **Remote APIs:** These are essential in distributed computing as they allow communication between software on different networked systems. Examples are Java's **Remote Method Invocation (RMI)** and Microsoft's **Distributed Component Object Model (DCOM)**.

- **Library or Framework APIs:** These are collections of predefined classes and methods for developers to achieve various tasks, from designing UIs to accessing databases. The .NET framework APIs for C#, and the Ruby on Rails framework API for Ruby, are examples.

- **Hardware APIs:** These interfaces allow software applications to interact with hardware at a lower level. They simplify the complexity of hardware, making it easier for developers to work with it. APIs for general computing with GPUs, such as CUDA and OpenCL, or those for specific hardware devices, such as printers, fall under this category.

APIs are vital in software development as they allow different software components to work together seamlessly. They enable the development of modular, scalable software systems, no matter the programming languages or platforms used. By allowing developers to leverage existing functionalities, APIs save the effort of creating everything from scratch, allowing them to focus on their software's unique features.

Providing an API also opens the door for other developers to build applications that interact with a software product, enhancing its functionality and value. This approach has proven successful for various software products, including operating systems such as Windows and iOS, and web platforms such as Facebook and X (Twitter).

In summary, APIs are the glue that holds various software components together. They are key to promoting interoperability, enabling code reuse, and fostering the creation of sophisticated software systems. They're a fundamental component of today's software development landscape.

Programming language APIs

Diving deeper into the realm of APIs, we encounter an array of programming languages that leverage these powerful tools to create robust and dynamic applications:

- **Ruby**: In the world of Ruby, the language's standard library comes replete with a wide range of APIs catering to various functionalities such as file **input/output (I/O)**, networking, data serialization, and multithreading, among others. Notably, the Rails gem stands out as an exemplary library for web development, offering developers a rich and comprehensive API to craft sophisticated web applications. Furthermore, the Ruby ecosystem boasts a plethora of gems specifically designed for interfacing with APIs from various web services, opening up an abundance of possibilities for developers.

- **C#**: Stepping into the realm of C#, developers find themselves immersed in the vast capabilities of the .NET framework. This powerful platform enables the creation of diverse applications, ranging from web and mobile applications to desktop software and even games. The .NET framework provides a vast collection of APIs, covering essential areas such as file handling, data access, networking, graphics, and more. One notable example is ADO.NET, a set of APIs dedicated to data access and manipulation. Another standout is **Windows Communication Foundation (WCF)**, which offers a suite of APIs for developing connected and service-oriented applications.

- **Go**: Venturing into the realm of Go, we discover a language celebrated for its simplicity and efficiency. Go's standard library offers a comprehensive set of APIs for a myriad of tasks, including file I/O, image manipulation, text processing, and network requests. As Go is particularly well suited for web server development, its robust support for HTTP facilitates the creation of APIs for web-based applications. Additionally, third-party libraries such as Gorilla Mux provide developers with powerful tools for routing and handling HTTP requests, streamlining the development process.

- **Rust**: For those seeking the perfect balance of performance and safety, Rust stands tall. The Rust Standard Library is home to a diverse array of APIs tailored to tasks related to data manipulation, file I/O, networking, and multithreading. Moreover, the Rust ecosystem is enriched by the Cargo package manager, which grants access to countless third-party libraries (crates) providing APIs for an extensive range of functionalities, further expanding the language's capabilities.

- **Swift**: When delving into the realm of Swift, we cannot ignore its role as the language of choice for iOS app development. Swift's API offerings cater specifically to the unique demands of iOS development, providing essential tools such as UIKit for crafting captivating UIs, Core Data for seamless data management, and Core Animation for dynamic and engaging graphics and animations. Additionally, Swift developers can seamlessly interact with existing APIs from the Cocoa framework used in Objective-C, further expanding their toolkit.

- **Kotlin**: Embracing the world of Kotlin, we encounter a language that enjoys full interoperability with Java, making it a powerful contender for modern software development. Kotlin developers can leverage the entire suite of APIs provided by the Java ecosystem, tapping into a wealth of existing functionalities. Furthermore, Kotlin boasts its standard library, which comes equipped with APIs designed for tasks such as collection management, I/O operations, and string manipulation. For Android development, Kotlin embraces APIs from the Android SDK, enabling developers to create exceptional mobile applications.

- **Scala**: As we journey into the domain of Scala, we discover a language that harnesses the power of the **Java Virtual Machine** (**JVM**) to its advantage. Scala developers can seamlessly tap into Java's extensive API ecosystem while simultaneously enjoying APIs tailored to Scala's functional programming features. Notably, Scala offers APIs for unique tasks such as actor-based concurrency, a hallmark of the language's capabilities. This combination of interoperability with Java and Scala's own distinctive APIs opens up endless possibilities for developers.

- **PHP**: In the realm of web development, PHP reigns supreme. This popular server-side scripting language empowers developers with APIs for diverse functionalities, such as file handling, database integration (MySQL and PostgreSQL), and HTTP request handling, among others. The **Standard PHP Library** (**SPL**) is a treasure trove of APIs, providing essential tools for data structures, file management, and more. With PHP's robust API support, developers can create dynamic and interactive web applications with ease.

To reiterate the importance of APIs, we must recognize the various types available. As mentioned previously, APIs can be classified as local (library-based), web-based (often utilizing HTTP), or specific to operating systems, databases, and any system or service exposing an interface for software components to interact with. Among these, web APIs hold immense significance, with a plethora of online services, including social media platforms, cloud providers, weather services, and more, offering APIs for developers to seamlessly integrate their applications with these powerful services. The availability of web APIs fosters a high degree of integration between different services, empowering developers to craft more sophisticated and feature-rich applications by leveraging existing resources.

Blockchain-related APIs

Blockchain, a revolutionary technology that came about with Bitcoin in 2009, is essentially a ledger system that's distributed across many computers and doesn't rely on a single, central authority. It keeps a careful record of all transactions. The setup of this system is such that if you want to change one transaction record after the fact, you have to change all the records that came after it. This makes the records virtually unchangeable, a quality known as **immutability**.

One of the great things about blockchain is that it has changed how we can interact with each other directly (peer-to-peer), without needing to trust a central authority. Blockchain isn't just for digital currency transactions such as Bitcoin, either. It's also used for more complex things such as managing a "smart contract," which is a program that runs on a blockchain, as we see with Ethereum.

However, while blockchain is very strong in its decentralized form, it still needs a way to communicate with the more traditional, centralized systems that most users and applications still work with. Undoubtedly, the broader blockchain ecosystem is significantly shaped by the critical role of blockchain-related APIs. These APIs simplify the complexities inherent in blockchain networks by encapsulating the underlying blockchain protocol operations and presenting them to developers through a more intuitive and user-friendly interface.

Let's dive deeper into what these APIs do: they act as a bridge allowing software applications to communicate directly with individual computers, also known as **blockchain nodes**, that form the entirety of a blockchain network. This bridge, so to speak, enhances the user experience by enabling users by way of **blockchain oracles** to directly access and interact with blockchain data, conduct various transactions, and implement smart contracts. Blockchain oracles serve as vital intermediaries that establish a connection between blockchain networks and external systems, thereby facilitating the seamless execution of smart contracts by incorporating real-world inputs and outputs into the process.

The abstraction offered by these APIs, which shields the complicated specifics of the blockchain protocol, is of enormous benefit to developers. It enables them to incorporate blockchain functionalities into their applications without requiring an in-depth understanding of the underlying blockchain protocol. This simplification doesn't merely make the developer's job easier; it also opens up a plethora of possibilities in the field of blockchain application development.

Furthermore, the advent of blockchain-related APIs has paved the way for the creation of applications that are blockchain agnostic, or, in other words, applications that can operate irrespective of the type of blockchain involved. Consider, for instance, a financial application that is programmed to handle transactions involving a diverse range of cryptocurrencies. Such an application would inherently make use of the respective APIs of Bitcoin and Ethereum, among others, while simultaneously providing its users with a uniform interface. This degree of abstraction, enabled by blockchain-related APIs, marks a significant milestone in fostering seamless and uniform interactions across different blockchain networks. There are three main types of blockchain-related APIs:

- Blockchain APIs (including blockchain oracles)
- Wallet APIs
- Contract APIs

Blockchain APIs

APIs related to blockchain data are instrumental when it comes to directly sourcing raw information from the blockchain. By utilizing these APIs, you can gain access to a wide spectrum of information. This encompasses details of transactions, intricate block data, statistics related to the network, and an array of other types of germane metadata. There are myriad use cases, but one of their main applications lies in the realm of number crunching or transaction verification within an app.

Not only can blockchain APIs aid in data extraction and analysis but they can also provide valuable insights into transaction patterns and network statistics. These could be pivotal in AI modeling and predictive analysis, thereby enriching the overall AI project and enhancing its accuracy and effectiveness. So, for individuals or organizations looking to combine the potential of AI with the transparency and immutability of blockchain, these APIs are not just helpful, they are indispensable.

Wallet APIs

When addressing the task of managing digital assets within the realm of blockchain networks, wallet APIs emerge as invaluable resources. They are equipped with the requisite utilities for all critical operations, paving the way for the seamless functionality of the blockchain-driven digital ecosystem.

To start with, wallet APIs facilitate the creation of new digital wallets. These wallets are essential for individuals and organizations to safely store and manage their digital assets within the blockchain environment. Without these wallets, we would find it challenging to operate in the increasingly prevalent digital asset space. The APIs simplify the process of creating these wallets, allowing even those without extensive technical knowledge to partake in the digital asset revolution.

Next, wallet APIs provide tools for inspecting a wallet's balance. This function is vital for users to keep track of their digital asset holdings. With the volatile nature of many cryptocurrencies and digital assets, it's crucial for users to have access to real-time, accurate data about their assets' value. Wallet APIs allow for this level of transparency and accessibility, supporting informed decision-making within the blockchain ecosystem.

Finally, Wallet APIs are indispensable for executing transactions within the blockchain network. If you intend to build an application that necessitates the transfer or management of assets on the blockchain, these APIs are absolutely fundamental. Transactions are the lifeblood of the blockchain, facilitating everything from cryptocurrency trades to the execution of complex smart contracts. Wallet APIs provide the essential tools for running these transactions, enabling the movement and management of digital assets with ease and security.

Smart contract APIs

If your work involves a blockchain that supports programmable contracts, contract APIs become an indispensable tool in your repertoire. They provide the crucial bridge for applications to communicate with smart contracts that are deployed live on the blockchain, enabling both interaction and observation.

Smart contracts, self-executing agreements inscribed into lines of code, are automated and eliminate the need for an intermediary, thereby enforcing the contract's execution under predefined conditions. These contracts play a significant role in the blockchain ecosystem, with their actions and effects rippling through the entire network.

One of the primary functions of contract APIs is to enable applications to call methods on these smart contracts. Methods are essentially functions embedded within a contract, dictating how the contract interacts with the broader blockchain. With the help of contract APIs, your applications can call upon these methods, triggering specific functions to attain desired outcomes within the blockchain context.

Apart from invoking contract methods, contract APIs serve a secondary yet equally important function. They allow applications to monitor the sequence of events that are initiated by these smart contracts. Smart contracts can initiate a cascade of events based on certain conditions or results. Such events can result in substantial changes across the blockchain, inducing a host of consequential actions.

Through contract APIs, you can keep your applications updated with these contract-induced events. They allow applications to not only keep a record of these events but also comprehend the circumstances that led to them and respond accordingly.

In summary, when working with a programmable contract-supporting blockchain, contract APIs are invaluable. They serve a dual purpose of enabling direct interaction with smart contracts and offering a surveillance mechanism for the events triggered by these contracts. Whether it's invoking contract methods or keeping an eye on contract-spurred events, contract APIs provide you with the ability to exert control and maintain transparency in the complex world of blockchain-based smart contracts.

Digital payment APIs

APIs in the realm of digital payments are crucial as they bridge the gap between financial institutions, merchants, payment processors, and other service providers, allowing them to interact, complete transactions, and tap into valuable data.

There are several categories of APIs in the domain of digital payments:

- **Cryptocurrency Payment APIs**: These APIs facilitate businesses to accept cryptocurrencies as a mode of payment. Apart from that, they can also be utilized for trading, acquiring price details, or for blockchain-associated functions.

- **Wallet APIs**: Through Wallet APIs, integrations with digital wallet services, such as Google Wallet or Apple Pay, become feasible. They permit users to conduct secure transactions in stores, apps, and online.

- **Banking APIs**: Enabling third-party developers to create applications and services around a bank's operations, these APIs may offer access to account information, transaction history, and fund transfers. This forms the crux of open banking where banks offer their APIs to third-party developers to foster a financial services ecosystem.

- **P2P Payment APIs**: Services such as Venmo or Square's Cash App offer peer-to-peer payment APIs, allowing money transfers between individuals, predominantly via a mobile app.

- **Payment Gateway APIs**: Payment Gateway APIs such as Stripe and Square are designed to integrate with e-commerce platforms and applications, enabling businesses to securely process payments from customers. They handle various forms of payment, such as credit/debit cards and digital wallets, ensuring transactions are encrypted and compliant with financial regulations.

- **Card Issuing APIs**: Businesses can issue virtual or physical cards through APIs provided by certain services (such as Marqeta). These cards can be programmed with specific controls, such as restricting expenditure to specific merchants or transaction types.

Each of these APIs offers a set of key functions:

- **Transaction Processing**: This forms the central functionality of most payment APIs, facilitating authorization and capture of payments, voiding transactions, and issuing refunds, among other services.

- **Security and Fraud Detection**: Many payment APIs come with built-in features for detecting and preventing fraud. They may flag suspicious transactions, provide risk scores, or support 3D Secure authentication.

- **Data Access and Analytics**: APIs often offer access to data that can be leveraged for analysis and reporting, including transaction history, payment status, and chargebacks.

- **Compliance and Regulatory Management**: Payment APIs can often assist in managing complex regional regulatory requirements such as **Anti-Money Laundering** (**AML**) processes or **Know Your Customer** (**KYC**) checks.

- **Customer Management**: Certain APIs offer customer management features, which could include storing customer details, managing customer profiles, or supporting loyalty programs.

Incorporating APIs in digital payments offers several benefits:

- **Innovation**: APIs foster the development of new, innovative financial services, opening new avenues for businesses to offer novel types of products and services.

- **Efficiency**: By streamlining and automating payment processing, APIs can make it faster and less error-prone.

- **Flexibility**: APIs provide businesses with the leeway to integrate payment processing in a manner that fits their unique needs, allowing them to decide which services to utilize and how.

- **Cost Savings**: Automation of payment processes via APIs can significantly reduce costs associated with manual processing.

However, utilizing APIs comes with its own set of challenges and considerations:

- **Security**: Since APIs are potential targets for hackers, it's essential to implement robust authentication, encryption, and other security measures.

- **Compliance**: The heavily regulated payment industry necessitates businesses to ensure compliance with all relevant regulations and laws.

- **Performance**: API performance can directly affect the user experience. Slow or unreliable transactions could result in lost sales.

- **Integration**: The process of integrating APIs can be complex, especially when dealing with multiple APIs or legacy systems.

Geolocation APIs

Geolocation APIs are a set of protocols, routines, and tools used by developers to build software applications that require geolocation data. Geolocation, in this context, refers to the identification or estimation of the real-world geographic location of an object, such as a radar source, mobile phone, or internet-connected computer terminal.

A geolocation API provides geographical location data based on various inputs, such as an IP address, a cell tower ID, or even data from a GPS satellite. They can provide the latitude and longitude coordinates of a location, and often additional information such as the city, state, country, postal code, time zone, and other related details.

This information can be used in a multitude of ways across different sectors – from delivering targeted content based on a user's location, calculating distances between two points, and tracking the location of devices in real-time, to more advanced uses such as predictive analytics, location-based advertising, and more.

There are several types of geolocation APIs, including the following:

- **IP Geolocation APIs**: These APIs take an IP address as input and return the estimated geographical location of that IP. They work by mapping IP addresses to a database of geolocation information. They're typically used to identify the location of web users without the need for GPS or other hardware-based location services.

- **Geocoding APIs**: These APIs convert addresses into geographic coordinates (latitude and longitude), a process known as **geocoding**. Some geocoding APIs can also perform reverse geocoding, which is the process of converting geographic coordinates into a human-readable address.

- **GPS-Based Geolocation APIs**: These APIs use **Global Positioning System** (GPS) data to pinpoint a device's exact location. These are commonly used in mobile applications and require hardware support from the device.

- **VPS-Based Geolocation APIs**: **Visual Positioning System** (**VPS**) APIs are a specialized subset of geolocation services that rely on computer vision and AR technologies to determine a device's precise location by analyzing the surrounding environment. These are often used in conjunction with GPS-based geolocation APIs and have diverse applications, encompassing indoor and outdoor navigation, AR experiences, and the precise localization of autonomous vehicles. Google integrates VPS into ARCore for more accurate outdoor and indoor AR positioning, while Niantic employs similar technology in its Lightship platform to create location-based AR gaming experiences such as Pokémon GO.

- **Cell Tower and Wi-Fi-Based Geolocation APIs**: These APIs estimate a device's location based on proximity to cell towers and Wi-Fi networks. They can provide location data when GPS isn't available and can often provide indoor location data, which GPS can struggle with.

Uses of geolocation APIs

Geolocation APIs are used in a wide array of applications:

- **Content Personalization**: Websites and apps can tailor content based on the user's location. This could include showing local news and weather or even changing the language of a website.

- **Location-Based Services**: Services such as food delivery, ride-hailing, and e-commerce, as well as games and social apps, use geolocation to provide location-based services and capabilities, calculate delivery routes, or estimate shipping costs.

- **Location-Based Advertising**: Advertisers can use geolocation to deliver targeted advertising based on a user's location.

- **Fraud Detection**: By identifying the geographic location of a user, businesses can detect unusual activity that might indicate fraud.

- **Analytics and Insights**: Businesses can use geolocation data to gain insights into their users, such as where their users are located or what local trends might be affecting their behavior.

The technology behind geolocation APIs

Geolocation APIs make use of several technologies and methodologies to achieve their goal, which is providing accurate geographical data:

- **IP-Based Geolocation**: Geolocation APIs use large databases that map IP addresses to specific geographical locations. These databases are populated using a variety of methods, including data from **internet service providers** (**ISPs**), regional internet registries, and even data from users themselves, who sometimes input their location data into websites or apps.

- **Geocoding**: To convert a physical address to geographical coordinates, or vice versa, geocoding APIs use databases that associate addresses with latitude and longitude coordinates. These databases are continually updated and refined to maintain accuracy as new buildings are constructed or old ones are demolished.

- **GPS**: GPS-based APIs receive signals from multiple GPS satellites (a process known as **trilateration**) and calculate the receiver's precise location on Earth. This method provides high accuracy and is used extensively in mobile and automotive applications.

- **VPS**: VPS APIs function by examining the visual characteristics of the environment as seen through a device's camera. These visual cues, such as unique landmarks or objects, are then cross-referenced with existing maps or models of the surroundings. This matching process allows VPS to accurately ascertain the device's current position and orientation in real time, facilitating applications such as AR, navigation, and precise localization for autonomous vehicles.

- **Cell Tower Triangulation**: APIs using this methodology calculate a device's location based on measurements of its distance from multiple nearby cell towers. This approach is often used in mobile applications where GPS data might not be available or accurate enough.

- **Wi-Fi Positioning**: Similar to cell tower triangulation, Wi-Fi positioning uses the signal strength of known Wi-Fi networks in the vicinity to estimate a device's location.

Geolocation API performance metrics

When selecting a particular geolocation API for an application, several performance metrics relevant across all the different types of geolocation APIs that should be considered are as follows:

- **Accuracy**: How precise is the location data provided by the API? Depending on the use case, this might be the most crucial factor. For example, a delivery service app would need high-accuracy location data, whereas a website displaying local weather may not.

- **Coverage**: Does the API provide data globally, or is it limited to certain regions?

- **Response Time**: How quickly does the API return results? In use cases where real-time data is necessary, a low-latency API would be preferable.

- **Rate Limiting**: How many requests can you make to the API per minute/hour/day? Depending on your application's scale, this could be an important consideration.

Integration of geolocation APIs

To integrate any type of geolocation API into a software application, developers use the API's endpoints (URLs). The developers make HTTP requests to these endpoints, typically using `GET` or `POST` methods, and include any necessary parameters. The API then returns a response, often in a standard data format such as JSON or XML.

For instance, here is how a developer might use a hypothetical geocoding API:

1. The developer sends a GET request to the API's geocoding endpoint, including the address to be geocoded as a parameter.

2. The geocoding API returns a JSON response containing the geographic coordinates associated with that address.

3. The developer's application parses this JSON response and uses the coordinates for its intended purpose, such as displaying the location on a map.

Challenges with geolocation APIs

Despite their wide range of applications and powerful capabilities, all the different types of geolocation APIs have these challenges:

- **Data Accuracy and Availability**: The accuracy and availability of geolocation data can vary based on the method used to gather it. For instance, IP-based geolocation can often only provide an approximate location.

- **Data Privacy and Security**: Handling geolocation data requires adherence to a range of laws and regulations, which vary by region. Developers must ensure that they are not infringing upon users' privacy rights when using this data.

- **Cost**: Many geolocation APIs are not free, and their cost can increase based on the number of requests made.

Understanding these challenges and addressing them appropriately can ensure the successful integration and utilization of geolocation APIs in any software development project.

In conclusion, the importance of geolocation APIs cannot be overstated in today's digital era. They are critical for a host of applications, from personalizing user experiences to aiding in complex logistics and transportation operations. However, developers must exercise caution to ensure they use these tools responsibly, keeping data privacy and security at the forefront.

Mapping APIs

APIs specializing in mapping offer an invaluable toolkit for developers. These APIs are a meticulously curated assortment of protocols and programming instructions, the purpose of which is to allow developers a simplified yet effective gateway to the features provided by a specific service or software component. The principal function of these APIs lies in facilitating digital map-related tasks such as creating, manipulating, and providing an interactive user experience.

There are numerous features and services that these mapping APIs typically encompass:

- **Geocoding and Reverse Geocoding**: This feature stands as a fundamental component of mapping APIs. It acts as a translation service between the language of human-understandable addresses and the numerical language of latitude and longitude coordinates that computers use. The facility to transform an address into geographical coordinates makes it straightforward to position that location accurately on a digital map. On the other hand, reverse geocoding permits the conversion of geographic coordinates back into a human-readable address. This feature is especially beneficial when there's a need to discern the physical location that a set of coordinates represents.

- **Navigation and Route Planning**: A boon, especially for sectors such as transportation and logistics, this feature enables developers to provide comprehensive, turn-by-turn directions between multiple points. In effect, it allows for the calculation of the most efficient pathways for vehicles to follow, whether that efficiency is measured in terms of time, fuel, or some other resource. The ability to optimize routing in this way can lead to notable enhancements in operational efficiency within these sectors.

- **Interactive Map Interface**: Mapping APIs provide developers with the capability to embed a fully interactive map within a web page or an application. This feature transforms the user experience, allowing end users to engage with the map. Users can zoom in for granular detail or zoom out for a broader view, navigate across different regions, or interact with markers to obtain additional information. Such an interactive UI can significantly enrich the user experience by making it more immersive and user-friendly.

- **Geospatial Analysis**: A sophisticated form of analysis, geospatial analysis is tied to a specific geographical location. The analysis can help uncover patterns or trends within geographic data. For instance, by employing this tool, we could determine areas with a high concentration of a particular demographic group. These insights can be instrumental in shaping strategic decisions in various domains, including (but not limited to) marketing, urban planning, and resource allocation.

- **Places API**: An essential element of mapping APIs, the Places API delivers data related to places. Within this context, *places* can refer to a wide variety of locations, including establishments, geographical regions, or prominent landmarks. The Places API does more than simply locate a specific business or point of interest in a designated area – it provides a wealth of details such as reviews, ratings, and contact information, enhancing the overall comprehension of the users about the place in question.

- **Street View**: A unique feature of mapping APIs, Street View offers users a chance to explore a location virtually from the street level. Frequently deployed in real estate or tourism-related applications, this feature allows users to feel as though they're physically present at a location. Whether users are considering buying a property or planning a vacation, the street-level perspective can offer critical insights to aid in decision-making processes.

In summary, mapping APIs offer an efficient and versatile toolkit for developers, enabling a variety of operations related to digital maps and geographical data. With their plethora of features, these APIs open up a realm of possibilities for creating dynamic and interactive applications, transforming the way both businesses and consumers interact with location-based data and services. Moreover, these APIs have wide-ranging implications for numerous sectors, contributing significantly to operational efficiencies and providing a more immersive, engaging user experience.

Messaging APIs

In the sphere of information exchange between systems, messaging APIs, or text message APIs (often referred to as SMS APIs) are vital tools. These APIs form a structured set of rules, standards, and protocols that serve the crucial function of enabling diverse applications to communicate seamlessly with each other.

The term **messaging APIs** isn't exclusively confined to **Short Message Service (SMS)** or **Multimedia Messaging Service (MMS)** interactions. It also has a broader application that covers various types of message transfers, such as emails, in-app messaging, and web messaging. However, in this particular context, our primary emphasis will be on SMS and associated messaging utilities, while keeping in mind that these principles could extend to numerous other communication modalities.

Turning our attention to the domain of SMS, we find that APIs act as potent facilitators for software applications, granting them capabilities to send, receive, and monitor SMS messages via an intermediary known as a **gateway**. In the business world, these APIs are utilized for an array of purposes, including (but not limited to) transactional alerts, various notifications, marketing communities, and paving the way for bidirectional communication with customers.

To illustrate, an SMS API might be employed by an e-commerce platform for dispatching shipment tracking updates, by a banking institution for broadcasting transaction alerts, or even by a medical office for issuing appointment reminders. Fundamentally, an API operates by liaising with an SMS gateway – a service that functions as a bridge, connecting the cellular network (which intrinsically handles SMS messages) and the internet. When your application transmits a request to convey a message via the API, this request is picked up by the gateway, which subsequently transmits it over the cellular network to reach the intended recipient.

Here are several key attributes of messaging APIs:

- **Dispatching SMS**: At its most fundamental, an API enables SMS transmission to one or several recipients.

- **Receipt of SMS**: Some APIs come equipped with the ability to accept an SMS sent to a predetermined number.

- **Delivery Reports**: To monitor the successful delivery of an SMS to its intended recipient, APIs can provide delivery reports.

- **Message Scheduling**: A select few APIs permit you to schedule the dispatch time of an SMS.

- **Automated Replies**: APIs can be programmed to transmit automatic responses based on predetermined triggers or keywords in the received SMS.

- **Message Formatting**: Some APIs might also provide the feature to format your messages – for example, allowing the inclusion of links, images, or structurally designed text.

When it comes to selecting a messaging API, numerous factors ought to be taken into consideration:

- **Coverage:** Does the API provider maintain good coverage in the regions where your recipients reside? *Coverage* in the context of a messaging API refers to the geographical reach of the API service. For global businesses or those with customers, clients, or users spread across multiple regions, it's important to ensure that the API provider offers robust service across all necessary regions. You'll need to ensure your provider has strong relationships with carriers in those countries or regions.

  For instance, if you have a substantial user base in Asia, you'll need an API that has reliable connectivity to mobile carriers in the region. Note that coverage can also impact the delivery rate and speed of your messages. Some providers may also have more robust features or better pricing in certain regions.

- **Reliability**: Can the provider guarantee a high delivery rate? Reliability is a measure of the API's uptime and its success rate in delivering messages. A good messaging API should have strong uptime guarantees (usually above 99.9%), meaning it's consistently available and operational.

  The delivery rate of the messages is another aspect of reliability. This indicates how often messages successfully reach their intended recipients. Inquire about the average delivery rates, how they handle message failures, and whether they have redundancies in place to cope with network failures. High reliability is often a result of superior infrastructure, efficient routing, retry mechanisms, and good relationships with network providers.

- **Cost**: How is the service priced? Is it on a per-message basis? Are there any concealed charges? The pricing model for messaging APIs can vary significantly from one provider to another. Some common models include pay-as-you-go (where you're billed for each message sent), monthly plans (which allow for a certain volume of messages per month), and tiered plans (where the cost per message decreases as you send more messages).

  It's important to understand the full cost of using the API. This includes looking for any hidden fees such as costs for using certain features, additional costs for sending international messages, or charges for receiving messages. A lower headline cost per message may be offset by these additional charges, so consider your usage patterns and the total cost of operation when comparing providers.

- **Ease of Use**: Is the API adequately documented and user-friendly? The ease of use of an API can be a significant factor, especially if you have a small team or limited development resources. This includes the quality of the API's documentation, the availability of **Software Development Kits** (**SDKs**) for your programming language of choice, and the overall simplicity of integrating the API into your systems.

  A well-documented API, complete with clear instructions, examples, and perhaps even a test environment, will be much easier to integrate and less likely to cause issues down the line. Look for providers that offer comprehensive, clear, and up-to-date documentation.

- **Security**: How does the provider safeguard the security of your messages and data? Security is paramount when you're sending potentially sensitive information via text messages. This includes not only the security of the messages themselves but also the security of your API credentials and the data stored by the API provider.

  Check the provider's security certifications (such as ISO 27001 or SOC 2), and look for features such as data encryption, secure transmission protocols (such as HTTPS), and secure handling of API keys. You should also check the provider's privacy policy to understand how they handle and store your data.

- **Customer Support**: Does the provider offer satisfactory customer support? The level of support provided by the API provider can be crucial, especially if you're operating a service with high availability requirements. Look for providers with strong support offerings, such as 24/7 availability, multiple contact methods (email, phone, chat, etc.), and proactive communication about issues or planned maintenance.

  Consider the provider's reputation for customer support, which you can often gauge through customer reviews and testimonials. Quick response times, knowledgeable staff, and effective problem resolution are all signs of excellent customer support.

  In conclusion, choosing a messaging API is a multifaceted decision that depends on your specific needs and constraints. By carefully considering each of these factors, you can select an API that will be reliable, cost-effective, easy to use, secure, and well supported.

Artificial Intelligence (AI) APIs

In the domain of AI APIs, two noteworthy categories come to the forefront: voice and audio APIs, and vision APIs. These APIs leverage cutting-edge machine learning and neural networks, serving as a transformative conduit to infuse intelligence into applications and services across diverse domains. Voice and audio APIs empower applications to interact with spoken language, providing features such as speech recognition and text-to-speech conversion. Simultaneously, vision APIs employ advanced computer vision techniques for the analysis and comprehension of visual content, facilitating tasks such as image recognition and object detection. These APIs play a significant role in ushering in an era characterized by AI-driven innovation, reshaping user interactions with technology, and unlocking the potential for more intelligent and intuitive applications.

Voice and audio APIs

A diverse array of APIs, specifically designed for audio and voice functionalities, significantly shape many contemporary digital platforms and services. They facilitate the interaction between software applications and voice or audio features, and their applicability spans a multitude of functions, including (but not limited to) voice recognition, voice analytics, text-to-speech conversion, and call tracking.

Here are a few of the predominant types of voice and audio APIs, elucidated alongside their primary functionalities:

- **Speech Recognition APIs**: Serving as the backbone of voice-enabled services such as voice assistants (for instance, Google Assistant or Alexa), voice-activated searches, and real-time transcription services, these APIs translate oral language into written script. Speech recognition APIs employ machine learning algorithms to convert oral language into written text, with their capability extending to understanding various languages, dialects, and accents. Their broad applications include voice assistants, transcription services, and **Interactive Voice Response (IVR)** systems. In transcription services, they facilitate real-time or post-meeting transcriptions for notetaking or accessibility purposes; in voice assistants, they convert the user's command into text for better comprehension.

- **Text-to-Speech APIs**: As the name suggests, **Text-to-Speech (TTS)** APIs transmute written text into verbal language. These are employed in various tasks such as reading out text messages, books, or articles and assisting users with visual impairments to interact with applications. Google's Text-to-Speech API, Amazon Polly, and IBM Watson Text to Speech are notable instances of TTS APIs. Contemporary TTS systems offer a range of "voices," adjusting for pitch, accent, speed, and even emotion, leading to a more natural-sounding output. They find widespread usage in reading apps, accessibility tools, and voice assistants. For instance, in accessibility tools, they read out interface elements or messages for visually impaired or dyslexic users; in voice assistants, they vocalize the assistant's response to the user.

- **Text Generation APIs**: Text generation APIs harness the power of sophisticated **Natural Language Processing (NLP)** models, such as those found in the OpenAI GPT model APIs, to create text that closely mimics human writing. These models have been trained on extensive datasets, allowing them to understand context, grammar, and semantics to produce coherent and contextually relevant text. They find their crucial application in chatbots, automatic language translation, and sentiment analysis. NLP APIs, powered by machine learning models, decipher, interpret, and manipulate human language. They extract entities (such as dates, people, or places), understand sentiments (positive, negative, or neutral), and determine the intent behind the language. They are used chiefly in chatbots and voice assistants to understand user input and decide the appropriate response. In addition, they aid customer service tools in analyzing customer feedback or comprehending email content.

- **Voice Biometrics APIs**: These APIs are employed to authenticate a person's identity by analyzing the unique characteristics of their voice. They are commonly used in voice-activated security systems, enabling the identification of individuals based on unique voice patterns. Auraya's EVA Voice Biometrics and VoiceItt's Voice Biometrics API are examples of such APIs. Voice Biometrics APIs identify or authenticate an individual based on unique characteristics in their voice, analyzing aspects such as modulation, tone, pitch, and so on. Primarily, these are employed in security and authentication, providing a hands-free alternative to traditional password systems. They also find use in customer services to swiftly identify the caller, thereby hastening the verification process.

- **Audio Analysis APIs**: These APIs examine audio files for various elements such as genre, mood, tempo, pitch, and more, and are often employed in music-related applications, and can be used as part of generative AI song-making.

- **Call Tracking and Analytics APIs**: These APIs record and examine phone calls, providing insights into call volumes, caller demographics, call outcomes, and more. They prove beneficial for businesses seeking to understand these aspects. They are used by businesses to understand customer behavior, track the performance of marketing campaigns, and gain insights for optimizing sales or customer service. Companies such as CallRail offer such APIs.

In summary, these are only the main examples of voice and audio APIs. As AI and machine learning continue to advance, the functionality and applications of these APIs are expected to broaden, paving the way for innovative methods of interaction and understanding through sound and voice.

Vision APIs

Vision APIs, also known as computer vision APIs, serve as specialized APIs that equip developers with tools and services designed for the processing and examination of visual content, encompassing images and videos. These APIs leverage AI and ML algorithms to execute a diverse array of functions pertaining to visual perception and comprehension. The following are some typical functions and applications associated with vision APIs:

- **Image Recognition/Generation**: Vision APIs excel at the identification of objects, scenes, and patterns within still images and video. For instance, they can discern the presence of a cat, an automobile, or a beach vista in an image. Included here performing image recognition and generation are generative AI APIs such as those from Midjourney, Runway, and OpenAI's DALL-E (DALLE-3 now available through ChatGPT Plus), which also have NLP capabilities.

- **Optical Character Recognition (OCR)**: These APIs possess OCR capabilities that enable the extraction of text from images, facilitating the transformation of printed or handwritten text in images into machine-readable formats.

- **Face Detection and Recognition**: Vision APIs proficiently detect and identify faces within photographs and videos, which proves valuable for applications such as tagging individuals in images or implementing facial authentication.

- **Image Segmentation**: These APIs excel at segmenting images into distinct regions or objects, facilitating comprehensive analysis and comprehension of visual content.

- **Anomaly Detection**: Some advanced vision APIs possess the capability to detect anomalies or atypical patterns within visual data, a functionality highly beneficial for quality control and security applications.

- **Content Moderation**: Vision APIs are proficient at autonomously filtering and flagging inappropriate or harmful content in user-generated images and videos, thereby contributing to the maintenance of secure online environments.

- **Visual Search**: Visual search capabilities empower users to initiate searches for products or information employing images as queries, a feature commonly integrated into e-commerce applications.

- **Style Transfer**: Vision APIs are adept at applying artistic styles to images, videos, or graphics, enabling users to generate visually distinctive content.

Vision APIs find extensive utilization across various industries, spanning tasks such as streamlining image analysis and elevating user experiences in mobile applications, e-commerce, healthcare, and beyond. These APIs empower developers to seamlessly incorporate advanced visual recognition and comprehension capabilities into their applications, without the necessity of constructing intricate computer vision models from the ground up. Leading companies such as Google, Microsoft, and Amazon include vision APIs within their suite of cloud services to facilitate the deployment of these functionalities.

Examples of API uses in the Metaverse

APIs play a crucial role in enabling communication and interaction within the Metaverse. The following are some illustrative examples of how APIs can be used in the Metaverse.

Social interaction APIs

- **Description**: Social interaction APIs facilitate communication and collaboration among users within the Metaverse.

- **Example**: A social API could allow users to send friend requests, create virtual events, and exchange messages or voice calls with their virtual friends. This API would handle authentication, privacy settings, and message routing, ensuring a seamless and secure social experience.

Virtual goods and asset marketplace APIs

- **Description**: These APIs enable users to buy, sell, and trade virtual assets and goods within the Metaverse.

- **Example**: An asset marketplace API could provide functionalities for listing virtual properties, digital art, wearables, and more. It would handle transactions, verify ownership, and ensure secure asset transfers between users.

AI APIs

- **Description**: These APIs bring advanced AI capabilities into the Metaverse, enhancing interactions and experiences.

- **Voice and Audio API Examples**:

Speech Recognition API (e.g., Google Cloud Speech-to-Text):

Description: Converts spoken language into written text with high accuracy

Example: In healthcare, transcribes medical dictations into electronic health records for precise record-keeping that could be housed digitally in the Metaverse.

TTS API (e.g., Amazon Polly):

Description: Transforms text into spoken language, offering a variety of voices

Example: Enables visually impaired users to have books and documents read aloud in their preferred voice; can be used in the Metaverse for visually impaired users.

Text Generation API (e.g., OpenAI's GPT3):

Description: Creates dynamic and engaging content

Example: In content marketing, generates blog posts on various topics to save time for content creators.

Voice Biometrics API (e.g., Auraya's EVA Voice Biometrics):

Description: Provides secure voice-based authentication

Example: In finance, allows customers to access their accounts by speaking their passphrase, enhancing security and user experience, and keeping the Metaverse more secure.

Audio Analysis API (e.g., Apple's Shazamkit):

Description: Identifies songs based on a short audio sample, such as a snippet of a song playing in the background.

Example: Can be used anywhere in the Metaverse to identify songs heard.

Call Tracking and Analytics API (e.g., CallRail):

Description: Records and analyzes phone conversations for improved customer service.

Example: In retail, helps businesses understand customer behavior and refine marketing strategies for more effective Metaverse campaigns.

- **Vision API Examples**:

Image Recognition/Generation API (e.g., Google Cloud Vision [Recognition]; Generative AI text-to-2D and 3D imaging and image-to-video [2D and 3D] companies [Recognition and Generation]):

Description: Recognizes objects and products in still images and video and generates novel imaging.

Example: Allows creators to make a 2D video from a single still image for viewing in the Metaverse.

OCR API (e.g., Amazon Textract):

Description: Digitizes text from images and documents.

Example: In businesses that digitize in the Metaverse, transforms paper documents into searchable digital files for improved accessibility.

Face Detection and Recognition API: (e.g., Microsoft Azure Computer Vision):

Description: Identifies individuals in photos and videos.

Example: In security applications for the Metaverse, enhances access control by recognizing authorized personnel based on facial features.

Image Segmentation API (e.g., TensorFlow):

Description: Separates objects within images for detailed analysis.

Example: In autonomous vehicles, help detect pedestrians, vehicles, and road signs for safer self-driving experiences.

Content Moderation API (Custom Models):

Description: Automatically filters inappropriate content in user-generated images and videos

Example: In social media platforms, prevents the dissemination of harmful or offensive content, maintaining a safe online environment

Multiplayer networking APIs:

- **Description:** These APIs facilitate real-time communication and synchronization among multiple users in shared virtual spaces.

- **Example:** A multiplayer networking API could enable users to participate in virtual games or events together. It would handle user presence, positional tracking, and data synchronization, ensuring a smooth and lag-free multiplayer experience.

Spatial audio APIs

- **Description:** These APIs provide realistic 3D audio experiences, enhancing immersion within the Metaverse.

- **Example:** A spatial audio API could simulate audio sources based on their virtual location. For instance, if a user is standing near a virtual fountain, the API would render the sound of flowing water realistically, creating a more immersive and interactive environment.

Avatar customization APIs

- **Description:** These APIs enable users to customize their avatars with various visual elements.

- **Example:** An avatar customization API could offer a wide range of options for users to personalize their virtual personas, including hairstyles, clothing, accessories, and facial features. The API would handle avatar rendering and appearance data, ensuring consistent representation across different experiences.

Real-time translation and communication APIs

- **Description:** These APIs enable real-time translation of conversations between users who speak different languages within the Metaverse.

- **Example:** A translation and communication API could automatically translate spoken or written text from one language to another, facilitating seamless communication among users from diverse linguistic backgrounds. This fosters inclusivity and breaks down language barriers within the virtual space.

Virtual education and training APIs

- **Description**: These APIs support virtual education and training experiences, making learning more engaging and effective in the Metaverse.

- **Example**: A virtual education API could offer interactive lessons, simulations, and quizzes for various subjects and skills. Educators and trainers can use the API to create personalized learning paths, track progress, and provide feedback to learners, leading to more immersive and effective educational experiences.

Personal data and privacy APIs

- **Description**: These APIs handle user data, consent, and privacy settings to ensure the secure and responsible use of personal information within the Metaverse.

- **Example**: A personal data and privacy API could allow users to control the data they share with different virtual platforms and applications. Users can specify their privacy preferences and manage data access permissions, giving them more control over their virtual identities and interactions.

Economic and financial APIs

- **Description**: These APIs enable virtual economies and financial systems within the Metaverse, supporting transactions, currencies, and economic interactions.

- **Example**: An economic API could introduce a virtual currency that users can earn through various in-world activities, trade with others, or convert to real-world currencies. This opens up opportunities for virtual businesses and entrepreneurship within the Metaverse.

Health and well-being APIs

- **Description**: These APIs promote well-being within the Metaverse by integrating health-related features and services.

- **Example**: A health and well-being API could provide virtual fitness classes, mental health resources, and personalized wellness plans for avatars. It could also track users' in-world activity levels to encourage a healthy balance between virtual and real-life engagements.

Cross-platform and cross-reality APIs

- **Description**: Cross-platform and cross-reality APIs are integral to the world of immersive technologies, including VR, AR, and MR. These APIs facilitate interoperability between different reality platforms and experiences, allowing users to seamlessly interact across various virtual and augmented environments. In this comprehensive discussion, I will delve into the significance of these APIs and provide multiple examples to illustrate their practical applications.

The significance of cross-platform and cross-reality APIs is described here:

- **Interoperability and Collaboration**: At the core of their importance, cross-platform APIs enable users from diverse reality platforms to interact and collaborate seamlessly. This fosters a sense of unity and inclusivity, as individuals can partake in shared experiences irrespective of their chosen reality platform.

- **Expansion of User Base**: The integration of cross-reality APIs into applications can significantly expand the user base. These APIs ensure that experiences are not restricted to a specific platform, attracting a more diverse and extensive audience.

- **Facilitating Content Sharing**: Cross-platform APIs streamline the sharing of content across different realities. Users can effortlessly access and engage with content created in VR, AR, or MR, enhancing the overall user experience.

- **Elevated Experiences**: Developers can craft more immersive and enriching experiences by leveraging cross-reality APIs. For instance, a gaming company might develop an API that enables players in VR and AR to participate in the same game, introducing depth and excitement to the gaming landscape.

- **Examples**:

  - **Virtual Concerts**: Envision a cross-platform API that enables users from various reality platforms to attend a virtual concert collectively. Those in VR headsets could immerse themselves in a virtual arena, while AR users could enjoy the concert within their physical surroundings, all sharing the same live performance.

  - **Collaborative Design**: In sectors such as architecture and design, cross-reality APIs can prove invaluable. Architects utilizing VR for 3D modeling could seamlessly collaborate with clients employing AR devices. Changes made in VR would instantly manifest in the AR model, facilitating real-time design discussions.

  - **Educational Enhancements**: Cross-reality APIs have immense potential in the field of education. Picture a history teacher conducting a VR-guided exploration of ancient civilizations, with students using AR to receive overlays of historical information in their physical surroundings.

  - **Cross-Reality Social Networking**: Social networking platforms can harness the capabilities of cross-reality APIs to connect users in diverse virtual or augmented settings. Friends could assemble in a virtual park, engage in shared games, or attend virtual art exhibitions, all regardless of their choice of reality device.

  - **Training and Simulation**: In scenarios involving training and simulation, cross-reality APIs can play a pivotal role. Military personnel, medical professionals, or first responders can train together in simulated environments, each utilizing their preferred reality platform.

- **E-Commerce Revolution**: Retailers can revolutionize the shopping experience by incorporating cross-reality APIs. Shoppers in VR can virtually try on clothing, while those in AR can visualize virtual furniture placements in their living spaces to assess suitability.

In summary, the importance of cross-platform and cross-reality APIs lies in their ability to facilitate seamless interactions and collaborations across a diverse array of reality platforms. These APIs are of paramount significance across various industries, spanning entertainment and education to design and social networking. As technology continues to evolve, the role of these APIs in shaping the future of immersive experiences will become increasingly pivotal.

Environmental and sustainability APIs

- **Description**: These APIs promote environmental awareness and sustainability within the Metaverse.

- **Example**: An environmental API could incorporate sustainability practices into virtual world design and mechanics. For example, virtual ecosystems could be designed to mimic real-world ecosystems, and users could participate in virtual environmental conservation efforts.

These examples illuminate the sheer breadth and depth of possibilities that APIs offer for the Metaverse. As technology continues to advance and more APIs are developed, the Metaverse will grow increasingly dynamic, diverse, and personalized, catering to an ever-broadening spectrum of user needs and experiences.

Summary

The role of APIs in constructing, expanding, and managing the Metaverse is pivotal and indispensable.

The Metaverse is imagined as a vast, interconnected digital cosmos, bringing together diverse platforms, applications, and technologies into a coherent entity. This integration is not possible without sophisticated hardware and software infrastructure, and APIs are integral to this infrastructural framework. They are essentially the protocols enabling various software systems to interact, exchange data, and share functionalities, making a multifaceted user experience feasible across the scope of the Metaverse.

APIs act as the conduits of the Metaverse, facilitating the unimpeded flow of data and functionality across its multifarious elements. Whether it is creating bridges between diverse gaming environments, ensuring data transfer between social media, VR, and AR applications, or integrating an e-commerce platform into a VR or AR environment, APIs make it possible. In simple terms, they transform the Metaverse from a fragmented collection of virtual entities into a coherent and interconnected universe.

Also, APIs have a key role in enhancing and integrating AI functionalities in the Metaverse. They help to incorporate AI-driven tools and systems, thereby enabling the creation of smart virtual environments that can evolve in response to user interactions. Such AI-powered enhancements can take many forms, from NPCs in games to recommendation algorithms in virtual stores or personalized virtual assistants.

APIs can also enable Metaverse to incorporate real-world data, augmenting its realism and usability. For example, a weather API could be used to mirror real-time weather conditions in certain parts of the Metaverse, or a news API might provide updates on the latest news in a virtual environment.

As technologies continue to evolve, APIs are essential to incorporate these developments into the Metaverse. They standardize software interaction, making it simpler to add, update, or replace elements of the Metaverse, thus allowing it to remain a dynamic and adaptive entity.

In summary, APIs are not merely adjuncts but are central to the development, expansion, and management of the Metaverse. They ensure platform interoperability, facilitate the integration of AI technologies, enable the inclusion of real-world data, and help the Metaverse adapt to technological advancements. As we stride into this new digital epoch, it is essential to focus on API development, integration, and management. APIs will continue to be the pillars supporting the growth and evolution of the Metaverse, facilitating the creation of an ever-evolving, interactive digital universe.

In our next chapter, *Chapter 8, Making and Using 3D Models and Integrating 2D Content*, we dive into how the creative explosion and use of 3D content and integrated 2D content are integral to the fabric of the Metaverse.

Making and Using 3D Models and Integrating 2D Content

Software tools for 3D imaging, avatar creation, and integration of 2D visuals for the Metaverse are vital. These software tools, together with AI software, enable the Metaverse to have content. Avatar creation, especially, is undergoing rapid change from what was originally cartoony to more realistic renders. In this chapter, we'll review software functionality for each software area, as well as specify who the main companies are.

In this chapter, we're going to cover the following main topics:

- Why software that enables 3D imaging, avatar creation, and integration of 2D visuals is vital to the Metaverse

- How 3D imaging and Generative AI are starting a revolution of easily created 3D objects, as well as the start of 3D video, and how this greatly affects content, social interaction, and business in the Metaverse

- The current status of avatar creation, and how avatars used in the Metaverse aid in social interaction and business

- How 2D video will be integrated into the Metaverse

- What the main companies are in 3D imaging, Generative AI, and avatar creation

3D revolution and 2D evolution

The Metaverse, often described as the amalgamation of video games, social media, and the vast realm of the internet, represents the frontier of digital immersion. At its core, the magic that breathes life into this expansive virtual world hinges on technological marvels such as 3D imaging, intricate avatar creation, and the adept integration of 2D visuals.

Entering the Metaverse is akin to stepping into an alternate dimension. Thanks to 3D imaging, this dimension doesn't merely exist as flat visuals; it offers depth, tangibility, and a sense of place. Whether you're navigating bustling digital cities, ascending virtual peaks, or exploring underwater realms, 3D imaging ensures a richness of experience. And as our technological prowess continues to grow, the lines between this virtual space and reality will blur further.

Avatars serve as the inhabitants of the Metaverse, acting as our digital proxies. The ability to customize these entities provides a means for self-expression, allowing users to craft either a faithful digital rendition of themselves or perhaps a more imaginative alter ego. Engaging with others as these avatars lends a layer of personal connection and makes social and business interactions within the Metaverse both meaningful and dynamic.

While 3D environments form the backbone of the Metaverse, the role of 2D visuals remains indispensable. These elements function as informational plaques, interactive displays, or even digital masterpieces that adorn this virtual world, providing context and a touch of familiarity amid the 3D expanse.

Beyond mere exploration, the Metaverse offers educational and entrepreneurial avenues. Interactive learning experiences can transport users to a 3D simulation of ancient Rome or a virtual biology lab, transforming traditional education. Avatars, with their personalized traits, enhance these interactive sessions, while 2D elements offer supplementary information and tools.

From a commercial perspective, the Metaverse is ripe with opportunity. The intricacies of 3D imaging have given rise to concepts like virtual real estate. The design and personalization of avatars have the potential to burgeon into a significant industry. Meanwhile, 2D visuals, strategically placed, can serve as innovative platforms for advertising and branding.

Furthermore, the Metaverse also stands as a beacon for cultural and historical preservation. Through advanced 3D recreations, users can experience and explore digital replicas of historical landmarks and artifacts. Avatars can be crafted to represent and celebrate diverse cultures, and 2D visuals can enshrine art, literature, and history.

The fusion of 3D modeling, avatar customization, and 2D visuals is instrumental in constructing the Metaverse's vibrant tapestry. As these technologies intertwine and evolve, we stand on the cusp of a digital revolution, poised to redefine our interaction with the virtual realm.

The dynamic dance of 3D imaging and generative AI

The Metaverse, blending both VR and AR, is an emerging digital universe where the boundaries of physical reality blur into a vast expanse of virtual landscapes. No longer a mere concept of science fiction, it's swiftly evolving as our next digital domain. Pioneered by advancements in 3D imaging and generative AI, these technologies are transforming our interactions within this space and are fundamental in sculpting the Metaverse's evolving architecture.

3D imaging – the artisan's tool in the digital era

In the digital era, 3D imaging stands as a pivotal bridge between the digital and physical worlds, reshaping the way we visualize, design, and interact with our surroundings. Originally a domain for dedicated artists and film studios, its relevance has grown exponentially, finding applications in diverse sectors such as entertainment, architecture, healthcare, and eCommerce. This artistic and technological convergence not only offers a platform for visual storytelling but also breathes life into intricate ideas, underscoring its role as an essential tool in today's interconnected age.

The evolutionary journey of 2D graphics

Our journey began in the formative years of computing when machines primarily relied on textual interfaces. Users had to navigate their way around computers using command lines, the primary method of communication between humans and machines. In this setting, every command input through a keyboard would undergo processing, resulting in the display of lines of text as output. Though this method was effective in certain ways, it was not particularly intuitive. Envision a world where each simple task required the exact recall and typing of specific commands, followed by waiting for a textual response. With a slender margin for error, any misstep wasn't easily overlooked.

Such an interaction mode, being heavily text-dependent, overlooked a fundamental aspect of human cognition: our innate ability to process visual information quickly and efficiently. Early computer aficionados and experts soon recognized that enhancing the user interface with visual elements wasn't just a desired upgrade – it was essential.

This epiphany ushered in the age of **graphical user interfaces**, commonly known as **GUIs**. This development marked a pivotal moment in computing history. No longer were interactions centered around abstract commands; now, they were about visual indicators and symbols. This new environment was where 2D graphics began to shine.

Though the first GUIs might appear basic to our modern sensibilities, they were groundbreaking in their era. They introduced the computing world to visual symbols, replacing textual commands. For example, the image of a floppy disk became synonymous with saving a file, while a trash can icon meant deletion. These graphical representations not only made navigation simpler but also democratized computing, making it accessible to a broader audience.

As time progressed, the evolution of 2D graphics was swift and significant. It became evident to developers and designers that a well-curated blend of colors, symbols, and shapes could craft an immersive experience for users. What started with basic icons soon evolved into detailed designs, vibrant color schemes, and lively animations, all within a 2D framework. These enhancements played a critical role in rendering software more user-centric, captivating, and visually attractive.

But 2D graphics wasn't just about function – it was also about form. The digital canvas beckoned artists and graphic designers, leading to the emergence of digital art. Leveraging specific software, these creatives began producing elaborate 2D designs, illustrations, and animations that went beyond the practical facets of GUIs.

The emergence of 2D graphics wasn't a mere step forward in the computer world – it signified a seismic shift in the human-computer relationship. With the assistance of 2D graphics, computers transformed from being tools reserved for the technically proficient to indispensable devices for individuals from various backgrounds. As we look ahead, while the nuances of 2D graphics will continue to evolve with technological progress, its crucial role in the annals of computing cannot be overstated.

A brief history of 2D imaging

The following timeline provides a snapshot of 2D imaging's journey. From the ancient imageries carved on stone walls to intricate digital creations, 2D imaging's progression has consistently mirrored technological and cultural advancements.

- **Primitive beginnings with cave paintings**

  Prehistoric times: Early humans captured stories, daily activities, and their belief systems through depictions on cave surfaces.

- **Hieroglyphics of ancient Egypt**

  Around 3200 B.C.: Egyptians combined logographic and alphabetic symbols to inscribe religious content on materials such as papyrus and wood.

- **Woodblock printing in ancient China**

  9th century: Using wooden blocks with carved designs, images were inked and transferred to fabrics or paper.

- **Renaissance's contribution to 2D art**

  Between the 14th and 17th centuries: Artists in Europe integrated mathematical concepts to introduce depth and perspective in their 2D artworks.

- **Birth of photography**

  Early 19th century: Innovations such as the daguerreotype process and the camera obscura allowed for real-world scenes to be captured on light-sensitive surfaces.

- **Animation's Golden Era**

  Early 20th century: 2D images were brought to life by animation trailblazers such as Walt Disney, creating an entirely new entertainment medium.

- **The advent of graphical computer interfaces**

  1960s: Pioneering work at places such as Xerox PARC and MIT explored graphical methods of computer interactions, laying the groundwork for today's GUIs.

- **Interactive 2D graphics in video games**

  1970s and 1980s: The explosion of home and arcade video gaming systems utilized 2D graphics to immerse players in virtual worlds.

- **The rise of desktop design**

  1980s: The introduction of software such as Adobe PageMaker democratized design, enabling sophisticated 2D layouts on personal computers.

- **Digital graphics for the web era**

  1990s: The growth of the internet catalyzed the development of 2D imagery for online ads, web designs, and digital publications.

- **Tools of modern graphic design**

  Late 1990s-2000s: Software advancements saw platforms such as Adobe Photoshop and GIMP emerge, offering intricate 2D image creation and editing capabilities.

- **2D designs tailored for mobile devices**

  2010s-today: The widespread use of smartphones and tablets ushered in a new wave of 2D design, focusing on icons and imagery for compact screens.

- **Cutting-edge digital artistry**

  2010s-today: Advanced tools, software, and devices such as styluses and tablets have empowered artists to redefine the possibilities of 2D art.

The foundational principles of design and user experience were established during the 2D era, a time marked by innovation and experimentation. While the current digital landscape is heavily characterized by 3D visuals and immersive experiences, the importance of the 2D era remains profound. Animators, designers, and creators around the world are still influenced by the aesthetics and lessons from this period. The rich tapestry of visual language crafted during the 2D era has provided a wellspring of inspiration for the subsequent advancements in the 3D domain.

Contemporary digital landscape – The dominance of 3D graphics

The canvas of digital advancement, ever-evolving, showcases the transition from 2D to 3D graphics as a cornerstone of contemporary innovation. Viewing this transformation through the lens of 2023 offers a riveting exploration of technology's forward momentum.

Tracing back to our 3D beginnings, let's reminisce about Ivan Sutherland's Sketchpad system from the 1960s – a veritable beacon illuminating the 3D frontier. As we moved into subsequent decades, tools such as CATIA not only emerged but dominated sectors, especially the aerospace and automotive industries.

Yet, it's 2023 that magnifies 3D's transformative potential. The entertainment industry's AR and VR segments are a testament to this, with 2023 projections from Grand View Research estimating the AR market to soar to an astounding $90 billion, while VR isn't far behind, poised to cross the $50 billion mark.

Video gaming, a sector already rich in its 3D exploits, further amplified its immersive prowess in 2023. Industry analyses forecast that with the integration of advanced 3D mechanics, real-time ray tracing, and AI-powered enhancements, the global gaming market could very well touch $230 billion by year-end.

Education, once tethered to traditional methodologies, embraced 3D with renewed vigor in 2023. A global educational survey revealed that almost 70% of educational institutions were integrating 3D tools into their curricula, motivated by studies showcasing students' enhanced engagement and comprehension with 3D modalities.

Post-pandemic, the real estate industry's evolution continued, with 2023 heralding a new era of 3D integration. Matterport's survey for the year indicated that close to 80% of property listings offered virtual 3D tours, a trend directly linked to increased property valuations and broader audience reach.

On the bleeding edge of technology, the AI consultancy sector witnessed profound shifts in 2023, especially concerning 3D simulations. In a significant reveal, NVIDIA announced that their "Hopper" architecture-based GPUs, launched in 2023, delivered performance enhancements up to 90x in 3D-intensive tasks, signaling the vast horizons yet to be explored.

To encapsulate, the odyssey from 2D to 3D graphics, though initiated decades ago, found significant milestones in 2023. With the year's insights interwoven, the 3D narrative emphasizes its ever-expanding influence, seamlessly blending the realms of virtuality and reality in our intricate digital world.

A brief history of 3D imaging

The following timeline offers a condensed look at the evolution of 3D imaging, highlighting its milestones and significant developments. Over the years, the technology has grown in complexity and capability, becoming an integral part of entertainment, gaming, medical imaging, and even AI applications.

- **Up to the 1960s – stereoscopy emerges**
  - **1838**: Charles Wheatstone introduced the concept of stereoscopy. Though it predates the 20th century, this was a fundamental step in understanding 3D imaging.
  - **1950s-1960s**: The 3D movie craze. Anaglyph images, which are pictures consisting of two slightly offset images, were viewed through two-color glasses to create a sense of depth.

- **1970s – the birth of computer-generated 3D graphics**

  - **1972**: The Utah teapot, one of the first computer-generated 3D models, was created at the University of Utah.

  - **1976**: The first 3D video game, *Spasim*, was developed. It featured wireframe 3D graphics.

- **1980s – the rise of 3D software and rendering**

  - **1982**: The movie *Tron* was released, boasting some of the earliest CGI graphics.

  - **1986**: Pixar released *Luxo Jr.*, a short film that showcased advanced 3D animation capabilities.

  - **1988**: Radiosity, a global illumination algorithm in 3D computer graphics, was introduced.

- **1990s – the expansion of 3D graphics and VR**

  - **1992**: The *Lawnmower Man* movie introduced mainstream audiences to the concept of VR.

  - **1995**: The release of *Toy Story*, the first full-length film made entirely using CGI.

  - **1999**: NVIDIA introduced the GeForce 256, branded as the world's first "GPU," accelerating 3D graphics rendering.

- **2000s – 3D imaging becomes mainstream**

  - **2001**: Microsoft's Xbox console was launched, featuring powerful 3D rendering capabilities.

  - **2005**: Google Earth was launched, providing a 3D representation of the world using satellite imagery.

  - **2009**: James Cameron's *Avatar* was released, setting new standards for 3D in cinema. 3D TVs and 3D Blu-Ray players started entering the market.

- **2010s – AR and VR**

  - **2013**: Oculus Rift, a VR headset, gained popularity after a successful Kickstarter campaign.

  - **2016**: Pokémon Go, an AR game, became a sensation, pushing AR to mainstream audiences.

  - **2018**: Apple introduced ARKit, making it easier for developers to incorporate AR into apps for iOS devices.

- **2020s – immersive experiences and real-time 3D imaging**

  - **2020-2021**: Rapid advancements in AI technology, such as **Generative Adversarial Networks (GANs)** and **Neural Radiance Fields (NeRF)**, enabled the creation of highly realistic 3D scenes and images.

- **2022**: Generative AI gained traction
- **2023**: Apple announced its long-awaited Mixed Reality headset, the Vision Pro with visionOS, which easily displays 3D visuals, alongside legacy 2D.

Crafting 3D images

3D visual creation, an interdisciplinary field, has seen marked growth with tech innovations, particularly in sectors such as animation, gaming, VR, and AR. The following are the major methods, software applications, and leading companies in 3D imagery:

- **Methods for crafting 3D images**

  - **3D modeling**: A technique where a three-dimensional object's mathematical representation is crafted using vertices, edges, and polygons.

  - **3D scanning**: A process that involves capturing the contours and dimensions of an existing object and converting it into a digital 3D model.

  - **Photogrammetry**: A technique that extracts 3D shapes from multiple photographs taken from different angles.

  - **Procedural generation**: An algorithmic method, it allows for the automatic creation of detailed and varied 3D models.

  - **3D rendering – post-modeling**: This method involves adding elements such as lighting, textures, and shading to bring a 3D model to life in an image or animation.

  - **Digital sculpting**: This is a method where 3D objects are sculpted and refined using specialized software, much like clay modeling but in a digital space.

  - **Parametric design**: This involves defining certain parameters and rules for a design, allowing software to auto-generate shapes based on those rules.

  - **NURBS creation**: Utilizes Non-Uniform Rational B-splines, a mathematical representation, to design and depict curves and surfaces in computer graphics.

  - **Polygon-based design**: This method crafts 3D forms by manipulating polygons. It's a prevalent technique, especially in video game development.

  - **Surface subdivision**: Through this approach, a polygonal model undergoes subdivisions to yield a more refined and smoother surface.

  - **Boolean manipulation**: This combines multiple 3D shapes through operations such as union, intersection, or difference to produce novel forms.

  - **Voxel-based design**: Represents 3D structures as a grid of volume pixels or voxels, analogous to pixels in 2D imagery.

- **Detail stimulation with mapping**: Techniques such as displacement and bump mapping create the appearance of intricate surface details without altering the model's geometry.

- **Inverse kinematics**: Primarily used in animation, this calculates joint rotations based on the desired final position, leading to more lifelike movement.

- **Ray trace rendering**: This simulates light interactions, including reflections and refractions, to create ultra-realistic graphics.

- **Radiosity illumination**: A technique that imitates the diffusive property of light bouncing off surfaces, enhancing environmental realism.

- **Stereoscopic imaging**: Enhances depth perception by presenting slightly varied images to each eye.

- **Mesh optimization via retopology**: By redesigning an existing 3D model's mesh, retopology optimizes the model for better efficiency.

- **2D image integration (UV mapping)**: This essential texturing method maps a 2D graphic onto a 3D model's surface.

- **Depth enhancement with ambient occlusion**: This method evaluates ambient light exposure at each point in a scene, contributing depth and dimension.

- **Particle system simulations**: Manages numerous tiny entities to depict intricate phenomena, such as fog, sparks, or fountains.

- **Stimulating liquids and gases**: Using Fluid Dynamics, this models the interactions and behaviors of fluids within a 3D environment.

- **Strand detailing**: Techniques to design and animate strand-like elements, including hair, fur, and grass, adding realism to characters and scenes.

- **Animating musculature**: This simulates the nuanced movement and deformation of muscles beneath a character's skin.

- **3D model animation infrastructure (rigging)**: Introduces bones and movement controls to a model, paving the way for animation.

- **Detail transfer via baking**: Transfers intricate details from detailed models to simpler ones, aiding in texture, light, and shadow optimization for quicker rendering.

- **Notable software for crafting 3D images**

  - **Blender**: An open source 3D creation suite, it caters to a broad range of 3D design needs, from modeling to rendering.

  - **Autodesk Maya and 3ds Max**: Renowned software from Autodesk, both are industry standards for 3D modeling, animation, and rendering.

- **ZBrush**: Specializing in digital sculpting, this software provides an intuitive platform for intricate 3D design.

- **Cinema 4D**: Valued for its user-friendly interface, it offers robust 3D modeling and animation tools.

- **Rhino**: Known for its powerful surface modeling capabilities, it's often used in industrial and architectural design.

- **Agisoft Metashape**: A leading software for photogrammetry processes.

- **Unity and Unreal Engine**: Beyond their primary use as game engines, both platforms also serve as tools for 3D design and rendering.

- **Mudbox**: A 3D sculpting and painting software developed by Autodesk, primarily used for creating highly detailed textures and organic models.

- **Modo**: A versatile 3D modeling, texturing, and rendering tool developed by Foundry. It's known for its user-friendly interface and powerful modeling features.

- **Houdini**: Developed by SideFX, Houdini is a procedural-based 3D software well-regarded for its powerful dynamics, effects, and simulation capabilities.

- **SketchUp**: Originally developed by @Last Software and later acquired by Trimble, SketchUp is an intuitive 3D modeling program that's widely used in architectural visualization, interior design, and civil engineering.

- **Revit**: Created by Autodesk, this software is primarily tailored for **building information modeling (BIM)**. It's essential for architects and building professionals to design and manage building projects.

- **LightWave 3D**: A comprehensive software that offers modeling, animation, and rendering capabilities. It has been used in various movies, TV shows, and games.

- **Octane Render**: A GPU-based cloud rendering software by OTOY, it's known for delivering high-quality real-time scene rendering.

- **Maxon Redshift**: A popular GPU-accelerated renderer known for its speed and efficiency.

- **Substance Designer and Substance Painter**: Developed by Allegorithmic (now a part of Adobe), these tools are specifically tailored for creating and painting 3D textures.

- **Marvelous Designer**: This software is tailored for creating 3D cloth simulations and designing virtual garments.

- **KeyShot**: A real-time ray-tracing and global illumination software used for creating 3D renderings and animations.

- **VRay**: Developed by Chaos Group, VRay is a high-performance renderer compatible with major 3D software such as SketchUp, 3ds Max, and Blender.

- **Fusion 360**: An Autodesk product that integrates CAD, CAM, and CAE in a unified cloud-based platform, it's ideal for product design and manufacturing.

- **Poser**: A 3D computer graphics software optimized for 3D modeling of human figures. It is known for its ready-to-use human figure models.

- **3D Coat**: A software that specializes in voxel sculpting, UV mapping, texture painting, and retopology.

- **Sculptrix**: A beginner-friendly digital sculpting software, it acts as an introductory platform to the likes of ZBrush.

- **Vue**: A software for creating, animating, and rendering natural 3D environments, often used for landscapes and vistas.

- **DAZ Studio**: A software designed for creating and animating 3D human figures. It's known for its vast library of pre-made characters.

- **Carrara**: A full-featured 3D software suite that offers modeling, animation, and rendering functionalities.

- **Prominent companies in the 3D imaging sphere**

  - **Autodesk**: A giant in the 3D design, engineering, and entertainment software market. They're behind tools such as AutoCAD, Maya, and 3ds Max.

  - **Adobe**: Known primarily for 2D design tools, Adobe has ventured into 3D with tools such as Dimension and acquisitions such as Allegorithmic (Substance Designer and Substance Painter).

  - **Maxon**: Makers of Cinema 4D, a popular 3D modeling, animation, and rendering software.

  - **SideFX**: The company behind Houdini, a high-end visual effects and 3D modeling software.

  - **Foundry**: Developers of Modo and other post-production tools.

  - **Blender Foundation**: Responsible for Blender, a powerful free open source 3D content creation suite.

  - **Chaos Group**: Creators of the V-Ray rendering software, which integrates with various popular 3D tools.

  - **OTOY**: They developed Octane Render, a popular GPU-accelerated cloud renderer.

  - **Pixologic**: The company behind ZBrush, a leading digital sculpting tool.

  - **Unity Technologies**: Makers of Unity, a game engine that's widely used for 3D game development as well as simulations and other applications.

  - **Epic Games**: Known for Unreal Engine, another major player in the game development and real-time 3D content creation arena.

- **Dassault Systèmes**: The company behind SolidWorks and CATIA, both widely used in industrial design and engineering.

- **Trimble**: Owners of SketchUp, a user-friendly 3D modeling tool that's popular in architectural visualization.

- **ANSYS**: Focused on engineering simulation, their software is used to predict how product designs will behave in real-world environments.

- **NVIDIA**: Beyond their graphics hardware, NVIDIA has been deeply involved in the 3D industry, developing technologies such as RTX ray tracing and AI-driven tools.

- **McNeel & Associates**: Developers of Rhino, a design software that uses NURBS mathematical modeling to produce precise graphics.

- **Luxion**: The team behind KeyShot, a real-time 3D rendering and animation software.

- **CLO Virtual Fashion**: Makers of Marvelous Designer, software for 3D clothing design.

- **Reallusion**: They develop tools for real-time 2D and 3D animation, such as Character Creator and iClone.

- **Matterport**: Specializes in creating 3D spatial data from real-world environments, often used in real estate and architectural visualization.

- **Agisoft**: Known for their photogrammetry software, they created PhotoScan (now called Metashape), which is used to process digital images to generate 3D spatial data.

- **NewTek**: Developers of LightWave 3D, a software used for modeling, animation, and rendering.

- **Geomagic**: Offers a suite of 3D imaging solutions for reverse engineering, product design, and quality control.

- **Mapbox**: While known more for mapping solutions, they offer tools and SDKs that leverage 3D spatial data.

- **Enscape**: Provides real-time 3D visualization and VR tools for architectural and urban planning needs.

Generative AI – the next quantum leap in digital design

Navigating the technological renaissance of the 21st century, generative AI stands out as a transformative force. We're on the cusp of not just enhancing chat engines but ushering in a groundbreaking computational evolution. Let's delve into the capabilities and potential of generative AI for visuality.

What generative AI can do now for visuals

The accelerated growth and transformation of AI technologies have witnessed a significant rise in its adoption across the business landscape. This surge isn't limited to just emerging startups attempting to mark their presence in this competitive realm; even well-established industry giants are making substantial strides to integrate generative AI methodologies into their operational structures. Among these cutting-edge AI techniques, **Generative Adversarial Networks (GANs)** stand out, along with **Transformers**. GANs have gained considerable traction because of their unique ability to generate new data instances that resemble a given set of training data. This capability has unlocked an array of diverse applications, ranging from image generation to enhancing digital media quality, creating realistic video game environments, and even in advanced research domains. As a result, businesses are keenly exploring and leveraging the potential of GANs to drive innovation and achieve competitive advantage in their respective industries. Transformers are a kind of deep learning architecture that was initially used for NLP and conversational AI, such as chatbots and digital assistants, but has now proven able to perform text-to-image generation, as used by OpenAI's DALL-E. DALL-E 2 uses a **diffusion model** that DALL-E does not. Diffusion models hold latent variables capable of removing noise and deciphering the underlying structure of an image. As these models become trained to understand the abstract concepts depicted in images, they gain the remarkable ability to generate an array of new variations for the same image.

Today, we are seeing the first uses of generative AI – ChatGPT, GPT-4, Stable Diffusion, Claude, Bard, and Ernie, along with many writing and other tools. As of August 2023, about 4,600 tools are already listed on the site Futurepedia.io. The investors who are pouring billions into companies such as OpenAI and Stability AI aren't investing that money to get a chat engine that gives lots of errors. So, what are they investing in?

The potential for a radically different way of computing – something closer to *Star Trek's* Holodeck than Microsoft Windows. Such a Holodeck will be fed by potentially dozens of generative AIs. What are generative AIs? They are AIs that use unsupervised learning algorithms with human inputting of text prompts to create new virtual photos, videos, text, code, or audio. Unsupervised learning can identify previously unknown patterns in data. Generative AIs can currently create 2D and 3D still images and transform an existing 2D video into a new one (using Runway's Gen-1 software). Soon, generative AI will be able to create long-form consumer-ready 2D and 3D video just from text prompts. And there is work currently being done to make Generated AI video capable of streaming.

Understanding the challenge of 3D generation in AI

Unlike 2D data, which is straightforwardly depicted using pixel grids, 3D models comprise point clouds, voxel grids, or polygon meshes. This necessitates the AI to grasp concepts such as depth, structure, and orientation – a daunting task compared to 2D representation.

The hurdles of ambiguity

- **From text to 3D**: Descriptions such as "a wooden chair with a cushion" can be visually represented in myriad ways. AI must discern specifics from such broad descriptors, an immensely complicated feat.

- **From image to 3D**: A 2D image only provides one viewpoint, leading to challenges in inferring the unseen aspects of the 3D subject. Depth perception, occlusion, and perspective further add layers of complexity.

- **From video to 3D**: Videos, although offering varied viewpoints through frames, bring their own set of challenges. Factors such as rapid motion, light changes, and blurriness can obscure the true 3D nature of objects.

Resource intensity of training data

To train a generative AI model adequately, you need vast amounts of matched training data. This would mean every text, image, or video input has an associated 3D model. Compiling such a dataset is both resource-heavy and time-intensive.

High computational demands

The creation of 3D models, especially detailed ones, demands immense computational resources. Processing and deciding upon millions of data points in the 3D space is resource-intensive.

Challenges in evaluating AI outputs

While 2D images might be assessed visually, 3D models demand scrutiny on multiple fronts: geometry, spatial coherence, texture, and more.

Constraints in real-world usages

For applications such as VR, gaming, or architectural visualization, the 3D output should be not only accurate but also optimized for real-time rendering, maintaining a balance between detail and computational efficiency.

The need for interactive systems

A promising avenue to manage the inherent ambiguities is devising interactive AI systems. Users can give feedback or adjust their inputs, but creating such systems brings forth challenges in user interface design, real-time model generation, and iterative refinement.

A reputable breakthrough by NVIDIA

The NVIDIA Omniverse team explored the capabilities of the GPT-4 multimodal model and ChatGPT. Their goal was to simplify the creation of 3D digital assets. They paired GPT-4 with DeepSearch, an AI tool from Omniverse that searches large 3D asset databases, even those without clear tags. This collaboration resulted in an extension that lets developers and artists easily fetch and integrate 3D assets using text commands.

Named "AI Room Generator," this extension highlights the power of generative AI in digital design. Users can quickly generate and place high-quality 3D entities with a few text prompts. Furthermore, these assets align with the **Universal Scene Description** (**USD**) SimReady standards, ensuring they're both visually relevant and physically accurate in simulations.

As for other breakthroughs, OpenAI, Stability AI, Anthropic, Cohere, Midjourney, and Runway, among other companies, are candidates for evolving into 3D generative visuals. The world of generative AI is continuously evolving, driven by advances in computational capabilities, innovative algorithms, and an expanding pool of diverse training data. As these models grow more refined and adept at deciphering the nuances of the 3D realm, we can anticipate a transformative shift in how we approach 3D visualizations.

Furthermore, the convergence of expertise from 3D artists, AI specialists, and other domain professionals is fostering fertile ground for innovation. Such collaborations aim to enhance AI's capability to minimize ambiguities from diverse inputs and ensure the creation of 3D models that are both functionally robust and aesthetically compelling.

There's also a burgeoning momentum behind tools that promote human-AI synergy, ensuring that the output aligns with both artistic vision and practical requirements. This collaborative spirit is a testament to the belief in AI's potential in the realm of 3D creation.

Taking stock of these positive trends, it's evident that generative AI is on track to not just meet the challenges of 3D visualization but to reshape industries ranging from entertainment and gaming to architectural design. In the not-so-distant future, the seamless integration of AI with 3D visualization promises to unlock unprecedented opportunities and redefine the boundaries of digital creativity.

The Metaverse transformation – a multi-faceted impact

3D imaging stands at the very heart of the Metaverse, furnishing it with the intricacy, richness, and lifelike depth that users seek. It's this technology that transforms virtual spaces into convincing replicas of our world, or even entirely new imaginative realms. By embedding such vivid details and authenticity, 3D imaging transcends the digital divide, offering a Metaverse experience that's more than a virtual escapade – it's an enriched continuum of our very reality, brimming with endless possibilities for discovery and creativity.

In this section, we'll examine how 3D imaging is transforming the Metaverse, accompanied by explicit examples.

Personal appearance and lifestyle

- **Personal avatars and digital representation**: In the Metaverse, users can generate avatars that closely resemble themselves using 3D imaging. This technology can be taken a step further to develop a "digital twin" for individual users. A practical application can be seen when someone wishes to experiment with fashion, allowing them to virtually try on different outfits, reflecting how they would appear on their real physique.

- **Virtual fashion parades and digital dressing rooms**: In the Metaverse, designers can showcase their collections in dynamic fashion displays. Attendees, using their avatars fine-tuned through 3D imaging to mirror their actual physique, can experiment with digital apparel fittings.

- **Cosmetic and fashion makeover studios**: Users can enter digital salons where they can experiment with different hairstyles, makeup looks, or fashion accessories on their avatars, all accurately presented using 3D imaging, before making real-world choices.

Professional collaboration and workspaces

- **Collaborative virtual workspaces**: The corporate world can leverage the metaverse for virtual meetings, especially in sectors such as AI and technology consulting. Visualizing a scenario where a 3D prototype of a product is presented in a meeting, attendees from different locations can gather in this virtual space, interact with the prototype, and provide instant feedback.

- **Virtual architecture and real estate innovations**: Prospective property investors or buyers in the Metaverse can benefit from 3D imaging. Architects can showcase virtual properties by creating detailed 3D replicas of structures, allowing visitors to navigate through these spaces for a comprehensive understanding.

- **Virtual film sets and production studios**: Film directors and production crews can explore and customize virtual sets created using 3D imaging. This would enable them to plan scenes, camera angles, and set designs even before physical production begins.

Digital commerce and consumer experiences

- **Revolutionized virtual commerce**: The eCommerce realm within the Metaverse is set to undergo a major transformation. Instead of static product images, shoppers can experience and engage with 3D product renditions. A tangible application could be a virtual furniture showroom where a shopper can view a 3D image of a chair, interact with it, and virtually arrange it with other items.

- **Showrooms for vehicles and aircraft**: Virtual halls showcasing 3D-imaged cars or planes can be a treat for vehicle aficionados. Not only can they inspect the minutiae of the latest models, but they can also simulate the experience of driving or flying them.

- **Personalized home décor and interior design workshops**: Interior designers could give tours of virtual home spaces, allowing clients to visualize and adjust decor elements, furniture, or color schemes in real time, all brought to life via 3D imaging.

Gaming, entertainment, and social engagements

- **Elevated gaming and virtual entertainment**: Game creators can incorporate 3D imaging to develop incredibly detailed game settings and characters. By scanning real-world scenarios, such as dense forests or historical ruins, developers can immerse players in these lifelike settings, offering an unparalleled gaming experience.

- **Music creation spaces and virtual concert halls**: Musicians could usher fans into a digital sound studio, where every instrument and piece of equipment is rendered through 3D imaging. Fans might participate in the melody-crafting journey or be spectators in a virtual concert with authentic acoustics and stage designs.

- **Adventure challenges and Metaversal escape rooms**: By harnessing 3D imaging, businesses can design riveting escape rooms or adventure quests. Users face intricate environments and riddles, ensuring a challenging and lifelike problem-solving adventure.

- **Social digital spaces for gatherings and engagements**: People, despite being separated geographically, can congregate in virtual locales such as parks or cafes, perfectly mimicked using 3D imaging. These settings are ripe for activities such as digital feasts, gaming nights, or simply catching up.

- **Metaversal dance and performance art studios**: Dancers and performers can use virtual spaces to teach, practice, or even host shows. The studio's surroundings, instruments, and even the ambiance can be crafted using 3D imaging, allowing for a highly interactive and dynamic experience.

Education and training

- **Educational platforms and immersive training**: 3D imaging can be a game-changer for educational modules in the Metaverse. Envision a medical training session where students can delve deep into a 3D representation of the human body, promoting a hands-on learning experience.

- **Paleontology and archeological digs**: Virtual dig sites, based on real-world archaeological locations, can be set up for enthusiasts. These digital grounds allow for the exploration and study of ancient relics or fossilized entities, offering a hands-on historical delve.

- **Surgical simulations and on-the-spot 3D rendering**: Within the surgical field, the integration of 3D imaging marks a groundbreaking leap forward. With this innovation, surgeons can craft accurate 3D depictions of a patient's internal structure before embarking on the actual surgery. This intricate visual model serves as a "digital twin" to the patient's inner body or the targeted surgical area.

- **Forensic training through virtual crime scenarios**: 3D imaging can facilitate the creation of detailed crime scenes in the Metaverse for training forensic students or professionals. This controlled setting is ideal for honing investigative skills or working on mock cases.

- **3D book narratives and story exploration**: Authors and publishers can create interactive book experiences where readers traverse 3D landscapes of the story, meeting characters and exploring settings, offering a more immersive reading experience.

- **Virtual culinary workshops**: Culinary enthusiasts could partake in virtual cooking sessions where 3D images of ingredients and tools are at their disposal. As an expert chef illustrates a technique on a 3D-imaged vegetable or piece of meat, participants can emulate the actions with their virtual counterparts.

Cultural and historical preservation

- **Digital preservation of culture**: The Metaverse can serve as a repository for historically significant sites and artifacts using 3D imaging. Even if the actual location succumbs to the ravages of time, its meticulously crafted 3D rendition will be available for posterity.

- **Museums and digital art galleries in the Metaverse**: Users can traverse a museum or gallery where exhibits, whether they're ancient relics or extinct species, are meticulously recreated through 3D imaging. This offers a risk-free, up-close interaction with global treasures.

- **Historical reenactments and time travel**: Users can step into different epochs, from ancient civilizations to significant historical events, all meticulously reconstructed using 3D imaging. This could serve as an educational tool or simply an immersive way to experience history firsthand.

Exploration and tourism

- **Space voyages and astronomical tours**: With 3D scans from institutions such as NASA, users can undertake virtual space journeys. They might wander the Martian terrain or embark on deep-space adventures based on real-life mission data.

- **Virtual excursions and tourism**: World-renowned sites, be it the Eiffel Tower or Machu Picchu, can be rendered in the Metaverse using 3D imaging. Users can embark on digital journeys, relishing the splendor of these landmarks without the need for physical travel.

- **Virtual underwater expeditions**: Delve deep into the digital oceans, navigating through 3D-replicated coral labyrinths, submerged caverns, or ancient ship remnants. This allows marine biologists and ocean aficionados to explore without the constraints of physical depths or equipment.

Creative and artistic ventures

- **Artistic digital creation and sessions**: In the Metaverse, artists could host workshops employing 3D-imaged art supplies. Attendees could witness the artistry up close, manipulating 3D materials to paint, sculpt, or craft alongside the host.

- **Virtual reality journalism and on-ground reporting**: Reporters can relay news stories from virtual replicas of real-world events or locations, allowing viewers to virtually "be" at the scene, understanding contexts and events more deeply.

Sports, wellness, and physical activities

- **Sports training and simulation**: Athletes might access a virtual training arena that's a 3D replica of a real-world sports venue. This digital space can serve as a ground for practice, strategy formulation, or replicating matches against rivals.

- **Digital therapy and virtual healing centers**: In a therapeutic Metaverse setting, patients can undergo rehabilitation sessions overseen by medical practitioners. Realistic tools and surroundings, forged through 3D imaging, aid in exercises or therapeutic regimens.

Nature, agriculture, and environmental studies

- **Botanical virtual ventures**: Using 3D scans of actual flora, digital forests or gardens can be established. Whether you're a botany student, a researcher, or just an enthusiast, you can study plant morphology or simulate the planting process in these digital green spaces.

- **Virtual farms and digital agricultural ventures**: Using 3D imaging, virtual farms come alive, offering researchers or farming hobbyists a chance to engage with crops, tools, and livestock. This provides a sandbox to simulate farming practices or delve into agrarian studies.

The digital landscape of the future is poised to be profoundly influenced by the integration of 3D imaging, especially within the Metaverse. Its applications are broad and multifaceted, impacting countless facets of our daily existence. Whether it's transforming professional domains, amplifying personal encounters, or paving the way for in-depth exploration and learning, 3D imaging plays a pivotal role.

Avatar creation

Designing avatars for the Metaverse encompasses a blend of artistic design, technological skills, and an understanding of societal norms. The Metaverse, essentially a vast shared digital realm born from the fusion of augmented reality and immersive digital environments, has elevated the importance of avatars as our digital counterparts. Let's delve into the multifaceted procedure and considerations involved in crafting these avatars for the Metaverse.

Envisioning and blueprinting

- **Avatar's role**: Determining if the avatar will act as an individual's digital twin, an imaginative character, or a representation of an idea is the initial step.

- **Design origins**: Inspirations can be drawn from diverse sources, such as myths, modern media, historical events, or individual narratives.

- **Preliminary designs**: Before diving into 3D, initial 2D sketches give a basic framework of the avatar's appearance.

- **Iterative feedback**: Collecting insights from potential users or other stakeholders is invaluable in refining the avatar's conceptual design.

Crafting in 3D

- **Choice of tools**: Depending on the avatar's intricacy, designers might use Blender, ZBrush, or Maya to sculpt the model, or generative AI to create it.

- **Formation of mesh**: At this juncture, the 2D concept transforms into a three-dimensional shape.

- **Mesh refinement**: Especially for platforms with constraints, it's pivotal to make sure the 3D model is streamlined in terms of its polygon structure.

Surface details and reflection properties

- **Texture implementation**: By flattening the 3D model into a 2D plane through UV mapping, artists can effectively add colors or patterns.

- **Texture crafting**: Tools such as Substance Painter come into play to design intricate surface details.

- **Light interaction**: Through shading, designers specify how the avatar's surface will reflect or absorb light, giving it a unique look.

Movement and life infusion

- **Framework of movement**: By adding a "skeleton," artists give the model the capability to move

- **Defining movement influence**: Weight painting ensures that the avatar moves naturally, by specifying which part of the model moves with which "bone"

- **Animating Life**: Avatars can either be fitted with preset motions or crafted to be compatible with custom animations

Onboarding to the Metaverse

- **Fine-tuning**: The avatar might require adjustments in form or texture details, depending on where it's being used

- **Initiating**: The tailored avatar is then brought into the desired Metaverse environment

- **Real-time interaction**: Many platforms offer avatars real-time capabilities such as voice modulation or expressive facial movement

Societal and cultural dimensions

- **Universal appeal**: Crafting avatars that cater to diverse body types, ethnic backgrounds, and features champions inclusivity

- **Avoiding pitfalls**: Steering clear of potential cultural stereotypes and ensuring respect for all groups is fundamental

- **User customization**: Giving individuals the tools to modify avatars enhances personal connection and immersion

AI's intersection with avatars

- **Automated movements**: AI algorithms can spontaneously create lifelike facial expressions and movements

- **Intelligent companions**: The next-generation avatars might double as AI aids, evolving based on user habits and preferences

- **Evolving avatars**: Advanced AI can enable avatars to grow and adjust based on the interactions they encounter

Crafting avatars for the Metaverse is an evolving art, enriched by technological advancements. When AI converges with these digital personas, they transition from mere representations to smart entities capable of interaction, evolution, and assistance in the virtual universe.

Entities that produce avatars

Generative AI, especially the use of GANs, has heralded a new era in digital design, particularly in the creation of virtual avatars, characters, and other visual representations. The following are some prominent entities and initiatives that have championed the sphere of virtual avatar synthesis, powered by the prowess of generative AI:

- **Apple's Vision Pro visionOS**: Just announced in June 2023, visionOS can create a hyper-realistic avatar using a neural network to use during Facetiming.

- **Artbreeder**: Using GANs, Artbreeder offers a platform for both individual and collaborative image creation. This platform is widely appreciated by artists, game developers, and illustrators to create unique avatars.

- **MetaHuman Creator by Epic Games**: MetaHuman Creator is a sophisticated app streamed via the cloud, aiming to elevate the realism and detail of real-time digital humans. A blend of generative techniques, including AI, drives this tool, making it a go-to for filmmakers, game developers, and artists.

- **DALL-E 2 by OpenAI**: DALL-E 2, celebrated for materializing images from textual cues, possesses vast potential in the avatar generation landscape. The prospect of delineating an avatar's features via textual inputs and having DALL-E illustrate the visual counterpart is truly groundbreaking.

- **Ready Player Me**: This avant-garde platform empowers users to design bespoke 3D avatars suitable for virtual ecosystems, as well as VR and AR experiences. Marrying AI with facial recognition, Ready Player Me can transfigure casual selfies into intricate 3D personas ready for digital immersion.

- **DAZ 3D**: DAZ 3D's renowned Genesis platform is used extensively for 3D avatar generation. With the incorporation of AI in recent updates, it offers enhanced realism and morphing capabilities.

- **NVIDIA**: NVIDIA is a leader in GAN research, with its StyleGAN series laying the groundwork for high-quality image generation. Though NVIDIA doesn't commercialize avatars directly, many third-party solutions rely on their foundational research.

- **Toonify**: Toonify specializes in transforming standard photos into cartoon-style avatars, leveraging the capabilities of StyleGAN.

- **Runway**: Runway provides an assortment of tools, GANs included, for creative endeavors. It's versatile and caters to avatar creation, style adaptations, and more.

- **Loom.ai**: Specializing in transmuting 2D photographs into animated 3D avatars, Loom.ai is revolutionizing the concept of digital doppelgangers. Their pioneering approach taps into deep learning to deduce 3D facial dynamics and nuances from singular or multiple snapshots, with potential applications spanning from immersive gaming to virtual simulations.

- **Synthesia**: Using Synthesia, users can generate videos and virtual characters. By simply entering text, the AI will create a video where the chosen virtual avatar communicates that text.

- **Crypto Avatars**: As the realms of blockchain and cryptographic collectibles expand, GANs have found a niche in sculpting one-of-a-kind avatars that are tradeable and ownable on decentralized platforms. These digital personas not only serve as unique crypto identities but also as gateways to a burgeoning virtual economy.

- **This Person Does Not Exist**: This popular website, built on NVIDIA's StyleGAN, produces a fresh high-resolution image of a fictional person with every refresh, highlighting the potential of GANs in avatar creation.

- **DeepArt**: DeepArt employs neural networks for artistic transformations, converting images into artwork that mimics famous painters or user-given styles, though it's not solely an avatar creator.

- **AI Dungeon by Latitude**: Leveraging GANs, AI Dungeon crafts characters and scenarios for its interactive text adventures.

- **ChatGPT Avatars by OpenAI**: OpenAI, the entity behind me, has ventured into generating virtual assistant avatars using GANs to enhance user interaction and experience.

- **Meta's Celebrity Digital Assistants**: Meta's initial foray into avatars had been met with ridicule due to their simplicity. However, in September 2023, Meta announced that they had enlisted celebrities such as Paris Hilton, Mr. Beast, and Kendall Jenner to become the faces of digital assistants. Notably, Meta plans to initially test the ability of users to create their own digital assistants, which could serve as avatars through select businesses before considering a broader implementation.

2D integration

The Metaverse represents a convergence of digitally enhanced physical reality with interactive virtual spaces. For it to be comprehensive, it's imperative to seamlessly incorporate traditional 2D digital content, such as videos, images, and text, into its predominantly 3D environments.

Why is 2D integration essential?

2D Integration is a key element in the modern digital world, particularly in the Metaverse. It connects the vast amount of existing 2D content with new 3D environments, creating a bridge between simple content creation and immersive experiences. By providing an accessible way to link different digital formats, 2D integration proves essential in shaping the future of technology, demonstrating its ongoing importance and practical value.

- **Legacy and historical relevance**: Given the vast reservoir of 2D content, such as historical data, media, and educational resources, it's crucial to make them accessible in the Metaverse without converting everything into 3D.

- **Simplicity of 2D content creation**: Despite the growing ease of creating 3D content, 2D remains the most accessible medium for general users, especially for documents, simple visual media, and presentations.

- **Combining multimedia dimensions**: Envisage attending a Metaverse-based virtual concert where 2D videos play on virtual screens within a 3D space, offering a multidimensional experience.

Technological techniques facilitating the integration

The blending of 2D content with 3D spaces in the Metaverse is facilitated by innovative technological techniques. These include virtual display mechanisms that replicate real-life screens, AR overlays for projecting 2D imagery onto actual surfaces, and touch-sensitive interactive boards that allow tactile interaction within 3D environments. Together, these technologies bridge the gap between conventional media and the immersive virtual world:

- **Virtual display mechanisms**: One straightforward approach is using virtual screens within the 3D Metaverse that can display 2D content, mimicking real-life TVs or computer displays.

- **AR overlays for real-world fusion**: In Metaverse regions that overlap with the real world, 2D content can be projected as AR overlays. For example, using AR glasses, a 2D video could be showcased on an actual wall.

- **Touch-sensitive interactive boards**: 2D content can be made available as interactive touch panels within the 3D space, allowing users to engage with them as they would with touchscreens in reality.

Potential challenges in integration

In the evolving field of digital content integration, the blending of 2D and 3D content brings forward several complex challenges. Ensuring accessibility to historical 2D content in 3D environments such as the Metaverse, enhancing interactivity between these dimensions, and maintaining the ease of 2D content creation are key concerns. Additionally, the obstacles of preserving an immersive experience, innovatively weaving multimedia dimensions for enriched experiences, and devising conversion tools contribute to the intricacies of this integration process:

- **Maintaining immersion**: It's a challenge ensuring that 2D content, though familiar, doesn't disrupt the immersion of a 3D space.

- **Enhancing 2D-3D interactivity**: Technical challenges arise when making 2D content both responsive and interactive within a 3D environment. An example is figuring out VR controls for zooming into a 2D image in a 3D space.

- **Development of conversion utilities**: While not all 2D content needs a 3D transformation, there's a surge in demand for tools facilitating this conversion, particularly for dynamic content.

- **Legacy and historical relevance**: Given the vast reservoir of 2D content, such as historical data, media, and educational resources, it's crucial to make them accessible in the Metaverse without converting everything into 3D.

- **Simplicity of 2D content creation**: Despite the growing ease of creating 3D content, 2D remains the most accessible medium for general users, especially for documents, simple visual media, and presentations.

- **Combining multimedia dimensions**: Envisage attending a Metaverse-based virtual concert where 2D videos play on virtual screens within a 3D space, offering a multidimensional experience.

The roles of AI in bridging AI and the Metaverse

In shaping the evolving Metaverse, AI plays a vital role, emphasizing its transformational capabilities. By streamlining the conversion of traditional 2D elements into 3D, optimizing the interaction between users and content, and providing personalized recommendations within the virtual world, AI enhances both the accessibility and dynamism of the Metaverse. These functions illustrate AI's indispensable contribution to forging a more intuitive and customized virtual experience:

- **Streamlining 2D to 3D conversion**: AI algorithms can facilitate the automation of converting 2D elements into their 3D counterparts, minimizing manual input.

- **Optimizing user interaction**: AI can discern user behaviors and intentions, refining the interaction between users and 2D content in the 3D environment.

- **Tailored content suggestions**: AI's capability to analyze 2D content allows for personalized recommendations and modifications within the Metaverse.

In essence, the seamless fusion of 2D and 3D in the Metaverse, propelled by AI's advancements, will shape the degree of interactivity and immersion users experience in future virtual realms.

Summary

The Metaverse's dynamism is heavily powered by 3D imaging and soon, more by generative AI techniques. The spaces we explored, rich in detail and depth, owe their realism to these cutting-edge 3D technologies. But it's not just the environments that captivate us – it's also the inhabitants. AI-generated avatars bring individualism and realism to the virtual space. These AI-generated avatars not only mirror real-world likenesses but can also encapsulate fantastical representations, allowing users to express themselves uniquely.

While 3D offers an immersive depth, the 2D visuals embed vital context and navigational cues. In essence, the vast 3D landscapes and the AI-enhanced avatars, which are punctuated by 2D elements, make the Metaverse both a wonder to explore and intuitive to navigate.

In our next chapter, *Chapter 9, Understanding User Experience Design and User Interface*, we'll examine the distinctions between UX and UI design for AR and VR in the Metaverse. We will delve into user concerns that are unique to AR and VR, explain the differences in UX and UI for both mediums, and discuss their application to the Metaverse overall.

Understanding User Experience Design and User Interface

In the realm of the Metaverse, the intricacies of **user experience (UX)** and **user interface (UI)** design manifest distinctively between VR and AR. Diving into this chapter, we'll unpack the unique user concerns associated with VR compared to AR, highlighting the fundamental reasons behind their UX and UI disparities. As we delve deeper, we'll also shed light on the overarching principles of UX and UI design as they seamlessly integrate and shape the comprehensive user journey within the expansive Metaverse ecosystem.

In this chapter, we're going to cover the following main topics:

- What is meant by UX and UI
- How user concerns are different for VR versus AR
- How UX and UI are different for VR versus AR
- How UX and UI apply to the Metaverse as a whole
- How good UX and UI can greatly improve business results

What are UX and UI?

Both UX and UI are pivotal in determining a user's interaction with digital platforms.

UX encompasses the full spectrum of a user's emotions and interactions related to a specific product or service. It isn't merely confined to the visual or interactive elements; instead, it delves into the entire journey, evaluating the intuitiveness and satisfaction levels of the product or service from the viewpoint of the user.

The UI refers to the medium through which a user interacts with a digital device or application. It encompasses all the visual elements, such as screens, icons, and buttons, that facilitate a user's engagement with the service or product.

The following are general features of both UX and UI.

UX

Within UX design, the careful interplay of user research, creative design, and iterative testing leads to digital interfaces that harmonize with user behaviors and needs. This involves crafting user personas, mapping their journeys, and employing usability and A/B testing methodologies to fine-tune designs based on real-world input. Empowered by diverse tools and technologies, UX designers navigate the intricacies of human-computer interaction, yielding experiences that seamlessly blend aesthetics and functionality, ultimately cultivating user satisfaction and loyalty.

Components

The pillars of UX design encompass vital components. Research informs user needs, while wireframing creates interface frameworks. Prototyping tests product models, user testing gathers feedback, and iterative design refines based on it. Information architecture organizes content logically, ensuring a seamless user experience. Let's look at these in more detail:

- **Research**: This involves understanding user behaviors, needs, and pain points.
- **Wireframing**: This is the process of creating a skeletal framework of the interface.
- **Prototyping**: Crafting a usable model of the product for testing purposes.
- **User testing**: Procuring feedback by letting actual users navigate the product.
- **Iterative design**: This is the constant refining based on the feedback received.
- **Information architecture**: This is the process of logically organizing and structuring the content.

Importance

UX holds utmost importance in design—tailoring products to user needs for enhanced satisfaction and usability. The absence of good UX can deter users, overshadowing functionalities. Effective UX boosts loyalty and garners recommendations.

- UX is vital as it ensures that products are tailored with the user's needs in mind, leading to better satisfaction and optimized usability.
- A product lacking in UX may dissuade users regardless of its functionalities.
- A good UX augments user loyalty and amplifies recommendations.

Process and methodologies

Key UX processes and methodologies encompass constructing user personas, mapping user journeys, usability testing for user-friendliness, and A/B testing to optimize product versions:

- **User personas**: Constructed profiles representing the target audience of the product
- **Journey mapping**: A depiction of a user's sequential interactions with the product
- **Usability testing**: A method to determine the product's user-friendliness with real users
- **A/B testing**: A technique to contrast two product versions to discern which one fares better

Tools and technologies

UX thrives on pivotal processes. User personas align designs with user needs. Journey mapping uncovers pain points. Usability testing assesses user-friendliness. A/B testing refines versions for resonance. These threads weave the fabric of intuitive, user-centric experiences.

- **Research tools**: Such as UserTesting, Google Forms, and SurveyMonkey
- **Wireframing and prototyping tools**: Such as Figma, Axure, and Balsamiq
- **Analytics tools**: Such as Hotjar, Google Analytics, and Mixpanel

UI

UI design is a dynamic blend of components that shape the visual fabric of applications and websites. Layouts orchestrate the arrangement of elements, while typography curates fonts for hierarchy and aesthetics. Colors, graphics, and imagery weave emotions into palettes, with buttons and controls guiding interactions. Feedback and animations provide cues for users. The UI profoundly influences user perception, with an intuitive design correlating to success and visual appeal impacting emotions and choices. Design tools such as Sketch, Adobe XD, and Figma, along with prototyping tools, fuel the creation process, while coding languages such as HTML, CSS, and JavaScript implement the visual vision.

Components

UI design is a synergy of elements. Layouts arrange screen components, typography shapes fonts for hierarchy, and colors evoke emotions. Graphics and imagery enrich interfaces; buttons and controls enable interaction; and feedback with animations guides users. These components unite to craft captivating user experiences:

- **Layouts**: The arrangement and organization of elements on a screen
- **Typography**: How fonts are utilized in terms of type, size, hierarchy, and color
- **Colors**: Selection and emotional implications of palettes

- **Graphics and imagery**: Including icons, photos, and illustrations that augment the visual interface

- **Buttons and controls**: Elements such as dropdown lists, checkboxes, and sliders

- **Feedback and animations**: Visual cues that guide users regarding their interactions

Importance

The UI wields immense influence on user experiences. It shapes perceptions, fosters success through intuitiveness, and molds emotions and decisions via visual allure:

- The UI profoundly impacts a user's perception of an application or website.

- A product with an intuitive UI is more likely to succeed.

- The visual appeal of the UI can shape users' emotions and choices.

Tools and technologies

UI design flourishes with versatile tools. Design tools such as Sketch, Adobe XD, and Figma spark creativity. Prototyping tools such as Marvel and InVision bring ideas alive. Coding, spanning HTML, CSS, JavaScript, and platform-specific languages such as Java and Swift, fuels implementation across platforms.

- **Design tools**: Examples include Sketch, Adobe XD, Illustrator, and Figma

- **Prototyping tools**: Tools such as Marvel, InVision, and Framer

- **Coding**: Languages such as HTML, CSS, and JavaScript, as well as platform-specific ones such as Java for Android or Swift for iOS and C# for Unity

To summarize, in UX design, the interplay of user research, creative design, and testing forges user-centric digital interfaces. Processes such as constructing user personas, mapping journeys, usability, and A/B testing refine designs based on real-world insights, fostering satisfaction and loyalty. UI design harmonizes elements such as layouts, typography, colors, graphics, buttons, controls, and animations to craft captivating digital experiences. Tools such as Sketch, Adobe XD, Figma, Marvel, and InVision, and coding languages such as HTML, CSS, JavaScript, Java, and Swift propel the creation and implementation of visually appealing interfaces.

In the next section, we review user concerns regarding UX and UI for VR versus AR. In the emerging Metaverse, the convergence of AR and VR necessitates a deep understanding of user concerns related to UI and UX. From comfort and privacy to engagement and aesthetics, our exploration delves into challenges and considerations. It highlights the need for addressing inclusivity, privacy, and contextual relevance, while also emphasizing the role of gesture-based interaction, typography adaptation, and seamless integration of virtual elements. Both VR and AR demand responsible design, combining multi-sensory engagement with ethical considerations. Our overview showcases the complex landscape of immersive experiences, with UX and UI as the cornerstone shaping the evolving Metaverse.

UX and UI features specific to AR and VR in the Metaverse are addressed after this, followed by how businesses can grow through having good VR and AR UX and UI design.

User concerns for VR versus AR

As the Metaverse emerges through the convergence of AR, VR, and virtually enhanced physical realities, understanding the intricate user concerns tied to UI and UX becomes vital. The challenges encompass not only the technological aspects but also the human elements such as comfort, health, privacy, and real-world integration. Developers, designers, and technology consultants focusing on AI must craft experiences that resonate with human values while pushing the boundaries of innovation. Ranging from accessibility and security to engagement and aesthetic appeal, this comprehensive exploration seeks to unearth essential insights and considerations for shaping immersive experiences in VR and AR, paving the way for the burgeoning and interconnected digital landscape known as the Metaverse.

VR UX-related user concerns

Addressing user concerns in VR UX is pivotal for a seamless experience. Inclusivity is achieved through cultural sensitivity and disability support, while ease of use involves language adaptation and age considerations. Realism and engagement are bolstered by personalization and compelling narratives, alongside immersive features such as environmental interaction and sensory synchronization. Comfort and health considerations encompass solutions for eye strain, including breaks and long-term impact management, as well as addressing motion sickness through customizable settings and symptom examination. By thoroughly exploring these user concerns, VR UX aims to create a holistic and satisfying virtual environment for all users.

Accessibility

Ensuring accessibility in VR UX is paramount. Inclusive design encompasses cultural sensitivity and disability support, while ease of use involves language adaptation and age considerations, catering to a diverse audience.

Inclusive design

- **Cultural sensitivity**: Taking various cultural norms into account in content creation
- **Disability support**: Using adaptive technologies such as voice commands or specialized controllers

Ease of use

- **Language adaptation**: Offering support for multiple languages to reach a broader audience
- **Age consideration**: Designing for all age groups, from young to old, with unique needs in mind

Realism and engagement

Realism and engagement are pivotal in VR UX. Emotional engagement thrives through personalization and compelling narratives, while immersion is achieved by enabling environmental interaction and sensory synchronization for a truly immersive experience.

Emotional engagement

- **Personalization**: Customizing experiences through avatars, environments, or experiences
- **Story depth**: Engaging users with compelling narratives

Immersion

- **Environmental interaction**: Facilitating real-life interactions with virtual objects and surroundings
- **Sensory synchronization**: Ensuring alignment between visual, auditory, and tactile feedback

Comfort and health

Comfort and health are paramount considerations in VR UX. Addressing eye strain involves implementing breaks and settings adjustments, while also examining potential long-term consequences. Solutions for motion sickness include customizable settings and symptom examination, aimed at enhancing user comfort and well-being during VR experiences.

Eye strain

- **Impact mitigation**: Implementing breaks or adjusting settings to reduce fatigue and discomfort
- **Long-term consequences**: Delving into the prolonged effects such as dryness or eye strain

Motion sickness

- **Solutions**: Utilizing settings to modify motion parameters and providing warnings for sensitive users
- **Symptoms examination**: Investigating underlying causes such as visual-vestibular conflicts or latency

VR UI-related user concerns

In the context of VR UI, addressing user concerns regarding visual clarity, design aesthetics, navigation, and control is of paramount importance. This overview offers a brief insight into the essential considerations for VR UI design, contributing to the creation of user interactions that are both immersive and compelling.

Visual clarity and design

Regarding VR UI, the focus on visual clarity and design entails the pursuit of aesthetic unity, involving the alignment of visuals with the theme and mood, as well as balanced design, which ensures a harmonious arrangement of visual elements. Readability is pivotal, necessitating accessible layouts and a thoughtful selection of contrasting colors and legible fonts for effective information display.

Aesthetic unity

- **Theme alignment**: Ensuring that visuals resonate with the overall theme and mood
- **Balanced design**: Crafting a harmonious balance among visual elements

Readability

- **Accessible layout**: Ensuring that essential information is readily available through effective layouts
- **Color and font selection**: Opting for contrasting colors and legible fonts

Navigation and control

Navigating through VR interfaces effectively involves catering to user feedback by employing multi-sensory reinforcement—utilizing auditory, visual, and haptic cues—and offering clear error guidance. Additionally, ensuring an intuitive user experience requires upholding consistent control schemes across different interactions and providing learning assistance, which includes tutorials and familiar controls to smoothen the learning curve.

User feedback

- **Multi-sensory reinforcement**: Employing auditory, visual, and haptic feedback
- **Error guidance**: Presenting clear error messages and directions

Intuitive use

- **Consistent controls**: Maintaining uniformity in control schemes across experiences
- **Learning assistance**: Easing the learning curve with tutorials and familiar controls

AR UX-related user concerns

In the domain of AR UX, addressing user concerns is paramount. This overview highlights key areas such as privacy and security, relevance and context, and accessibility and usability, all of which play pivotal roles in creating immersive and responsible AR interactions.

Privacy and security

In AR, prioritizing privacy and security is essential. This entails ethical usage for public interaction and consideration of social factors. Effective data handling involves implementing security protocols such as encryption and authentication, while transparency assurance is achieved through clear communication about data collection and obtaining user consent.

Public interaction

- **Ethical usage**: Following guidelines for responsible AR application
- **Social considerations**: Recognizing how AR affects others, such as privacy in public spaces

Data handling

- **Security protocols**: Applying encryption and authentication
- **Transparency assurance**: Clarifying data collection and obtaining consent

Relevance and context

In AR, ensuring relevance and context is essential. Real-world fusion encompasses object recognition for interacting with real-world objects and spatial alignment to maintain physical consistency. User customization involves adapting content based on location, time, and preferences, creating a personalized and contextually rich experience.

Real-world fusion

- **Object recognition**: Facilitating interaction with real-world objects
- **Spatial alignment**: Adhering to real-world physics and spatial considerations

User customization

- **Contextual content**: Adapting dynamically to location, time, or social context
- **Preference alignment**: Tailoring content to individual interests and behaviors

Accessibility and usability

In AR, prioritizing accessibility and usability is key. User guidance involves providing resources such as FAQs, tutorials, and onboarding guides. A cross-platform experience requires responsive design and device compatibility across systems and devices for a seamless user experience.

User guidance

- **Available assistance**: Including FAQs, video tutorials, or chat support
- **Onboarding**: Offering comprehensive guides for newcomers

Cross-platform experience

- **Responsive design**: Adapting to different screens and orientations
- **Device compatibility**: Ensuring uniformity across various devices and systems

AR UI-related user concerns

In AR, user concerns are centered around effective interaction, responsiveness, visual harmony, and aesthetics. Interaction involves subtle animations, timely responses, accurate gesture recognition, and intuitive interaction indicators. Visual aspects encompass design coherence, brand consistency, recognizable symbols, managing overlaps, and maintaining aesthetic integrity. These collective considerations shape AR experiences, ensuring user engagement and visual appeal.

Interaction and responsiveness

In the realm of UI, addressing concerns related to interaction and responsiveness is crucial. Visual interaction involves the use of subtle animations for seamless transitions and actions while ensuring prompt responsiveness ensures timely reactions to user input. Moreover, precision in gesture and touch sensitivity enhances the accuracy of recognizing gestures and touches, further complemented by intuitive interaction indicators that contribute to a fluid and engaging user experience.

Visual Interaction

- **Animation guidance**: Utilizing subtle animations for transitions and interactions
- **Prompt responsiveness**: Ensuring timely responses to user actions

Gesture and Touch Sensitivity

- **Control accuracy**: Implementing precise recognition of gestures and touches
- **Interaction indicators**: Designing visual cues to highlight interactive elements

Visual harmony and aesthetics

When it comes to creating a seamless and visually appealing AR UI, addressing design coherence and visual clarity is paramount. This entails maintaining brand consistency and incorporating recognizable symbols for design coherence. Visual clarity involves managing overlaps between AR elements and reality to prevent clutter, all while upholding aesthetic integrity that aligns with real-world context. These factors collectively contribute to a captivating AR UI.

Design coherence

- **Brand consistency**: Adhering to brand guidelines and visual identity
- **Universal symbols**: Using recognizable icons and symbols

Visual clarity

- **Managing overlaps**: Preventing clutter by controlling overlap between AR elements and the real world
- **Aesthetic integrity**: Maintaining a cohesive visual style in conjunction with real-world context

In the developing fields of VR and AR, users' needs and preferences are multifaceted. In VR, comfort, health, engagement, and accessibility are key. Designers and developers must consider everything from motion sickness to cultural sensitivity. Meanwhile, AR emphasizes privacy, security, real-world integration, and customization, requiring attention to everything from ethical guidelines to spatial alignment.

Both realms highlight the critical role of UI, serving as the medium of interaction between users and these revolutionary technologies. From intuitive navigation and aesthetic design to responsive interactions, each aspect must be finely tuned.

As technology continues to evolve, this in-depth user-centric analysis is a guide for all professionals, shaping the world of immersive technology. This detailed understanding ensures that technology not only enhances our reality but also resonates with our human values and expectations, forming the foundation of the ever-expanding Metaverse.

User experience design

The burgeoning Metaverse presents unprecedented challenges and opportunities in UX design. As a realm that goes beyond the constraints of traditional digital experiences, it demands a revolutionary approach to creating human-centric, responsive, and adaptable interfaces. From the nuanced handling of gesture-based communication to the thoughtful integration of three-dimensional typography and seamless alignment of virtual elements, the UX design for the Metaverse must be intricate and forward-thinking. Ensuring user comfort, personalizing engagement, and upholding ethical standards form the core of this complex design ecosystem. This transformation from conventional interfaces to an encompassing virtual universe signifies a bold new era in design, one that strives for immersive, inclusive, and responsible interaction within virtual spaces. The following subsections have more details regarding UX design for the Metaverse.

Utilizing gesture-based inputs for human-centered interactions

Gesture-based inputs revolutionize human-centered interactions. In the Metaverse, interpreting voice commands, subtle gestures, and empathetic virtual agents mimics natural communication. Incorporating diverse interaction modes enhances accessibility and inclusivity for users with varying abilities:

- **Innovating interactions through human-like communication:** In the Metaverse, interpreting nuanced voice commands, subtle gestures, and the integration of empathetic virtual agents enable a natural, real-world communication experience.

- **Accessibility through inclusion:** By incorporating various interaction modalities such as voice, gesture, and eye-tracking, the Metaverse ensures an environment that accommodates users of diverse physical abilities.

Navigating typography's challenges in three-dimensional domains

In the context of three-dimensional environments, addressing typography challenges requires adapting for various viewing angles, adjusting font sizes dynamically, refining contrasts, and strategically placing text to ensure clear communication within the Metaverse. Typography is very relevant with regard to the Metaverse, as evidenced by Google's new font, AR One Sans, developed to be used specifically in AR environments (announced in October 2023).

Ensuring clarity across multiple perspectives: Typography in the Metaverse should adapt to varied viewing angles and distances, with considerations for dynamic resizing, contrast adjustments, and intelligent spatial placement.

Creating a tailored experience that resonates and adapts

Crafting a personalized and adaptable user experience involves maintaining consistency while accommodating various virtual environments, merging coherence with novelty, and harnessing data-driven insights to anticipate individual needs, providing predictive support and customized content within the evolving digital landscape:

- **Delivering a consistent but adaptable user experience:** Creating a familiar yet flexible experience across diverse virtual landscapes fosters a sense of continuity, blending freshness with coherence.

- **Anticipating and adapting to individual needs:** Personalized experiences leverage data and user behavior to not only adapt to user preferences but also to anticipate their needs, offering predictive assistance and tailored content.

Embedding virtual elements seamlessly within environments

Seamlessly incorporating virtual elements into environments demands the innate integration of virtual objects and information within the Metaverse, ensuring they appear intrinsic rather than superficial enhancements.

Achieving intrinsic integration of virtual objects: In the Metaverse, virtual objects and information must be naturally integrated, appearing as intrinsic parts of the environment rather than superficial add-ons.

Strategically orchestrating engagement through sensory cues

Enhancing engagement through sensory cues requires an immersive design that builds emotional bonds in the Metaverse, along with utilizing multi-sensory signals such as visual, auditory, and haptic feedback for intuitive navigation guidance without imposing directives:

- **Building emotional bonds through immersive design**: Engagement in the Metaverse is enriched through storytelling, aesthetic appeal, and interactive designs that foster emotional connections.

- **Utilizing multi-sensory signals for navigation guidance**: A blend of visual, auditory, and haptic feedback creates intuitive navigation paths, guiding user attention without imposing directives.

Designing a multi-dimensional spatial experience

Designing a multi-dimensional spatial experience encompasses integrating sensory elements for realism and promoting natural movement within the Metaverse, achieved through physics-based interactions and spatially intelligent navigation aids:

- **Enhancing realism through sensory integration**: Realism goes beyond graphics, involving sensory feedback, physics-based interactions, and contextual responsiveness to create a tangible virtual environment.

- **Promoting natural movement and exploration**: Navigation within the Metaverse should mirror human movement patterns, facilitated by physics-based dynamics and spatially intelligent navigation aids.

Addressing user comfort and ethical considerations

Addressing user comfort and ethical concerns involves prioritizing health and ergonomic well-being through informed design, alongside ensuring robust privacy and security measures to establish trust within the evolving Metaverse:

- **Prioritizing health and ergonomic comfort**: Attention to ergonomics and prevention of fatigue and motion sickness must be central to the design, guided by health research and user feedback.

- **Implementing robust privacy and security measures**: Trust in the Metaverse is built on transparent privacy policies, user control over personal information, and secure technological infrastructures.

The UX design of the Metaverse presents a multi-dimensional tapestry that intertwines technology, psychology, art, and ethics. From seamless integration to personalized engagement, and natural navigation to responsible design, it requires a holistic and innovative approach. The challenges are met with opportunities to craft immersive, inclusive, and ethical virtual experiences. As designers continue to explore and innovate, the Metaverse promises to be an evolving and exciting frontier, redefining the boundaries of virtual interaction.

VR versus AR UX comparison

UX design for VR and AR in the Metaverse has distinct differences, stemming from their inherent technological characteristics, use cases, and user expectations. The following is a detailed comparison.

Engaging with content

- **VR UX**: Interaction involves a blend of real-world activities and digital elements. The UX challenge is integrating the digital seamlessly without hindering real-world tasks.

- **AR UX**: Users navigate a wholly digital realm, making immersion a crucial UX aspect. The challenge lies in making the digital environment appear tangible and engaging.

Safety precautions

- **VR UX**: Here, motion sickness and user disorientation are concerns. Smooth transitions and logical movements are integral to the UX design to reduce such issues.

- **AR UX**: Given the combination of digital and real-world interaction, UX should have provisions for safety alerts to ensure users don't collide with real-world obstacles.

Interacting with the surroundings

- **VR UX**: In this enclosed digital space, maintaining a consistent environment is paramount. The world is predefined, ensuring users have a controlled experience.

- **AR UX**: The UX must be adaptive, as AR is highly responsive to the physical environment. Features such as object detection, spatial orientation, and light analysis become essential.

Device capabilities and limitations

- **VR UX**: Depending on whether the VR device is tethered or independent, there are mobility and processing challenges. UX designs should factor in these aspects.

- **AR UX**: Given the portable nature of AR devices, such as glasses, there are constraints such as battery duration, computational power, and screen clarity. UX strategies should be tailored to accommodate these.

Designing for space

- **VR UX**: Movement pertains only to the digital domain. UX elements should guide users within this space, using depth, audio cues, and perspectives.
- **AR UX**: Users traverse their physical environment, so digital content should adjust to their movement and surroundings. It's about merging reality with digital augmentations effectively.

Content layout

- **VR UX**: The entire visual range is available for digital content. Depth, spatial sound, and viewpoint need to be effectively utilized to craft a genuine experience.
- **AR UX**: Digital content is projected onto the real world, so it should neither obscure essential real elements nor be too faint. It's about striking the right balance.

Distinct applications

- **VR UX**: Emphasizing fully immersive scenarios, VR is ideal for games, simulations, training, and virtual sightseeing. The aim is to make these digital settings as realistic as possible.
- **AR UX**: It's often employed for directions, data overlays, gaming, and virtual help. UX designs should focus on enhancing these tasks in the real context.

Interacting socially

- **VR UX**: This provides complete digital communal areas. Here, avatars, activities, and communication occur in entirely virtualized spaces, and the UX should foster easy interactions.
- **AR UX**: Social engagements occur in both the digital and real world. Thus, features may include digital personas, annotations, or collaborative AR zones.

The UX design principles for VR and AR can be substantially divergent, owing to the distinctions in their application and the way they integrate digital and physical elements to create engaging experiences. While both of these technologies strive to blend the virtual and real worlds for enriched experiences, the subtle differences in their interaction paradigms necessitate unique approaches to design. In the context of the swift advancement of the Metaverse and related technologies, it becomes imperative for UX designers, particularly those focusing on VR and AR, to remain well-informed and flexible. By doing so, they are better equipped to meet and adapt to the ever-changing needs and expectations of users.

UI design

Navigating the intricate landscape of the Metaverse, which seamlessly combines AR and VR, poses unique challenges and opportunities in the spheres of UI and UX. This rapidly evolving space challenges traditional design methodologies and prompts a new era of innovation. Let's delve deeper into these refined aspects of Metaverse UI:

- **Dimensional UI in spatial contexts**: Moving beyond the confines of standard flat screens, the Metaverse immerses users in vivid 3D worlds. This shift necessitates designers to innovate where and how interface components are showcased, ensuring that they harmoniously fit within these expansive virtual spaces.

- **Interactions governed by gesture-based inputs**: The Metaverse bids adieu to conventional tools such as mice and keyboards, steering toward more instinctual forms of communication. From simple gestures to voice prompts and even motion-sensor inputs, the realm demands a design that's attuned to these nuanced methods of interaction.

- **Embedding UI elements within the virtual scene**: Rather than superimposing UI elements, the Metaverse environment allows them to become intrinsic parts of the virtual milieu. An illustrative example could be a game's energy gauge manifested as a tangible entity within its virtual ecosystem.

- **Strategically orchestrating user engagement**: Owing to the extensive scopes of VR and the interlaced nuances of AR, maintaining and directing user focus becomes paramount. Employing a mix of visual cues, such as strategic lighting, guided motion, or auditory markers, can play a pivotal role in sustaining user immersion.

- **Typography's evolution in depth-filled domains**: Conveying text within the 3D spaces of the Metaverse requires a fresh approach to typographical design. Variables such as font size, contrast ratio, and relative placement need recalibration to ensure clarity and legibility across different vantage points.

- **UIs that resonate and adapt:** In the dynamic Metaverse, the interface's capacity to morph based on user surroundings or preferences becomes a crucial design element. This ensures that as users traverse varied virtual terrains, their interaction experience remains consistent yet customizable.

Navigating the expansive 3D worlds of the Metaverse requires a radical transformation in UI design, including instinctual interactions through gesture-based inputs such as voice commands and motion detection. The design must also include the intrinsic embedding of UI elements within the virtual space, such as specific game components, and the adaptation of typography to ensure clarity across various perspectives. Key to this evolution is a strategic orchestration of user engagement through visual or auditory cues and the development of interfaces that are both resonant and adaptable to different virtual terrains. Together, these elements create a user experience that is consistent yet customizable, moving beyond the confines of traditional flat-screen interfaces and into a new era of immersive virtual interaction.

VR versus AR UI comparison

In the ever-evolving landscape of the Metaverse, both VR and AR play defining roles. These two technologies, however, offer distinct user experiences and, by extension, require different approaches to UI design. The following is a comparison of their unique UI attributes.

The base of interaction

- **VR UI**: Within a completely virtual realm, VR immerses users. The entire environment, from nearby objects to far horizons, is a crafted digital experience. This allows UI designers greater freedom to innovate, yet also tasks them with fabricating wholly intuitive systems.

- **AR UI**: Digital elements supplement the user's actual surroundings in AR. To maintain a harmonious experience, these elements need to coexist seamlessly with the physical environment, either anchored or contextually floating based on the viewer's perspective.

Spatial interactions and design

- **VR UI**: Within the controlled bounds of VR, spatial dynamics are consistent. This makes it easier to design UI components that predictably fit within set scenarios.

- **AR UI**: The AR experience is intimately tied to real-world spatial arrangements. As such, its UI components should be adaptable to various environments, considering factors such as real-world objects, lighting, and distances.

Navigating digital space

- **VR UI**: Interaction modes in VR can range from hand controllers and gaze tracking to advanced gestures. It's essential for the UI to offer intuitive guidance in these immersive realms.

- **AR UI**: AR usually leans on voice commands, touch (on handheld devices), or physical gestures for interactions. Simplifying these interactions is crucial, ensuring users aren't overwhelmed by a new system.

Visual elements and integration

- **VR UI**: Designers have the latitude to imagine innovative UI components in VR—think 3D menus or interactive holograms. Still, they should incorporate clear visual cues to assist users.

- **AR UI**: Given AR's overlay on reality, its UI design calls for a minimalist approach. By using features such as shadows and strategic placement, AR UI elements can blend more naturally with the real world.

Hardware implications

- **VR UI**: Primarily designed for VR headsets, this UI has to factor in aspects such as device resolution, field of view, and tracking capabilities.

- **AR UI**: Devices such as smart glasses, tablets, and smartphones are common AR platforms. For these, UI design should consider factors such as screen dimensions, battery efficiency, and camera functions.

Comfort and ergonomics

- **VR UI**: The immersive nature of VR necessitates UI designs that counter motion sickness and disorientation. Stable horizons and smooth scene transitions can improve comfort.

- **AR UI**: With AR keeping users rooted in their physical surroundings, the primary focus becomes ensuring the visibility and accessibility of UI components without inducing strain.

Functional intent

- **VR UI**: With many VR applications centered on entertainment or social interaction, UI designs can afford to be experimental and engrossing

- **AR UI**: Given its more practical applications such as data overlay or navigation, the AR UI should be efficient and straightforward

To summarize, while VR UI thrives on creating an enveloping digital universe, AR UI emphasizes harmonizing digital and physical worlds. Tailoring the design process to each medium's unique constraints and potential is crucial for delivering compelling experiences in the Metaverse.

Business growth through VR and AR UX/UI design

The advent of VR and AR within the Metaverse represents an uncharted frontier for innovation, creativity, and business expansion. Dividing our exploration into four critical sections, we focus on the distinct business benefits related to proficient UX and UI design in both VR and AR. These sections encapsulate the facets of enhancing engagement through immersive UX, driving revenue generation, optimizing operational efficiency, and navigating risk mitigation with ethical considerations. Furthermore, the application of strategic UX and UI design in brand building, innovation in products and services, sustainability, cost savings, and market expansion is detailed. Through this analysis, we aim to illuminate the multifaceted opportunities that both VR and AR offer, enabling businesses to leverage these technologies for success and long-term growth within the Metaverse ecosystem.

VR UX business benefit

VR UX is more than a technological advancement; it's a multifaceted toolkit for reimagining business success. Through immersive 3D showrooms and exclusive subscription content, it opens new revenue channels that don't just sell products but also build long-term relationships (more detail on particular use cases like this for VR, as well as AR, can be found in *Chapter 12, 3D and 2D Content Forms and Creation*). Engagement is at the heart of VR UX, with intuitive navigation and guided tutorials that welcome newcomers and enhance user satisfaction. This approach not only boosts retention but also expands market reach. Ethical considerations, such as user safety and accessibility, resonate with broader societal goals, ensuring a responsible and inclusive brand image. But the transformation doesn't stop there. With virtual collaboration and employee training, VR unleashes innovation and efficiency, tapping into global talent and reducing costs. In a landscape where competition is fierce and consumer expectations are ever-changing, VR UX stands as a strategic ally, positioning businesses at the forefront of their industry and setting the stage for sustainable growth and differentiation.

Revenue generation through effective VR UX

Unlocking revenue potential with effective VR UX is a dynamic strategy. Through subscription-based models, recurring revenue is stabilized while fostering loyalty with exclusive content. Additionally, virtual showrooms enhance sales by engaging customers with 3D exploration, leading to informed purchases and improved conversion rates. This synthesis of strategic VR UX approaches holds the key to substantial revenue growth and business transformation.

Subscription-based experiences

- **Stabilizing income through recurring revenue channels**: Premium content subscriptions not only ensure continuous revenue streams but also promote customer loyalty and long-term relationships.

- **Exclusive content for loyalty building**: Providing unique content via subscriptions encourages repeat engagements and fosters trust among customers.

Increased sales via virtual showrooms

- **Engaging shopping and higher conversion rates**: 3D exploration in virtual showrooms enhances user experience, leading to informed purchases and improved sales.

- **Rich product interactions for informed decisions**: Detailed 3D views promote understanding and satisfaction, enhancing conversion rates.

Enhancing engagement with immersive UX

Immersive UX revolutionizes engagement. Guided tutorials ease adoption and empower users, while intuitive navigation broadens the audience and boosts retention. This strategic approach reshapes the VR experience, enhancing engagement and satisfaction.

Guided interaction with tutorials

- **Facilitating adoption through reduced learning curves**: Easing users into VR with tutorials enhances confidence and quickens the adoption process.

- **Satisfaction via user empowerment**: Tutorials create self-assurance, encouraging further engagement and satisfaction within VR spaces.

Intuitive navigation for comfort

- **Expansion of audience reach**: By focusing on comfort, VR appeals to a diverse audience, including newcomers, broadening market reach.

- **Boosting user retention through ease of use**: Comfortable and intuitive navigation keeps users returning, improving retention rates.

Risk mitigation and ethical considerations

Navigating the landscape of VR design involves addressing risks and ethics head-on. Safety takes precedence with guidelines to minimize health risks and prevent legal disputes, safeguarding both user well-being and business reputation. Simultaneously, a commitment to accessibility amplifies inclusivity, resonating with societal goals and expanding the user base. This dual approach underscores the pivotal role of ethical considerations in shaping a responsible and thriving VR environment.

Emphasis on user safety

- **User well-being through minimized health risks**: Clear safety guidelines prioritize user health, avoiding potential problems related to VR usage.

- **Protection against legal disputes**: Safety protocols reduce legal liabilities, safeguarding the business reputation and mitigating associated costs.

Promoting inclusivity with accessibility

- **Aligning with societal goals and inclusivity**: Prioritizing accessibility in design resonates with broader social responsibility and attracts a more diverse audience.

- **Reaching a wider user base through accessible design**: Focusing on varied abilities ensures a broader appeal, enhancing the user base.

Operational streamlining with VR UX

Enhancing operations through VR UX is paramount. Virtual collaboration fuels innovation and global talent access, while immersive training drives skill development and cost savings. This strategic integration not only streamlines operations but also shapes customer experiences and ethical positioning, setting the stage for success in the competitive VR landscape.

Remote team collaboration

- **Innovation through team cohesion**: Virtual spaces enhance creativity and team connection, leading to groundbreaking outcomes.

- **Leveraging global talent**: Virtual collaboration allows access to diverse skill sets and perspectives, boosting operational efficiency.

Efficient employee training

- **Skill development with realistic simulations**: Immersive training environments foster skill enhancement and retention, aligning closely with real-world scenarios.

- **Significant cost savings**: Virtual training negates the need for physical resources, offering a cost-effective solution.

The strategic integration of VR UX within business frameworks offers multifaceted advantages. From revenue generation through immersive shopping experiences and subscriptions to engagement enhancement via intuitive navigation and tutorials, VR UX shapes the customer journey. Additionally, focusing on ethical considerations such as safety and accessibility ensures responsible market positioning, while leveraging VR for collaboration and training streamlines operations. This holistic approach to VR UX design sets a foundation for sustainable success in the competitive VR landscape, catering to both user needs and business goals.

VR UI business benefit

The VR UI is a pivotal force in business, with tangible impacts on sustainability, cost savings, innovation, and brand loyalty. Through energy-efficient practices and ethical sourcing, the VR UI embodies environmental responsibility. Its reusable and adaptive components directly translate to development efficiency and reduced costs. By catering to diverse cultural contexts and accessibility needs, it enables market expansion. Furthermore, the VR UI's immersive capabilities foster new opportunities in virtual tourism and healthcare, while its unique virtual branding tools strengthen customer relationships. In essence, the VR UI's multi-dimensional influence is redefining success and growth in the modern business landscape, proving itself as an indispensable tool for future endeavors.

VR UI's contribution to sustainability and social responsibility

The multifaceted contributions of the VR UI span various domains, from sustainability and social responsibility to brand loyalty and innovation. In terms of sustainability, the VR UI displays a commitment to environmental goals through energy-efficient rendering and ethically sourced assets. Meanwhile, its impact on brand loyalty is exemplified by spatial navigation, avatar customization, and interactive elements that foster memorable brand spaces and exclusive virtual events. In the broader context, the VR UI not only enhances user experience but aligns with responsible business practices and drives cost savings. The integration of cultural considerations and accessibility features widens market reach, while its role in healthcare and virtual tourism pioneers new avenues. This comprehensive approach signifies VR UI's pivotal role in shaping business success through innovation, sustainability, and brand allegiance.

Responsible design aligning with environmental goals

- **Energy-efficient rendering**: By embracing green rendering techniques, the VR UI illustrates a commitment to environmental stewardship.

- **Ethical sourcing of UI assets**: Ensuring responsibly sourced visual components aligns with ethical business practices.

VR UI's role in cost savings and market expansion

VR UI's impact is twofold: by tailoring content to cultural contexts and embracing accessibility, it widens market appeal. Efficient development is achieved through reusable components and adaptive design, reducing costs and maximizing reach.

Reaching diverse audiences through cultural consideration

- **Localization and cultural icons**: Tailoring content to specific cultural contexts enhances global appeal.

- **Accessibility options**: By embracing diverse accessibility features, the interface appeals to a broader user base.

Efficient development through user-centric design

- **Reusable UI components**: Leveraging repeatable and modular UI elements fosters development efficiency and cost reduction.

- **Adaptive UI**: Designing a versatile UI that adapts to various devices amplifies reach and streamlines the development process.

Innovation and new opportunities in products and services through VR UI

VR UI design is revolutionizing industries and experiences. In virtual tourism, 360-degree views with guided exploration and multilingual interfaces enrich immersion. In healthcare, gesture recognition, haptic feedback, and real-time analytics are transforming therapies and progress tracking. This tech-human fusion heralds an era of limitless innovation and transformation.

Emerging market in virtual tourism

- **360-degree views and virtual guides**: Panoramic imagery paired with guided exploration invigorates the virtual tourism experience. NeRFs and 3D Gaussian splats could be creating these highly detailed 360 views and virtual guides for commercial use, given a little more finessing. Next up are videos made this way, which are much more difficult to create and highly anticipated.

- **Language selection**: Multilingual capabilities in UI design enhance accessibility to various linguistic communities. There are a number of new AI companies that are developing near real-time multilingual voice capabilities that are featured in videos, with Elevenlabs leading the bunch.

Therapeutic applications

- **Gesture recognition and haptic feedback**: Integrating tactile feedback and intuitive gestures enhances the therapeutic application of VR

- **Real-time analytics and progress tracking**: The ability to monitor treatment and patient progress in real time facilitates innovative healthcare solutions

Building brand loyalty and identity through VR UI

In the dynamic landscape of VR, businesses are leveraging seamless spatial navigation and personalized avatars to enhance brand recognition and user connections. Incorporating 3D elements, interactive menus, and visual consistency across virtual spaces further strengthens brand affinity. Beyond aesthetics, VR UI components drive eco-friendly practices, cost efficiency, and innovation in sectors such as healthcare and tourism. This strategic integration underscores the pivotal role of VR UIs in shaping comprehensive business success.

Memorable brand spaces

- **Spatial navigation**: Navigational ease within virtual brand spaces fosters stronger brand recognition.

- **Avatar customization**: Personalized avatar options contribute to a unique and memorable brand experience.

Exclusive virtual events for brand association

- **3D elements and interactive menus**: Engaging three-dimensional designs paired with interactive menus reinforces brand affinity.

- **Visual consistency**: Uniform visual cues such as logos and colors solidify a cohesive brand image across virtual spaces.

From environmental considerations to brand loyalty building, the VR UI's multifaceted components extend well beyond mere aesthetics. The strategic alignment of energy-efficient processes and ethical sourcing manifests a broader societal responsibility, while the careful assembly of reusable and adaptive design elements yields significant cost savings.

Simultaneously, innovative opportunities in healthcare and virtual tourism showcase the capacity of VR UI to pioneer new market avenues. With the integration of culturally relevant content and diverse accessibility features, the scope for market expansion becomes virtually boundless.

In the competitive world of VR, the sophistication and flexibility of VR UI are integral not just to the user experience but also to the holistic business strategy, promoting innovation, sustainability, and robust brand allegiance. It is the delicate balance of these aspects that heralds the future of VR and its pivotal role in shaping business success.

AR UX business benefit

The transformative power of AR UX in the business world is only beginning to be realized, and its potential for future growth and impact is immense. Already enhancing operational efficiency through visual aids, expert collaboration, and manufacturing support, AR UX is poised to revolutionize these processes further. The seamless integration and personalized interactions currently offered will evolve, opening new frontiers in user engagement. Sales tactics, including interactive ads and virtual shopping experiences, will become even more effective and targeted. Ethical considerations and risk mitigation will continue to mature, reflecting the dynamic legal and social landscape. The current benefits are substantial, but the future of AR UX promises even greater advancements, positioning it as an essential tool for innovative, responsible, and successful business strategies moving forward.

Operational efficiency gains through AR UX

The convergence of AR and UX design is reshaping industries. AR UX's potential, including streamlined operational efficiency, real-time manufacturing assistance, and personalized user experiences, is profound. Ethical considerations and sales-driving capabilities further underscore its impact. This innovative tool is changing how businesses operate, engage users, and foster growth.

Remote technical guidance

- **Visual aids for troubleshooting**: Streamlining complex tasks with AR aids saves resources, increasing efficiency.

- **Expert support and collaboration**: Remote expert guidance via AR minimizes downtime, enhancing repair or maintenance procedures.

Real-time assistance in manufacturing

- **Quality inspection support**: Leveraging AR to spotlight defects improves accuracy and speed in quality checks.

- **Guided assembly instructions**: Using AR to provide workers with detailed instructions reduces errors and training time, boosting productivity.

Seamless integration for enhanced UX

In the evolving field of UX, achieving seamless integration and contextual interaction is crucial for heightened engagement and personalization. By recognizing environments and adapting to user behavior, tailored experiences are created, fostering loyalty. The smooth fusion of virtual and physical elements relies on spatial awareness and visual cohesion, enhancing interaction and overall satisfaction.

Contextual interaction for personalization

- **Environment recognition**: Customized experiences through recognition of surroundings foster user engagement.

- **User behavior analysis**: Tailoring interactions based on user behavior and preferences leads to heightened satisfaction and loyalty.

Smooth blend with physical reality

- **Spatial awareness**: Proper alignment of virtual objects with real-world scenarios ensures an authentic and immersive experience.

- **Visual cohesiveness**: A smooth integration between AR elements and physical reality enhances user interaction and enjoyment.

Driving sales and revenue with AR UX

In the ever-evolving world of AR, harnessing user experience is essential for driving sales. Localized advertising involves creating interactive AR ads for engaging brand experiences and geo-targeted promotions, optimizing marketing impact. Guided shopping experiences take center stage, with detailed product information overlays and virtual try-ons that significantly influence purchasing decisions. These AR UX strategies pave the way for increased sales and revenue.

Localized advertising

- **Interactive ad experiences**: Creating engaging AR ads that users can interact with promotes brand awareness and potential sales growth

- **Geo-targeted promotions**: Personalized, location-based advertising maximizes marketing effectiveness and conversion rates

Guided shopping experiences

- **Product information overlay**: Detailed AR product information influences positive purchasing decisions
- **Virtual try-on**: Customers virtually trying on products fosters confident buying, lowering return rates

Ethical usage and risk mitigation

In the context of AR, maintaining ethical usage and mitigating risks are critical. User safety considerations include legal compliance and health-conscious design, while ethical data handling involves secure management and transparent communication. These measures are essential for fostering user trust, safety, and legal adherence in AR applications.

User safety considerations

- **Legal compliance**: Adherence to regulations regarding AR use reduces legal risks.
- **Health and comfort guidelines**: Design that prioritizes health, such as avoiding eye strain, enhances safety.

Ethical data handling

- **Secure data management**: Strong data security measures protect user privacy and mitigate legal concerns.
- **Transparent data usage**: Clear communication about data storage and usage sustains user trust and legal compliance.

The AR UI has emerged as a multifaceted tool with profound implications for various aspects of business. From building and cementing brand trust through interactive experiences and personalized support to pioneering innovative applications in education and healthcare, the AR UI is shaping new paradigms.

Its capability to bridge the physical and virtual worlds offers businesses the chance to connect with consumers on deeper, more meaningful levels. The immersive experiences crafted through AR can resonate emotionally with users, creating long-lasting bonds and loyalty.

Furthermore, the transformative power of AR in the fields of medical training and educational innovation underlines its potential to disrupt traditional methodologies. By offering realistic simulations and interactive learning environments, AR is redefining the way professionals are trained and educated.

The global appeal and consistency achievable through thoughtful AR UI design cannot be overlooked. Recognizing cultural nuances and maintaining a uniform brand experience across platforms can be key determinants of global success.

Additionally, AR's potential to unlock new market avenues, from virtual tourism to wellness applications, provides businesses with uncharted territories to explore and capitalize on.

In a world where technology continually evolves and consumer expectations shift, leveraging the AR UI as a strategic asset offers businesses a path to differentiation, engagement, and growth. The intricate blend of creativity, technology, and understanding of human interaction that AR UI demands positions it as a pivotal element in modern business strategy.

AR UI business benefit

AR UI is transforming modern business by offering innovative solutions across various sectors. It enables companies to strengthen brand trust through real-time support, foster global appeal with adaptable design, and open new markets in areas such as virtual tourism and healthcare. By bridging virtual and physical worlds, AR UI creates engaging experiences that resonate with customers, offering a unique competitive edge. This technology not only deepens customer connections but also paves the way for future business innovation and success.

Brand trust and identity building with AR UI

In the ever-changing world of AR, establishing brand trust and identity is a central goal. This involves providing immediate AR customer assistance for real-time query resolution, enhancing customer satisfaction and potential referrals. Interactive guides and augmented manuals further bolster brand perception by offering seamless product usage instructions, which nurtures customer loyalty and trust.

Strengthening trust through support

- **Immediate AR customer assistance**: Leveraging AR for real-time support in handling queries and problems amplifies customer satisfaction, leading to sustained patronage and potential word-of-mouth referrals.

- **Interactive guides and augmented manuals**: Developing AR-based user instructions not only eases product usage but also forms a positive perception of the brand, enhancing customer loyalty and trust.

Interactive branding and immersive experiences

- **Creating engaging brand interactions**: Through AR, brands can provide interactive and tangible experiences with products, building memorable impressions that resonate with customers.

- **Emotional connections through virtual storytelling**: The artful use of AR to narrate the brand's story can create an emotional connection, forging a deeper bond with the brand.

Innovative AR UI in service and product development

AR is reshaping education, training, and medical procedures. AR simulations revolutionize learning with hands-on skill practice, while interactive educational materials deepen understanding. In the medical field, AR enhances surgeries through real-time data overlay and empowers medical staff with risk-free training simulations, ultimately driving advancement.

Revolutionizing education and training with AR UI

- **Hands-on simulations for skill building**: AR simulations allow students to practice and refine skills, transforming conventional learning and opening opportunities in vocational and higher education.
- **Immersive learning environments**: By integrating AR into educational materials, students can interact with complex concepts, promoting deeper understanding and retention.

Enhancing medical procedures and training through AR UI

- **Improving surgical outcomes with real-time information**: Utilizing AR to overlay critical data during surgeries can lead to more precise and safer procedures.
- **Robust medical staff training**: AR-powered simulations offer medical professionals an opportunity to practice in risk-free environments, improving skills and confidence.

AR UI's impact on global appeal and brand consistency

The AR UI transforms brand consistency and global appeal. With unified design across platforms, AR UI ensures seamless interaction and a cohesive brand image. Culturally sensitive AR interfaces resonate with diverse markets, while localized content enhances engagement and expands the brand's reach.

Fostering consistency across various platforms through AR UI

- **Adaptable and cohesive interface design**: A unified AR UI across multiple devices ensures seamless interaction, enhancing user satisfaction and contributing to a coherent brand image.
- **Cross-platform brand expression**: Consistency in AR UI design across platforms offers customers a familiar experience, regardless of device, further strengthening the brand's identity.

Culturally inclusive design with AR UI

- **Embracing cultural sensitivities**: Designing AR interfaces that resonate with local customs and values can foster acceptance in diverse markets, expanding the brand's global footprint.
- **Customized content and language localization**: Adapting content to fit regional preferences through AR increases user engagement and satisfaction, contributing to success in various markets.

AR UI's role in pioneering new market opportunities

AR UI drives innovation in emerging markets. Virtual tourism and real estate benefit from immersive AR tours, while interactive real estate showings enhance sales. In healthcare, AR-assisted rehabilitation and wellness applications offer engaging therapeutic and mental health support.

Emerging markets in virtual tourism and real estate

- **Virtual exploration of destinations**: AR can offer immersive tours of travel destinations, opening new avenues in the tourism industry.

- **Interactive real estate showings**: Providing prospective buyers with augmented real estate experiences can enhance sales and customer satisfaction.

Therapeutic and wellness applications of AR UI

- **AR-assisted rehabilitation**: Tailoring therapeutic AR experiences for rehabilitation can lead to more engaging and effective recovery processes.

- **Wellness and mental health support**: Using AR for mental wellness applications provides a new dimension in health care and personal well-being support.

The AR UI has emerged as a multifaceted tool with profound implications for various aspects of business. From building and cementing brand trust through interactive experiences and personalized support to pioneering innovative applications in education and healthcare, the AR UI is shaping new paradigms.

Its capability to bridge the physical and virtual worlds offers businesses the chance to connect with consumers on deeper, more meaningful levels. The immersive experiences crafted through AR can resonate emotionally with users, creating long-lasting bonds and loyalty.

Furthermore, the transformative power of AR in the fields of medical training and educational innovation underlines its potential to disrupt traditional methodologies. By offering realistic simulations and interactive learning environments, AR is redefining the way professionals are trained and educated.

The global appeal and consistency achievable through thoughtful AR UI design cannot be overlooked. Recognizing cultural nuances and maintaining a uniform brand experience across platforms can be key determinants of global success.

Additionally, AR's potential to unlock new market avenues, from virtual tourism to wellness applications, provides businesses with uncharted territories to explore and capitalize on.

In a world where technology continually evolves and consumer expectations shift, leveraging the AR UI as a strategic asset offers businesses a path to differentiation, engagement, and growth. The intricate blend of creativity, technology, and understanding of human interaction that the AR UI demands positions it as a pivotal element in modern business strategy, promising a future filled with innovation, connection, and success.

Summary

This chapter has placed a spotlight on the critical role of UI and UX within the Metaverse, particularly as it pertains to VR and AR. Through an exploration of the unique attributes and requirements of VR and AR, we've seen how these technologies demand specialized approaches to design, considering both the immersive nature of VR and the blending of real and virtual in AR.

Beyond mere aesthetics, this chapter has emphasized the tangible business implications of sound UX and UI design. The connection between engaging user experiences and business success in the Metaverse is clear. Effective design can lead to enhanced user retention and satisfaction, ultimately translating to positive business outcomes.

In summary, this investigation into the nuanced realm of UX and UI within the Metaverse, focusing on the specific contexts of VR and AR, offers practical insights and lessons. It underscores the vital role of thoughtful design, not just for creating compelling virtual experiences but also for achieving real-world success. For those venturing into the virtual landscape, the principles and understanding gleaned from this chapter are foundational to thriving in the exciting and ever-evolving world of the Metaverse.

In our next chapter, *Chapter 10, New Ways of Social Interaction*, we explore new forms of social interaction in the Metaverse using AR and VR technologies. Three Metaverse scenarios are presented: a limited contact scenario for friends and family, a personalized contact scenario based on preferences, and an open exploration scenario. The next chapter is the start of four chapters in which we present use cases that actively portray how AR and VR are used in the Metaverse.

Part 3: Consumer and Enterprise Use Cases

In *Part 3*, we will immerse ourselves in the evolving landscape of social interaction, work, arts and entertainment, and retail within the Metaverse, emphasizing a wide range of practical use cases that leverage AR and VR and best business practices. We will start with an examination of emerging social media technologies, envisioning scenarios spanning from limited contact to personalized interactions, and shedding light on how individuals will engage and cooperate in this digital realm. Ways of working within the Metaverse are then discussed, covering both virtual and physical tasks. Furthermore, we anticipate a Metaverse brimming with 3D and 2D games, streaming entertainment, and artistic expression, accentuating the distinctions between AR and VR experiences. The prospect of shopping within the Metaverse is highly anticipated, promising tailor-made retail encounters. In essence, *Part 3* provides an encompassing grasp of how the Metaverse is positioned to transform social dynamics, work environments, arts and entertainment, and retail experiences through a diverse array of practical scenarios.

This part has the following chapters:

- *Chapter 10, New Ways of Social Interaction*
- *Chapter 11, Virtual and Onsite Work*
- *Chapter 12, 3D and 2D Content Forms and Creation*
- *Chapter 13, Retail Experiences*

10

New Ways of Social Interaction

In this chapter, we will journey into the Metaverse to investigate how social interaction is being transformed by AR and VR. This chapter will unfold through three critical scenarios: the Friends and Family Metaverse focuses on maintaining close relationships in a digital environment, the Personalized Metaverse explores how individual preferences shape social experiences, and the Exploration Metaverse opens a broader range of social possibilities. With insightful use cases interspersed throughout, this chapter aims to arm you with a comprehensive skill set for navigating this new frontier, from personal interactions to business efficiencies.

In this chapter, we're going to cover the following main topics:

- How people in a friends and family circle could interact in the Metaverse
- How people could interact in a personalized version of the Metaverse
- How people are projected to explore a completely open version of the Metaverse
- How social interaction in the Metaverse differs when using AR versus VR
- Best business practices that are exemplified in each use case in this chapter

In an era where digital realms are fast becoming significant venues for human interaction, this chapter turns its attention to the growing Metaverse. Supported by groundbreaking technologies in AR and VR, we will take a comprehensive look at the myriad ways in which social interaction is being redefined in this expansive digital landscape.

The chapter kicks off with a section on the *Friends and Family Metaverse*, a virtual environment engineered to enrich connections with those closest to you. Here, we will examine the dual role of technology: how it can enhance these close relationships, and how it also poses challenges that necessitate thoughtful management to maintain meaningful interactions in a digital context.

From there, we will turn our attention to the *Personalized Metaverse*, where the digital landscape is a living, adaptive entity, attuned to individual user preferences. This responsive space dynamically alters social interactions, modes of content engagement, and even the mechanics of business transactions.

This portion of this chapter is dedicated to unpacking the profound implications of such a tailored digital experience on social dynamics and engagement.

We will round out this chapter by venturing into the *Exploration Metaverse*, a digital space unshackled by conventional limitations, offering endless possibilities for spontaneous social encounters and discoveries. This concluding section will outline both the exhilarating prospects and potential hazards of operating within an open and boundless virtual world.

Threaded throughout this chapter are detailed use cases that not only ground the conceptual discussions in real-world applications but also serve as actionable guides. These are valuable for anyone keen to navigate these emerging landscapes effectively, both on a personal level and from a business optimization standpoint.

By the end of this chapter, you will be equipped with an array of skills that cover everything from maintaining meaningful relationships in a digital setting to navigating a responsive and adaptive virtual space to understanding how to engage safely and productively in a limitless virtual ecosystem. Additionally, you will grasp how AR and VR technologies distinctly impact these various facets of digital interaction, and you'll learn how businesses can adapt these insights for improved operational effectiveness.

Friends and Family Metaverse

In our increasingly digital world, the burgeoning Metaverse provides a captivating new arena for us to reevaluate our definitions of social engagement. Enabled by the technological frameworks of AR and VR, this digital universe is reshaping the norms of human interaction. In this section, we will scrutinize how these revolutionary technologies are altering our social fabric, specifically looking at the role they play in sustaining close relationships.

Far more than just a 3D video chat, the Friends and Family Metaverse is a carefully crafted ecosystem aimed at elevating the quality of our digital interactions with those we hold dear. This environment gains significance against the backdrop of evolving societal trends like remote work and long-distance relationships, where digital means have become the main vehicle for sustaining connections. Within this digital haven, technology assumes a dual role:

- **Augmentation**: Technological advancements serve to enhance the emotional depth and scope of our interactions. Be it a haptic-feedback-enabled virtual hug or a communal space where family members can enjoy a movie "together," the objective of the Friends and Family Metaverse is to lessen the emotional gaps often experienced in digital communication.

- **Management**: While technology offers us amplified means of bonding, it also ushers in a host of challenges that necessitate cautious management. Among these are concerns surrounding data security, the potential for emotional detachment through excessive automation, and the complexities of maintaining a balanced life in an age of pervasive digital connectedness.

By providing use cases here, we aim to illuminate both the transformative opportunities and inherent risks technology brings to the table, particularly as it pertains to the reconfiguration of our social dynamics in this digital ecosystem.

Use case 1 – virtual family reunions

Imagine you're part of a family spread across multiple continents. Scheduling a video call that suits everyone's time zone is a Herculean task, let alone the complex logistics of arranging a physical reunion. Enter the Friends and Family Metaverse, where *virtual family reunions* become a feasible and enriching experience.

The setup

Each family member puts on their VR headset and enters a pre-designed virtual space that mimics the family's ancestral home. The environment is hyper-realistic, replete with family photos, heirlooms, and even the familiar smell of grandma's cookies, thanks to olfactory simulators.

Interactivity

Members can pick up objects, show them to others, or even initiate group activities like virtual board games. The kids can run off to a virtual playground while the adults "sit" in the living room, discussing life updates.

Technical innovation

Real-time voice and emotion recognition tools analyze speech patterns and facial expressions to add layers of emotional depth to the conversation. These technologies then adjust the virtual environment accordingly, perhaps dimming the lights when a sentimental topic arises, thereby amplifying the emotional resonance among participants.

Challenges

As enchanting as it may seem, this scenario also raises questions about data security. How is emotional and behavioral data being stored and used? Could a hacker interrupt this intimate family setting? Companies providing these services must be hyper-vigilant about security protocols.

Use case 2 – augmented reality playdates

As technology becomes increasingly ubiquitous, even children's playdates are evolving. While nothing can replace the experience of physical play, AR offers an alternative when that's not possible.

The setup

Using an AR headset or glasses, children can engage in a "playdate" with their friends who are miles away. In their respective living room, they can see a combination of their real-world environment and virtual elements.

Interactivity

They can build a virtual castle together, go on a treasure hunt, or even have a superhero battle, all while being physically apart but virtually together.

Technical innovation

Advanced AI algorithms can monitor children's safety during these interactions, blocking any inappropriate content or external intrusions. Moreover, parental control settings allow for the adults to ensure the interaction remains both safe and fun.

Challenges

Again, the issue of data privacy arises. Moreover, there's the ethical concern of introducing advanced technology to children at a young age. What are the implications for their social development and understanding of reality?

Use case 3 – senior engagement and companionship

Addressing loneliness among the elderly, this use case leverages VR in the Friends and Family Metaverse to foster meaningful connections and emotional well-being. It explores technical and health-related challenges while offering an immersive way for seniors to engage with loved ones from home.

The setup

Elderly individuals often suffer from loneliness and may not always have family or friends available for regular visits. A VR-based setup in the Friends and Family Metaverse could enable more interactive and emotionally fulfilling connections.

Interactivity

Wearing a simple-to-use VR headset, a senior citizen could "attend" a family gathering, play with their grandchildren in a virtual park, or visit an old friend in a digital café – all from the comfort of their armchair.

Technical innovation

The crux of this innovative solution lies in the seamless integration of user-friendly VR headsets for seniors. These VR devices serve as a gateway to the Friends and Family Metaverse, providing an intuitive and immersive platform for meaningful interactions. The design of these headsets prioritizes simplicity and accessibility, allowing even the less tech-savvy elderly individuals to navigate the virtual world with ease.

Challenges

The digital divide could make it challenging for less tech-savvy senior citizens to engage with such platforms. Additionally, the long-term effects of VR on seniors, who may already have health conditions, would need to be studied meticulously.

Use case 4 – long-distance romantic relationships

Long-distance relationships face unique challenges in maintaining emotional closeness. This use case explores how the Friends and Family Metaverse can elevate virtual *date nights* using immersive VR or AR experiences, offering new layers of intimacy while also considering data privacy and emotional health implications.

The setup

Maintaining a long-distance relationship is emotionally challenging, even with existing digital communication methods. The Friends and Family Metaverse could offer a *date night experience* that virtually transports couples to a romantic setting, whether it's a beach sunset or a Parisian café.

Interactivity

Couples can wear VR headsets or use AR devices to enter a pre-selected or custom-designed environment. They can dine, dance, or take a moonlit stroll along the virtual beach. Real-world delivery services could be synchronized to deliver actual meals to both participants, making the virtual dinner date more tangible.

Technical innovation

Biometric sensors in wearables can capture data such as heartbeat or body temperature, which could be shared in a consensual, opt-in manner to increase the sense of closeness. For example, a small increase in your partner's heartbeat when you *hold hands* in the virtual environment might add a layer of emotional intimacy.

Challenges

Once again, data privacy is a concern, especially given the intimacy of the data being potentially captured and shared. Emotional health is also a consideration; while virtual experiences can enhance connection, they also risk becoming a substitute for real-world interaction, which could have various psychological implications.

Use case 5 – multiplayer VR and AR gaming

In the context of multiplayer gaming, VR and AR technologies are beginning to offer more personalized and interactive experiences. This use case explores how friends can engage in Metaverse-based games that employ AI and data analytics to create unique challenges and environments. While these advancements enhance engagement and immersion, they also introduce ethical challenges around data privacy and the addictive potential of such games.

The setup

Friends who enjoy gaming can participate in multiplayer games designed in a Metaverse framework, providing much more than just a casual gaming experience.

Interactivity

Unlike traditional online games, a Metaverse-based adventure game could incorporate elements from each player's data to create unique challenges and storylines, making the gameplay more engaging and personalized.

Technical innovation

Incorporating AI-driven **non-player characters** (**NPCs**) can elevate the complexity and adaptability of such games, making the experience incredibly immersive and responsive to each player's actions.

Challenges

Ethical considerations concerning data privacy, especially if personal data is used for game customization, need to be addressed. The addictive potential of such immersive games is another concern.

Negative implications of the Friends and Family Metaverse

The Friends and Family Metaverse stands as an emblematic representation of our collective aspiration to transcend the limitations of physical reality in our pursuit of meaningful human connection. However, it's crucial to navigate this landscape with a nuanced understanding of both its possibilities and pitfalls for the Friends and Family Metaverse carries profound implications that traverse the technological, social, and ethical areas. Let's explore some of the implications in detail.

Technological implications

Some technological implications to take into consideration include the following:

- **Advanced UI and UX**: Designing environments that cater to multi-generational users is a Herculean task. The UI and UX must be intuitive for all, from tech-savvy teenagers to older adults who may not be digital natives.

- **Interoperability**: For a unified family experience, the platform must support a variety of devices and software, including both AR and VR headsets, smartphones, and traditional computers.

- **Real-time adaption**: Employing machine learning algorithms to adapt the environment in real-time based on user behavior and preferences is both a remarkable feat and a challenge.

- **Resource intensity**: High-quality AR and VR experiences often require robust computational power and fast internet speeds, which could be limiting factors for some households.

Social implications

Here are some social implications to take into consideration:

- **Enhanced connectivity**: The ability to interact in a three-dimensional, immersive environment transcends the current limitations of video calls, adding depth to long-distance relationships.

- **Learning and skill-building**: As seen in the Wisdom Wells use case, the Metaverse could become a potent tool for intergenerational knowledge transfer and skill acquisition.

- **Emotional well-being**: Having a dedicated space to share emotions or celebrate milestones can contribute positively to emotional health but also poses challenges regarding emotional safety.

- **Social dynamics**: The Metaverse might amplify existing family dynamics, for better or worse. For example, a dominant family member could potentially dominate virtual interactions as well.

Ethical implications

Here are some ethical implications to take into consideration:

- **Data privacy**: Gathering behavioral data to customize experiences raises significant privacy concerns, especially with minors involved.

- **Mental health**: The fine line between simulation and reality may have yet-to-be-understood psychological impacts. For example, could living in an idealized virtual world lead to dissatisfaction with the real world?

- **Accessibility**: There's the risk of creating a "digital divide" between families who can afford the necessary technology and those who can't.

- **Commercialization and exploitation**: Companies behind these platforms could use the intimate knowledge of family dynamics for targeted advertising, which could be seen as intrusive.

Business implications

The business implications to take into consideration include the following:

- **New revenue streams**: Monetizing the Friends and Family Metaverse opens up new avenues, from subscription models to in-world purchases.

- **Integration with existing services**: Platforms such as social media or eCommerce websites could seamlessly integrate with the Metaverse, offering a more cohesive user experience but also more opportunities for data collection.

- **Regulatory challenges**: Different types of data and interactions may be subject to various laws, from data protection regulations to age-restricted content.

By examining these implications, we can get a clearer picture of the monumental changes the Friends and Family Metaverse could bring about. Balancing the vast opportunities with the potential pitfalls will be the key challenge as this technological frontier continues to evolve.

Personalized Metaverse

In the captivating domain of the *Personalized Metaverse*, the boundaries of conventional digital interaction are not just stretched but fundamentally redefined. Unlike static online platforms or even the more rudimentary forms of VR, the Personalized Metaverse presents a living, breathing digital universe. It continually reshapes itself based on intricate algorithms that learn and adapt from each user's behavior, preferences, and interactions. With the advent of sophisticated AR and VR technologies, this metamorphic digital landscape offers an unprecedented level of customization and adaptability.

But what makes the Personalized Metaverse truly groundbreaking is its seamless fusion of the physical and digital worlds. By incorporating real-time data and contextual information, it creates a multi-dimensional experience that is both immersive and deeply individualized. Whether it's altering a virtual storefront to reflect your favorite brands, personalizing AI avatar interactions based on your past engagements, or even modifying its virtual architecture to mirror your aesthetic preferences, the Personalized Metaverse goes beyond passive consumption. It empowers users to co-create their digital environment in real time, offering a dynamic feedback loop that constantly recalibrates to offer the most relevant and engaging experiences.

Yet, the potential benefits come with an array of challenges. From ethical quandaries such as the fine line between personalization and manipulation to technical obstacles around data security and user privacy, the implementation of a Personalized Metaverse raises complex questions that demand nuanced solutions. Additionally, businesses face an unprecedented challenge of how to ethically and effectively leverage this dynamic digital realm to enhance operational efficiencies while also providing value to consumers.

In this section, we will present use cases that exemplify the vast potential and complex challenges associated with the Personalized Metaverse. Through these use cases, you'll get to explore the intricate mechanics of how such a responsive digital environment can both enhance and complicate our social interactions, shopping experiences, content engagement, and much more. These examples serve not just as a glimpse into what the future could hold but as a framework for understanding the evolving dynamics between humans and technology in this brave new world.

Use case 1 – personalized virtual culinary experiences

In an immersive digital space where culinary art meets cutting-edge technology, imagine entering a virtual kitchen that not only knows your name but also your flavor profiles, dietary restrictions, and cooking skills. Welcome to the world of personalized virtual culinary experiences, where the marriage of AR sensory technology and IoT kitchen gadgets creates a deeply personalized, multi-sensory cooking journey. But as wondrous as this innovation is, it brings along its own set of challenges, from the need for accurate flavor representation to the imperative of ensuring no allergies are triggered.

The setup

Going beyond a simple virtual kitchen, this digital space is a culinary playground personalized to you. Tailoring itself to your skill level, dietary needs, and even your preferred types of cuisine, the environment presents you with recipes you're likely to enjoy and succeed in making. Using augmented reality overlays, it can show you the ideal chopping techniques, suggest spice blends, or even transform the digital workspace to mirror a famous chef's kitchen, if you desire.

Interactivity

The experience reaches a new level of engagement through cutting-edge AR sensory devices. Imagine "tasting" a dash of saffron or a slice of truffle in your virtual dish, as olfactory and taste sensors simulate the flavors and aromas for you. The AI-driven culinary instructor not only provides step-by-step guidance but is capable of recognizing when you make an error – or a creative twist – and adapting the recipe in real time, offering immediate feedback.

Technical innovation

The line between the virtual and the real world blurs as IoT devices in your actual kitchen are synchronized with the virtual environment. Your real-world oven preheats at the exact moment the virtual recipe calls for it. Smart refrigerators could inventory your ingredients and suggest them in the virtual world. Or your actual blender could be set to the precise speed and timing as indicated in the virtual recipe, ensuring seamless integration between your real-world actions and virtual instructions.

Challenges

As mouth-watering as this virtual realm seems, there are hard-to-navigate risks involved. The system must guarantee that the simulated flavors and ingredients do not pose allergy risks or inaccurately represent taste, which requires robust verification processes. Mistakes here could be more than just a culinary mishap – they could lead to real-world health issues. Thus, creating fail-safes and accurate representation becomes a critical aspect of this tantalizing technological innovation.

Use case 2 – holistic mental health support

Incorporating both VR and real-time biofeedback mechanisms, the holistic mental health support system creates a tailored therapeutic environment that dynamically adapts to individual needs. Here, the virtual AI therapist can adjust its approach based on a multitude of user-specific cues, from voice tone and facial expression to physiological indicators such as heart rate. While the system demonstrates a leap in personalized mental health care through technology, it raises pivotal challenges – chief among them, ethical considerations about AI-driven therapy and concerns over data privacy.

The setup

In a tailored mental health space, users encounter a calming and supportive environment designed around their specific therapeutic needs, such as anxiety management, cognitive behavioral therapy, or mindfulness.

Interactivity

A virtual AI therapist conducts sessions, while real-time biofeedback mechanisms adjust the surroundings, pacing, and therapeutic approach depending on the user's physiological responses.

Technical innovation

Advanced algorithms could analyze voice tone, facial expressions, and even physiological indicators like heart rate to adapt the therapeutic experience in real time.

Challenges

Ensuring that AI-driven therapy is ethically administered and doesn't replace the nuanced care that a human therapist can provide is crucial. Data privacy is also a significant concern.

Use case 3 – specialized education and training

In the realm of specialized education and training, envision an advanced, AI-driven ecosystem that does more than merely instruct – it intuitively understands each learner, dynamically adapting not just the content but the entire learning environment to suit individual needs and preferences. From elementary students to career professionals, this revolutionary educational space leverages AI, AR, and VR technologies to deliver unprecedented levels of customization, transforming the learning landscape into an adaptive, multi-sensory experience that not only educates but also engages and inspires. As exhilarating as this prospect is, it also raises complex challenges, from equitable access to data privacy, that must be judiciously addressed.

The setup

In this use case, the educational ecosystem is no ordinary online classroom or e-learning platform. It's an advanced, fully integrated environment that leverages cutting-edge AI algorithms to offer a customized learning experience that caters to individual learners at an unprecedented level of detail. Not only does it adapt to different learning styles – be they auditory, visual, kinesthetic, or a combination thereof – it also adjusts the pacing and complexity of the curriculum in real time to match the learner's own pace and skill level.

Interactivity

Here, the word "tutor" is an understatement for the AI-driven educational guides that facilitate the learning process. These AI entities are more like dynamic educational companions capable of understanding the emotional and cognitive state of the learner. They adapt their teaching style and even their "personality" to offer the most effective instruction. Imagine a physics lesson where not only the content but also the classroom itself morphs to visually demonstrate the laws of motion or a music lesson where the virtual environment shifts to mimic a concert hall as you practice. The AI tutors guide users through these variable settings, offering real-time feedback and adapting the sensory experience for maximum understanding and retention.

Technical innovation

To power this level of customization, the system relies on a network of machine learning algorithms that analyze data at multiple dimensions. These algorithms examine learner interactions, quiz results, time spent on each topic, and even facial expressions if the system is accessed via VR/AR with facial recognition capabilities. This data is used to identify not just what the learner knows but how they learn best. Accordingly, the algorithms dynamically adjust not just the subject matter but also the method of delivery and even the environmental settings in which the learning occurs.

Challenges

The promises of such an educational Metaverse are not without its pitfalls. One significant challenge is ensuring that this kind of personalized education is available to students across socio-economic strata. As education becomes more customized and efficient, there's a risk that only those who can afford such personalized services will benefit, thus widening existing educational and social disparities. Ethical considerations also extend to data privacy, as this system would inevitably collect an immense amount of data on the learner's behavior, preferences, and performance, which must be handled with the utmost care to prevent misuse.

Use case 4 – adaptive eCommerce platforms

In the rapidly evolving landscape of eCommerce, where online shopping has largely been defined by scrolling through static web pages and product listings, consider a revolutionary leap forward. This use case delves into a cutting-edge eCommerce platform that is far from static; it's an immersive, adaptive environment accessible through both VR and AR interfaces. Picture yourself navigating through a virtual or augmented mall where the store layouts, product displays, and even the sales representatives are dynamically responsive to your individual preferences and past shopping behaviors. Instead of you seeking out products, this highly intelligent ecosystem intuitively rearranges its virtual or augmented space in real time, offering you a shopping experience tailored specifically to your interests, needs, and even your physical location. Beyond just convenience, this technology aims to redefine how we interact with e-commerce platforms, making shopping not just a transaction, but a highly personalized, interactive experience.

The setup

Picture yourself entering an eCommerce platform that isn't just a series of static web pages, but an immersive, adaptive environment accessible through both VR and AR interfaces. Here, the virtual mall adjusts not just its store lineup, but also its offers 3D architecture and even the behavior of its sales representatives based on your past interactions and shopping behavior.

Interactivity

In the VR version, as you traverse this dynamic space wearing your headset, storefronts rearrange themselves in real time, featuring products specifically catered to your interests. In the AR version, imagine your smartphone or AR glasses overlaying 3D representations of products and offers onto your real-world environment, also dynamically curated to your preferences. Sales representatives in both versions are AI avatars, sophisticated enough to not just recognize you but also understand your prior shopping history to make highly personalized recommendations.

Technical innovation

In this seamless amalgamation of physical and digital retail, machine learning algorithms work incessantly in the background. They adapt the virtual or augmented space based on various metrics such as your clicks, length of gaze at specific items, and even your movement patterns. In the AR setting, geolocation data might also be considered to offer you contextual promotions if you are near a physical counterpart of the virtual store.

Challenges

Such high levels of personalization bring forth ethical and practical dilemmas. The balance between personalization and manipulation becomes blurred in this dynamic setup. If the system is too adept at showing you what you're likely to buy, it raises concerns about consumer autonomy. Additionally, there are significant implications for data security, as the system would need to store and analyze enormous amounts of highly personalized data to function effectively.

Use case 5 – ultimate fantasy fulfillment

Step into a realm where the limitations of reality are left at the door and your deepest desires take center stage. This use case explores an intricate sandbox environment, combining cutting-edge AR, VR, and AI technologies, designed to bring your ultimate fantasies to life. Whether you've dreamed of life as a rockstar, envisioned exploring unknown planets, or anything in between, this groundbreaking platform not only allows but actively adapts to make those dreams a tangible experience. Here, even the laws of physics and societal norms can be reshaped to fit your imagination. This is more than a game or a simulation; it's an unprecedented journey into the realms of possibility, all while grappling with complex ethical challenges surrounding the consequences – or lack thereof – of living out your deepest fantasies.

The setup

Imagine a deeply personalized, multi-sensory kaleidoscope of experiences tailored to your tastes, desires, and curiosities. In this bespoke sandbox environment, your unique dreams are the blueprint for limitless possibilities. In the VR realm, you might discover a labyrinth of crystalline caves that glow with a light that only you know signifies peace, or you could compose symphonies with echoes that evoke your favorite memories. But your utopia also extends into your real world through AR. The walls of your own home could metamorphose into cascading waterfalls based on your favorite nature settings, complete with ethereal fish that resemble your cherished childhood drawings. Your daily commute becomes a fantastical journey, as mythological creatures you've always been fascinated with join you, playfully racing alongside your train or car. You might even discover that a mysterious artifact from your VR adventures has been translocated onto your actual kitchen table, each design detail mirroring your aesthetic. This duality of worlds is tailored exclusively for you, offering a realm where your imagination sets the rules and reshapes the boundaries between the virtual and the actual.

Interactivity

The world around you isn't just passive scenery; it engages with you in a deeply interactive and personal dialogue, like a co-author writing your life's most captivating chapters. In the virtual world, your every touch might bloom into handcrafted gardens that only flourish with flora you've shown interest in. In the AR world overlaying your reality, personalized AI entities – be they mentors, companions, or storytellers – interact with you in meaningful ways based on your past conversations, interests, and even emotional states. They could recommend books you've been yearning to read, or even offer a listening ear, sharing comforting words in the voice of a loved one when you're feeling down. Real-time machine learning algorithms dissect the minutiae of your interactions, from your vocal inflections to your biofeedback, fine-tuning the experience to align perfectly with your emotional and psychological state at any given moment. The sum of these parts creates an interactive tapestry that continually adapts, making you not just an observer but an integral part of your unfolding narrative.

Technical innovation

Beyond merely acting as a stage for your escapades, the underlying technology operates like a symphonic orchestra finely tuned to your every whim. Cutting-edge AR algorithms can seamlessly intertwine your fantasy settings with the mundane reality of your living room or workplace, animating your everyday environments with awe-inspiring wonders. VR, on the other hand, transports you to an expansive, dynamic universe where the textures of each grain of sand or the hues of alien skies adapt to resonate with your aesthetic inclinations. Meanwhile, advanced AI engines labor diligently behind the scenes, continually refining their understanding of you. They integrate everything from your eye movement tracking data to your geolocation, all the while adjusting both AR and VR elements in real time to create a personalized, highly contextual experience. By leveraging quantum computing or specialized machine learning techniques, the system can even predict your needs before you articulate them, arranging serendipitous moments that feel both exciting and uncannily well-suited to your tastes.

Challenges

The highly personalized and vivid nature of the environment raises ethical and psychological questions. Could blurring the lines between reality and fantasy lead to negative psychological effects such as escapism or narcissism? The platform would also accumulate a significant amount of personalized data, demanding stringent data security measures. Furthermore, the risk of creating echo chambers, where users are shielded from differing viewpoints, poses challenges to social harmony. Addressing these complexities would require a multi-disciplinary approach involving ethicists, psychologists, and technologists.

Negative implications of the Personalized Metaverse

The Personalized Metaverse represents a quantum leap in how humans interact with technology and each other, offering tailored experiences that adapt to individual preferences, behaviors, and needs. While this promises an unprecedented level of engagement and satisfaction, the implications are multi-dimensional and complex, touching upon ethical, psychological, and social considerations.

Ethical implications

The ethical implications to take into consideration are as follows:

- **Data privacy**: A personalized experience means collecting and analyzing a vast amount of user data, which raises questions about data ownership and privacy.
- **Informed Consent**: Users might not be fully aware of how their data is being used or manipulated to customize their experience.

Psychological implications

Psychological implications to take into consideration include the following:

- **Reality distortion**: When the digital world adapts to individual preferences, the boundary between reality and VR can become blurry, possibly impacting real-world perceptions and behaviors.

- **Emotional well-being**: Tailored experiences could end up reinforcing your beliefs and behaviors, risking the creation of echo chambers that could be detrimental to mental health.

Social implications

Here are the social implications to take into consideration:

- **Inequality**: Not everyone will have equal access to high-quality personalized experiences due to technology gaps, potentially exacerbating social inequalities.

- **Cultural impact**: As experiences become more personalized, shared cultural touchpoints may decrease, gradually affecting societal cohesion and common understanding.

Economic implications

The following are the economic implications to take into consideration:

- **Business models**: Hyper-personalization could revolutionize advertising and commerce, but it could also raise questions about consumer autonomy.

- **Job market**: As the Metaverse evolves, new roles and skills will be in demand, while others may become obsolete, affecting the labor market dynamics.

Legal implications

Here are some legal implications to take into consideration:

- **Intellectual property**: The infrastructure needed to support hyper-personalization at scale is enormous, potentially leading to environmental concerns related to energy usage.

- **Regulation**: As personalization algorithms become more complex, ensuring that different systems can work together cohesively becomes a challenge.

Technical implications

Finally, here are some technological implications to consider:

- **Scalability**: Hyper-personalization could revolutionize advertising and commerce, but it could also raise questions about consumer autonomy.

- **Interoperability**: As personalization algorithms become more complex, ensuring that different systems can work together cohesively becomes a challenge.

Exploration Metaverse

The *Exploration Metaverse* is an expansive, virtually limitless frontier that integrates the latest in VR, AR, and AI technologies to create an unparalleled environment for human curiosity and social spontaneity. Imagine stepping into an intricately designed VR world where not only can you socialize in awe-inspiring landscapes, but you can also manipulate the physics or climate of that world in real time, enabled by AI algorithms.

In the AR counterpart, the boundaries between the physical and digital worlds blur, allowing you to overlay fantastical elements or social hubs onto your everyday reality. From bustling digital bazaars and social forums that pop up like whimsical street fairs to serendipitous encounters with AI-driven characters that teach you something new, this Metaverse is the epitome of dynamic interactivity.

However, the same technological marvels that make this space so compelling also expose it to ethical quagmires, data security risks, and the complex challenges that come with virtually limitless freedom. This concluding section aims to delve deeply into both the exhilarating opportunities and sobering pitfalls that come with navigating an uncharted, unrestricted, and intricately personalized digital universe.

Use case 1 – serendipitous social encounters across virtual and physical realities

In the Exploration Metaverse, a cutting-edge blend of VR, AR, and generative AI coalesce to create a social tapestry that's as unpredictably enriching as real life but with an unparalleled depth of digital interactivity. Imagine engaging with a generative AI philosopher who not only adapts their dialogue based on your level of interest in existentialism but also creates unique philosophical ideas in real time for you to ponder. In AR, your local park comes alive with information and experiences layered onto the physical world, including a chance encounter with a like-minded nature enthusiast, guided by AI that generates personalized topics for conversation. These social possibilities, while thrilling, also present a myriad of ethical, privacy, and data security challenges.

The setup

As you navigate through breathtaking VR landscapes, from eldritch forests to surreal cityscapes, or walk through your local neighborhood with a superimposed AR environment, generative AI ensures that your world is far from static. In VR, the landscapes themselves could be generated on the fly based on your mood or interests. In AR, public art installations you encounter are not just static sculptures; they're generative art pieces that evolve based on public sentiment and individual viewer interaction. The characters, both AI-driven and human, populating these environments are similarly dynamic, appearing based on complex algorithms that assess a multitude of factors such as your mood, recent online searches, or even your current biometric data.

Interactivity

Generative AI ramps up the level of engagement. In VR, you're not just an observer; you could become a co-creator of your experience. For instance, in a virtual music jam session, generative AI not only matches your skill level but also composes new tracks in real time for you to jam along with. In AR, your everyday activities can become far more engaging. Taking a cooking class? Generative AI could invent new recipes on the fly based on your past culinary experiments, your flavor preferences, and current trending flavors, effectively turning your kitchen into a real-world sandbox for culinary creativity.

Technical innovation

Beyond the integration of VR and AR, generative AI serves as the linchpin that elevates the user experience to unprecedented levels. It uses machine learning algorithms trained on enormous data sets to create new, personalized content in real time. Imagine a neural network so sophisticated that it understands your emotional responses through a blend of facial recognition, voice modulation analysis, and biometric feedback, then uses this information to generate real-time experiences that are precisely tuned to your current emotional state and intellectual interests.

Challenges

While the level of personalization and interactivity made possible by generative AI is incredibly compelling, it also presents significant ethical and technical challenges. Data privacy takes on new dimensions when the AI isn't just analyzing but generating personalized content. The line between creation and manipulation blurs, raising questions about user autonomy and consent. Further, the generative nature of AI, which may create scenarios or dialogues on the fly, presents novel challenges in ensuring that the generated content adheres to societal norms and ethical guidelines.

Use case 2 – randomized life simulations

What if you could live a multitude of lives in a single lifetime? The Exploration Metaverse presents *randomized life simulations*, an audacious venture that allows you to do just that. This isn't merely a game; it's an exploration of life's possibilities through the lenses of sociology, psychology, and history. Within the bounds of VR and AR, complemented by highly sophisticated AI algorithms, you can be reborn into different families, social classes, countries, or even different eras altogether. It's a contemplative space where you can examine the consequences of choices and circumstances, by stepping into alternate lives and living them in accelerated timelines. These simulations aim to offer not just escapism or entertainment but also introspective value and social empathy. However, the venture also grapples with ethical complexities, from the potential for misrepresentation to the trivialization of real-life hardships.

The setup

Upon entering this use case scenario, users find themselves in a "Life Lobby" where they can select the variables that will define their alternate life. Do you want to experience life as an 18th-century artisan, a modern-day tech entrepreneur in Silicon Valley, or perhaps a fisherman in a small coastal village? AI algorithms then model a world based on intricate data ranging from socio-economic factors, historical accounts, and cultural norms to simulate life in your chosen setting. Users can opt for a completely randomized setup where even the choice of time period, location, and socio-economic background is left to the system's algorithms, adding an element of surprise and unpredictability.

Interactivity

Once you've been "reborn," you navigate through life in fast-forward mode, kind of like watching a movie where you are the star, but you can pause, rewind, or change the script at any time. You'll be presented with critical life decisions at every stage, from choosing a school to deciding on a career, building relationships, and even ethical dilemmas. The AI models different outcomes based on your choices and even allows for chance events like winning a lottery or facing a natural disaster, all rendered in richly detailed VR or overlaid in your real world via AR.

Technical innovation

This scenario leverages generative AI to model the complexities of life itself. It utilizes real-time big data analytics to recreate socio-economic environments, behavioral AI to simulate interactions with other AI-generated individuals, and even machine learning to adapt the experience based on user behavior and choices. What's more, NLP allows for interactive dialogues that take into account regional dialects, idioms, and even historical forms of speech to deliver an authentically immersive experience.

Challenges

The simulation faces immense ethical and philosophical questions. Is it ethical to allow users to experience lives fraught with hardship and struggle, potentially trivializing the very real difficulties faced by individuals in those situations? Misrepresenting or simplifying the challenges associated with particular life paths could also perpetuate stereotypes or misinformation. Ethical guidelines and possibly advisory boards might need to be implemented to ensure that the simulations are not only accurate but also respectful of the intricacies and sensitivities associated with different life scenarios.

In essence, the randomized life simulations in the Exploration Metaverse offer a groundbreaking method for self-exploration and the cultivation of empathy, made possible by the harmonious integration of VR, AR, and cutting-edge AI technologies. However, navigating the ethical landscape will be as complex as the technological one, requiring careful thought and ongoing oversight.

Use case 3 – Temporal Tourism

Welcome to *Temporal Tourism*, the Exploration Metaverse's most ambitious journey through time. Ever wanted to sit in the forum of ancient Rome, participate in the American Revolution, or explore the red dunes of a future Mars colony? Now you can, and all you need is a headset and a thirst for discovery. Combining the magic of VR and AR with AI's incredible data-processing capabilities, Temporal Tourism offers a unique blend of education, entertainment, and experiential living. The platform allows you to virtually live through different epochs, be it a day, a week, or an extended stay, exploring the nuances of life and culture in vivid detail. While this represents a pioneering step in interactive education and entertainment, it also raises several challenges, particularly around historical accuracy, cultural sensitivity, and educational integrity.

The setup

Upon entering the Temporal Tourism hub, users are greeted with a timeline that spans from the dawn of civilization to speculative futures. After selecting a time period, they are transported into a meticulously designed environment replete with details, from the architecture to the fashion and customs of the era. Users can even choose the role they want to adopt during their journey. Would you like to be a Roman senator, a knight in medieval Europe, or perhaps a scientist in a future space colony? Each setup is not just a mere backdrop; it's a living, breathing world simulated in the Metaverse, driven by advanced AI algorithms that take into account everything known about that period or plausible future scenarios.

Interactivity

Once "inside" the chosen time period, you're free to explore and participate. In ancient Rome, you might find yourself engaging in a real-time debate in the Senate, your arguments assessed by AI-driven orators skilled in classical rhetoric. In a future Mars colony, you could engage in extraterrestrial agriculture experiments or explore Martian geography. Activities aren't just scripted events; they are dynamic experiences that evolve based on user interaction, made possible by generative AI. This enables not only factual learning but also emotional and social understanding of different eras or speculative futures.

Technical innovation

What sets Temporal Tourism apart is the blend of advanced VR capabilities for environmental immersion, AR for incorporating elements into real-world settings, and AI for ensuring historical accuracy and dynamic interactivity. High-fidelity 3D scans of historical artifacts can be incorporated into the VR environment for added authenticity. Additionally, machine learning algorithms constantly update the environment based on new archaeological findings or scientific breakthroughs. For future scenarios, speculative models built in consultation with experts in various fields create a plausible environment where the user can explore tomorrow's possibilities today.

Challenges

The concept of Temporal Tourism is not without its ethical and technical challenges. There are issues surrounding the accuracy and representation of historical or cultural contexts. Getting something wrong or unintentionally glossing over a sensitive issue could have repercussions, especially when dealing with periods or cultures with living descendants who may take offense. Furthermore, the balance between educational value and entertainment might be hard to maintain. Users could potentially misinterpret the experience as entirely factual when some elements might be speculative or simplified for the sake of gameplay.

Use case 4 – Pocket Universes

Imagine an Exploration Metaverse so vast and so diverse that it can house an infinite number of microcosms, each with its own set of rules and realities. Welcome to the world of *Pocket Universes*, where users are not merely spectators but creators, architects of their realms, limited only by their imagination. Seamlessly combining AR, VR, and AI technologies, Pocket Universes offers an unprecedented degree of freedom, enabling you to create a world governed by your laws of physics, social norms, and even morality. However, while this frontier pushes the limits of digital creation and personal expression, it also ventures into ethically complex territories and psychological uncertainties.

The setup

Upon entering the Pocket Universes interface, users are greeted with a set of creation tools that would make a deity blush. Beyond standard environmental features such as landscape and climate, you can set the gravitational constant, light spectrum, and even time dilation factors. Do you want a universe where water flows upwards, or where individuals age backward? Go for it. You can even encode social norms, laws, and forms of government into your universe, using an AI-assisted rule-setting feature that ensures your choices are coherent and functional.

Interactivity

Visitors to your pocket universe have the option to "sync" with the local laws and rules temporarily. This way, they can fully engage in activities and social interactions unique to your universe. For example, if you've designed a world with non-Euclidean geometry, visitors might engage in mind-bending architectural projects or solve puzzles that are impossible in our universe. The use of generative AI allows the creation of highly dynamic and responsive scenarios, adapting to the actions and decisions made by visitors, thereby creating an evolving experience that is different each time you visit.

Technical innovation

The real marvel of Pocket Universes lies in the dynamic algorithms and ethical AI monitoring that make these self-contained worlds possible. A state-of-the-art physics engine generates a unique set of physical laws for each pocket universe, all while maintaining computational efficiency. Additionally, ethical AI monitors oversee the universe to ensure that no activities occur that violate a predefined set of universal ethical principles, such as causing intentional harm or promoting discrimination. These monitors can also flag content for review if it strays into controversial or morally ambiguous territory.

Challenges

While the concept opens up endless opportunities for creativity, it also raises considerable ethical concerns. What limitations, if any, should be imposed on the kinds of worlds and rules that users can create? There's also the matter of psychological impact; what are the long-term effects of spending time in a world where fundamental aspects of reality are drastically different? The autonomy given to creators raises concerns about the creation of pocket universes that may be used for promoting harmful ideologies or activities. The challenge lies in balancing the freedom to create with the responsibility to maintain a socially and psychologically safe environment.

Use case 5 – dynamic marketplaces

Commerce takes on a revolutionary new form in the Exploration Metaverse through the concept of dynamic marketplaces, blending the spontaneity of a traditional bazaar with the cutting-edge capabilities of AI, AR, and VR technologies. Imagine strolling through a virtual forest and suddenly stumbling upon a marketplace selling artisanal products crafted in that very environment. Or, picture yourself in an AR overlay of your local mall, where a digital pop-up store appears to offer exclusive deals. While these transient, dynamic marketplaces create an electrifying commercial atmosphere, they also grapple with a host of challenges ranging from regulatory compliance to transactional security.

The setup

In a dynamic marketplace, businesses aren't constrained by physical locations or set hours of operation. Using AI algorithms, they can set up digital storefronts that materialize in different parts of the Exploration Metaverse based on real-time data. For example, a virtual fashion outlet could appear near a popular social hub during peak hours, and then relocate to a quieter space for exclusive, invitation-only showcases. Each storefront can be customized to fit the aesthetic or cultural milieu of its temporary location, allowing businesses to localize their offerings instantaneously.

Interactivity

Once inside these AI-generated marketplaces, users experience a level of interactivity that transcends traditional online shopping. Not only can you see and virtually touch the products, but thanks to AR and VR sensory technology, you can also smell and taste, say, a cup of coffee before buying it. The marketplaces facilitate a range of economic interactions, from straightforward purchases to complex negotiations and barter systems. Users can engage with AI-driven sales representatives, who can negotiate prices or offer personalized deals based on the user's past shopping history and preferences.

Technical innovation

The dynamic nature of these marketplaces is enabled by a blend of sophisticated technologies. Machine learning algorithms analyze vast datasets in real time to determine the most opportune locations and times for these marketplaces to appear, considering factors such as user density, prevailing social activities, and even current events that might drive consumer interest. These marketplaces also make use of generative AI to create unique and personalized shopping experiences, generating products or services that may appeal specifically to the individual user based on their data profiles.

Challenges

Despite the technological marvel and convenience offered, several challenges loom large. Regulatory compliance becomes complex in a setting where businesses can spawn in multiple virtual jurisdictions, each potentially with its own set of commercial laws and regulations. Ensuring secure transactions in such a fluctuating environment is another concern, requiring robust cybersecurity measures that can adapt as quickly as the marketplaces themselves. There are also ethical concerns surrounding data privacy and the usage of personal information for hyper-targeted selling.

Negative implications of the Exploration Metaverse

The Exploration Metaverse, as envisioned, would serve as a groundbreaking blend of technologies, including VR, AR, and AI. With use cases ranging from social encounters and creative endeavors to corporate strategy and problem-solving, the Metaverse would influence multiple sectors of human activity. However, its intricate web of technological marvels also paves the way for a multitude of implications, both positive and negative. Let's look are some of the major implications to consider.

Data privacy and security

In the Metaverse, data privacy and security are critical. Extensive data collection, third-party risks, and constant surveillance are concerns. Businesses also face risks such as intellectual property theft and corporate espionage. Strong security measures are essential in this virtual realm.

Let's look at this in more detail.

Information exposure

Here are some areas that should be taken into consideration:

- **Detailed data collection**: As users interact in various settings within the Metaverse, they will generate a wealth of behavioral and interactional data. The scope of this data is far beyond typical browsing history or purchase transactions.

- **Third-party risks**: Data could be accessed or purchased by third parties for targeted advertising, manipulation, or even malicious activities.

Surveillance concerns

The following areas should be considered:

- **Ubiquitous monitoring**: With advanced AI algorithms, almost all user activities can be tracked, from interactions with AI entities to conversations with real users.

- **Algorithmic profiling**: Tracking could be used to build comprehensive profiles that may inadvertently pigeonhole or unfairly label users based on their behavior.

Corporate data risks

The following areas should be considered:

- **Intellectual property theft**: Businesses operating within the Metaverse may inadvertently expose sensitive data or intellectual property to unauthorized parties.

- **Espionage**: There may be risk factors such as corporate spying or insider threats that are harder to manage in a virtual setting.

Ethical quandaries

In the Metaverse, ethical challenges abound. Engaging in moral scenarios can cause emotional stress, with decisions having real-world consequences. Concerns also arise about misrepresentation and bias, where AI algorithms may perpetuate stereotypes. Virtual escapism poses risks of detachment from reality and dissatisfaction with real experiences as users chase idealized virtual lives. These ethical dilemmas call for careful consideration in the evolving Metaverse.

Moral dilemmas

The following areas should be considered:

- **Ethical complexity**: Scenarios or games that involve making moral or ethical decisions could inadvertently cause mental or emotional stress.
- **Real-world consequences**: Decisions made within such scenarios could have real-world implications such as directing funds to charities, which might manipulate users' choices.

Misrepresentation and bias

The following areas should be considered:

- **Algorithmic discrimination**: AI algorithms could inherit the biases of their human creators, leading to unfair or stereotypical portrayals.
- **Cultural sensitivities**: Depictions of historical events or various cultures could be oversimplified or inappropriately represented, sparking controversies.

Virtual Escapism

The following areas should be considered:

- **Detachment from reality**: Users might prefer the idealized versions of themselves in the Metaverse, causing them to neglect real-world responsibilities and relationships.
- **Fantasy overload**: A constant influx of idealized experiences could cause dissatisfaction or disillusionment with real-world experiences.

Social impact

The Metaverse introduces significant social considerations. Economic barriers and digital literacy could create disparities, while reduced physical interaction and community fragmentation might lead to social isolation. Prolonged exposure to virtual and augmented realities could potentially alter individuals' perceptions and desensitize them to real-world emotions, emphasizing the need for an inclusive and balanced approach to Metaverse development.

Virtual divide

Areas that should be taken into consideration include the following:

- **Economic barriers**: High costs for cutting-edge hardware and software could exclude lower-income populations from participating.
- **Digital literacy**: Those with limited digital skills could face challenges in navigating or benefiting from the Metaverse.

Isolation

The following areas should be considered:

- **Reduced physical interaction**: The convenience and allure of virtual interaction might discourage traditional, physical social activities.

- **Community fragmentation**: While the Metaverse may bring together like-minded individuals, it may also segregate communities based on niche interests, preventing holistic social interaction.

Reality distortion

Here are some areas that should be taken into consideration:

- **Perception shifts**: Prolonged time in enhanced virtual or augmented realities might lead to changes in how individuals perceive and interpret the real world.

- **Emotional numbness**: Overexposure to heightened experiences could potentially desensitize individuals to real-world emotions or consequences.

Regulatory standards

Regulating the Metaverse is a complex task. The undefined jurisdiction, copyright issues, and global user base make legal processes challenging. Enforcing decisions and defining universal ethical standards is difficult. Content monitoring is complex, raising dilemmas about user freedom and preventing harmful content. Addressing these regulatory challenges is crucial for a well-structured and ethical Metaverse.

Legal framework

The following areas should be considered:

- **Undefined jurisdiction**: Legal processes would be complicated by the fact that the Metaverse could be hosted in multiple countries, each with its laws and regulations.

- **Copyright and IP issues**: The virtual world could potentially be rife with unauthorized replications of real-world intellectual property.

Global jurisdiction

The following areas should be considered:

- **Enforcement difficulties**: With users from around the globe, it would be challenging to enforce any sort of legal decisions or penalties.

- **Ethical universalism**: Defining universally acceptable ethical standards and norms for a diverse user base is a monumental challenge.

Content monitoring

The following areas should be considered:

- **Governance complexity**: Moderating the vast array of interactions and content generated within the Metaverse would require a Herculean effort.

- **Censorship dilemmas**: Striking a balance between user freedom and the prevention of harmful content poses ethical and operational challenges.

Psychological health

The Metaverse raises concerns about psychological health. Addiction risks are significant due to its immersive nature, potentially impacting real-world responsibilities and relationships. Disengagement from real-world interactions may lead to social skill deficits and life imbalances. Virtual experiences in the Metaverse can evoke intense emotions and blur the boundary between virtual and real worlds, posing potential psychological challenges. Addressing these concerns is essential for ensuring the well-being of Metaverse users.

Addiction

The following areas should be considered:

- **Immersion risks**: The highly interactive and immersive nature of the Metaverse could contribute to addiction issues, akin to those seen in video gaming.

- **Life impact**: Addiction could lead to a neglect of real-world responsibilities and relationships, impacting employment, education, and family life.

Disengagement

The following areas should be considered:

- **Social withdrawal**: Those who find virtual interactions easier or more rewarding than real-world relationships might begin to disengage, leading to a lack of social skills or increased social anxiety.

- **Life imbalance**: Excessive time in the Metaverse could detract from the attention given to real-world commitments, including professional and educational endeavors.

Psychological strength

The following areas should be considered:

- **Emotional intensity**: Virtual experiences, especially those mimicking real-world situations, could evoke strong emotions or exacerbate existing psychological conditions.

- **Reality blurring**: For some individuals, the boundary between the Metaverse and the real world might blur, potentially leading to confusion, disorientation, or a warped sense of reality.

These potential negative implications suggest a need for comprehensive governance, regulation, and ethical considerations as the Exploration Metaverse evolves. While some of these challenges can be addressed through technological advances, many will require interdisciplinary approaches involving law, psychology, ethics, and social science.

Best business practices

Examining the Friends and Family Metaverse, Personalized Metaverse, and Exploration Metaverse, several overarching best business practices emerge that can help guide businesses to navigate these complex, rapidly evolving virtual landscapes. While each area has its unique considerations, some principles provide a foundational approach across all three domains.

Let's consider the practices that should be followed

Data governance and ethics

- **Transparency and consent**: Always inform users about data collection practices across all three Metaverse types and seek explicit consent.

- **Security protocols**: Data encryption and robust authentication mechanisms should be uniformly applied.

User experience and engagement

- **Seamless integration**: Whether you're focusing on personal connections, personalization, or exploration, the user interface and experience should be consistently smooth and intuitive.

- **Quality content**: In all three Metaverse models, the quality of content, be it social interactions, personalized features, or exploration possibilities, is paramount.

Scalability and performance

- **Resource allocation**: Be prepared to scale resources up or down, depending on user engagement in each segment of the Metaverse.

- **Low latency**: Ensure quick load times and real-time interactions, regardless of the type of Metaverse environment.

Business models and monetization

- **Flexibility**: Different monetization strategies may be applicable, depending on whether the focus is on friends and family, personalization, or exploration.

- **Transparency in transactions**: Clear, straightforward financial transactions should be the norm across all Metaverse environments.

Compliance and regulation

- **Legal framework**: Given the varied activities that each type of Metaverse supports, adhering to both local and international laws is crucial.

- **Ethical guidelines**: Develop and adhere to ethical standards that go beyond mere compliance with legal requirements, particularly when personalization and data collection are involved.

Inclusivity and accessibility

- **Global reach**: All Metaverse models should be designed with global access in mind, including language localization and cultural considerations.

- **Universal design**: The Metaverse should be accessible to people with varied abilities and needs.

Community building and management

- **Moderation and oversight**: Regardless of the type, community moderation is essential for maintaining a positive and safe environment.

- **User empowerment**: Users should have the ability to report issues, block other users, and have some control over their experiences.

Business intelligence and analytics

- **Cross-domain analytics**: Use analytics tools that can generate insights across the different Metaverse types to create a more cohesive strategy.

- **User feedback loops**: Consistent mechanisms for capturing user feedback should be implemented across all areas.

Sustainability

- **Environmental impact**: Aim for eco-friendly server hosting and encourage sustainable practices within the Metaverse.

- **Social responsibility**: Develop corporate social responsibility initiatives that can be integrated into each Metaverse type.

Interoperability

- **Cross-platform consistency**: Users often switch between different types of Metaverse environments; the experience should be consistent.

- **Data portability**: Allow users to carry their profile data, achievements, and assets between the different Metaverse types.

By considering these overarching best practices, businesses can create a more unified strategy that allows for flexibility while maintaining core values and objectives. Implementing these practices will not only enhance the user experience but also build trust, which is crucial for long-term success in any Metaverse setting.

Summary

In this chapter, we embarked on an in-depth journey into the burgeoning world of the Metaverse, guided by the transformative impact of AR and VR on social interactions. Through three distinct scenarios – Friends and Family Metaverse, Personalized Metaverse, and Exploration Metaverse – we've tackled the complexity and nuances of social dynamics in these digital realms. Each section provided a detailed look at how these novel virtual spaces alter, challenge, and expand our traditional understanding of social interaction, whether it's preserving intimate relationships or enabling unprecedented exploratory engagements.

As we've seen, the *Friends and Family Metaverse* helps us explore the pros and cons of maintaining close relationships in a digital setting. While technology offers tools to enhance the intimacy and immediacy of our connections, it also raises challenges that require careful navigation. Likewise, in the *Personalized Metaverse*, we saw how AI-driven environments can morph in real time to align with our preferences, ushering in a new era of hyper-personalized social and commercial interactions. However, this unprecedented personalization also brings with it ethical and psychological implications that are critical to understand.

Our foray into the *Exploration Metaverse* allowed us to conceptualize a realm of virtually limitless possibilities. Yet, it's precisely this boundlessness that makes it fraught with ethical, psychological, and social challenges that users and policymakers must be prepared to address. Interspersed throughout this chapter were real-world use cases that served as practical guides for both personal and business endeavors. These cases were carefully curated to provide actionable insights into how you might navigate this new, expansive digital frontier effectively.

In summary, this chapter has aimed to equip you with a holistic skill set for understanding and navigating the new ways of social interaction in the Metaverse. We delved into how different types of social interactions are being affected – sometimes dramatically – by advancements in AR and VR technologies. The goal has been to arm you with knowledge, insights, and best practices that will allow you to engage productively and safely in these exciting new digital worlds.

For business leaders and entrepreneurs, the implications of the Metaverse are far-reaching and not to be overlooked. The emerging digital landscapes present groundbreaking opportunities for customer engagement, product development, and even internal operations. As AR and VR technologies continue to evolve, businesses have the chance to pioneer new models of commerce, collaboration, and customer service. However, this new frontier also brings a host of challenges, from data privacy to ethical considerations, that will require thoughtful and well-informed strategies. As you consider integrating these technologies into your business roadmap, it's essential to approach them not merely as trendy buzzwords, but as fundamental shifts in how consumers and professionals will interact, communicate, and transact. Understanding the nuanced dynamics of social interaction within the Metaverse will be key to leveraging its full potential and could very well serve as a competitive differentiator in a rapidly evolving digital landscape.

Our next chapter, *Chapter 11, Virtual and Onsite Work*, explores the game-changing impact of the Metaverse on various types of work, including office and virtual work, walk-around jobs, and factory work, along with training and diagnostics. Employing AR and VR technologies, this chapter will present in-depth use cases that show how these digital realms will revolutionize traditional workspaces and work processes. These use cases will provide you with actionable insights into best practices that can significantly improve business efficiency and effectiveness. The skills that will be taught in this chapter range from understanding the critical role the Metaverse will play in reshaping different kinds of work, to distinguishing between the utility of AR and VR in virtual and onsite work, to learning essential business practices for success in this new digital landscape.

11
Virtual and Onsite Work

As technology continues to advance, the Metaverse emerges as a groundbreaking development, poised to revolutionize the way work is conducted in diverse sectors. This chapter provides an in-depth exploration of the Metaverse's transformative impact on office and virtual work, walk-around jobs such as retail and field operations, and factory-related tasks, including training and diagnostics. Leveraging AR and VR, we will present use cases that not only highlight the changing dynamics of these work environments but also offer best practices for optimizing efficiency and effectiveness in this new virtual paradigm. Whether you're an organizational leader contemplating how to harness the capabilities of the Metaverse, or an individual keen on future-proofing your skills, the practical insights and competencies gained from this chapter are indispensable for navigating the future of work.

In this chapter, we're going to cover the following main topics:

- As displayed in use cases, how the Metaverse will radically change office and virtual work, walk-around jobs, factory work, training, and diagnostics
- How virtual and onsite work in the Metaverse differ when using AR versus VR
- Best business practices that are exemplified in each use case in this chapter

In a world where technology keeps evolving at breakneck speed, we're standing on the edge of something truly thrilling. This chapter will take you on an exhilarating journey into the Metaverse, a digital realm that's set to redefine the very essence of work. But this isn't just a revolution for one industry; it's a game-changer that'll touch every aspect of our professional lives, from the familiar office spaces to those hands-on jobs out in the field.

At the heart of this transformation are AR and VR, technologies that have grown far beyond their entertainment origins to become the cornerstones of this digital renaissance. Together, they create a canvas of endless possibilities, breaking free from the constraints of the physical world and inviting us to rethink work itself.

But here's the exciting part – we're not mere spectators in this show; we're active participants in this captivating technological drama. The Metaverse isn't just a vague concept; it's a tangible force reshaping how we communicate, collaborate, and get things done at workplaces worldwide. It's challenging us to adapt, grasp its intricacies, and harness its immense potential.

Our journey will take us through the Metaverse's transformative influence across various aspects of work, each with its unique set of opportunities and challenges. From traditional corporate boardrooms to the bustling aisles of retail, from the open expanses of field operations to the precision of factory floors, we'll experience firsthand how the Metaverse is reshaping these environments.

But it's not just about technology; it's about understanding how AR and VR work together to create immersive and efficient workspaces. It's about discovering best practices that light the path forward in this digital revolution.

Through real-world examples, we'll uncover how organizations are seamlessly integrating the Metaverse into their operations, gaining insights into strategies that maximize returns and benefits for both businesses and their workforce.

So, whether you're a forward-thinking organizational leader exploring the potential of the Metaverse or an individual eager to stay ahead in an ever-evolving job market, the insights within this chapter are your indispensable guide. They offer more than just a peek; they provide a profound understanding of the future of work in the Metaverse, equipping you to navigate this transformative era with confidence and flair. Welcome to the dawn of a new work paradigm, where the digital and physical converge to redefine the boundaries of what's possible.

Metaverse – redefining office and virtual work

In the ever-evolving landscape of work, a profound transformation is underway, and this chapter delves into its core. Here, we will explore how the Metaverse, an expansive digital realm, is fundamentally altering the dynamics of office and virtual work. This shift extends far beyond the boundaries of conventional workspaces, reshaping the very essence of how we collaborate, communicate, and thrive professionally.

Now, picture a world where professionals from different corners of the globe gather in virtual co-working spaces within the Metaverse. Equipped with AR and VR devices, they transcend geographical constraints and collaborate on projects with unprecedented efficiency. This virtual co-working experience, powered by generative AI, not only enhances productivity but also fosters a sense of community among remote workers.

Yet, this transformation isn't just about remote connections; it's about immersive experiences that redefine virtual meetings. Envision traditional video conferences transforming into immersive meetings, where participants leverage generative AI to create lifelike avatars, don AR and VR headsets, and find themselves seated around a virtual conference table in picturesque digital environments. These meetings, with AI-driven language translation, not only enhance engagement but also alleviate the fatigue associated with prolonged screen time.

As we delve deeper, consider a multinational corporation harnessing the Metaverse to enable cross-continental collaboration. Teams in different time zones work together in shared virtual spaces, breaking down geographical barriers and time constraints. With the aid of generative AI, language barriers vanish as AI-driven translators facilitate seamless communication. This fosters innovation and allows for 24/7 productivity, reshaping our approach to global teamwork.

Now, think about a tech company providing customer support through AR and VR, empowered by generative AI. When customers encounter technical issues, support agents don AR glasses to remotely access their devices in the Metaverse, troubleshoot problems, and guide them through solutions. Generative AI-driven chatbots provide instant assistance, enhancing customer satisfaction and problem resolution.

In the realm of employee well-being, visualize companies promoting wellness by offering virtual programs in the Metaverse, enriched by generative AI. Employees can participate in yoga sessions and mindfulness classes, or even take virtual walks in stunning digital landscapes. AI-driven health and wellness recommendations tailor experiences to individual needs, reducing stress and promoting work-life balance in a novel way.

These scenarios provide glimpses into how the Metaverse, augmented by generative AI, is reshaping office and virtual work. It's a world where professionals harness the power of AR, VR, and AI to transcend boundaries, enhance collaboration, and redefine the future of work itself. Welcome to an era where office and virtual work converge.

Use case 1 – virtual team collaboration

Organizations are harnessing AR and VR technologies to create immersive virtual office spaces within the Metaverse, facilitating collaboration among remote employees. In these virtual realms, employees wear AR/VR headsets, represented by customizable avatars, allowing for natural interactions. Meetings occur in lifelike conference rooms with real-time document sharing and spatial audio for a lifelike feel. Collaboration extends to face-to-face meetings, brainstorming on interactive whiteboards, and real-time discussions, enhancing the sense of presence. AI-driven technologies enhance interaction efficiency, but challenges include ensuring connectivity, minimizing technical glitches, addressing VR meeting fatigue, and maintaining data security. These virtual offices redefine remote work, offering a transformative approach to collaboration.

The setup

In a bid to foster seamless remote collaboration, organizations have established immersive virtual office spaces accessible through AR/VR headsets. In these virtual realms, each employee is represented by a customizable avatar, allowing for personalization and identification within the virtual workspace. Virtual meetings are held in lifelike conference rooms equipped with advanced tools for real-time document sharing and collaborative activities. The integration of spatial audio technology ensures that conversations sound natural, as the voices of participants emanate from the direction of their avatars.

Interactivity

Virtual team members engage in a wide range of collaborative activities, surpassing the limitations of traditional video conferencing. Face-to-face meetings take on a new dimension as team members, represented by their avatars, interact in a highly immersive environment. Brainstorming sessions come to life on interactive whiteboards, enabling real-time ideation and concept development. Real-time discussions unfold through avatar-based communication, fostering a sense of presence and interconnectedness. Moreover, the virtual workspace allows team members to physically interact with shared digital content, enhancing the collaborative experience.

Technical innovation

Cutting-edge AI-driven technologies significantly enhance the naturalness and efficiency of interactions within the virtual workspace. Voice recognition powered by AI ensures that verbal communication feels intuitive and responsive, bridging the gap between physical and virtual interactions. Gesture control empowers users to navigate the virtual environment effortlessly, further blurring the line between the physical and digital realms. Advanced haptic feedback devices enable users to engage in tactile interactions, such as handshakes or high-fives, creating a profound sense of physical connection despite the virtual setting.

Challenges

While virtual team collaboration in the Metaverse offers immense potential, several challenges must be addressed to ensure its effectiveness. Seamless network connectivity is paramount to prevent disruptions during critical discussions and meetings. Efforts to minimize potential technical glitches and optimize the performance of AR/VR hardware and software are ongoing. User fatigue in prolonged VR meetings is a concern, necessitating innovative solutions to maintain engagement and comfort. Additionally, robust data privacy and security measures must be in place to safeguard sensitive information shared within virtual spaces, addressing concerns related to data breaches and unauthorized access.

Use case 2 – virtual training and onboarding

Forward-thinking organizations are at the forefront of leveraging the potential of VR and AR technologies to craft immersive team-building experiences and social events within the expansive Metaverse.

The setup

These pioneering companies have embarked on a journey to redefine workplace interaction by meticulously designing virtual spaces within the vast digital landscape of the Metaverse. These spaces are accessible through an array of AR-equipped devices and VR headsets, each thoughtfully constructed to cater to a diverse spectrum of team-building activities and social gatherings. From serene virtual retreats nestled in idyllic settings to dynamic virtual convention centers, these environments offer a plethora of options to suit every team's unique preferences and needs.

Interactivity

What truly sets these virtual experiences apart is the unprecedented level of interactivity they afford. Employees, represented as highly customizable avatars, step into these digital realms to engage with their colleagues. These interactions transcend the limitations of traditional video conferencing, elevating the mundane to the extraordinary. Participants actively partake in a multitude of captivating team-building exercises, from navigating intricate virtual escape rooms and solving complex collaborative challenges to indulging in friendly competitions. The immersive capabilities of AR and VR amplify the sense of presence, enabling employees to feel as though they are physically together, regardless of their actual locations. It is in these shared virtual adventures and lighthearted contests that lasting bonds are nurtured, camaraderie flourishes, and innovative ideas are sparked.

Technical innovation

At the core of these remarkable experiences lies a fusion of technological innovation and human connection. Fueled by advanced AI algorithms, event planning becomes an intelligent and dynamic process. These algorithms take into account each participant's individual preferences, team dynamics, and historical engagement data to curate a personalized menu of activities. What sets this apart is the dynamic nature of these events – they adapt in real time, ensuring sustained engagement. Spatial audio technology further heightens the authenticity of interactions, allowing team members to converse as naturally as if they were in the same physical space. Avatars, thoughtfully designed to express a range of emotions, play a pivotal role in enhancing connections and conveying non-verbal cues crucial for effective communication.

Challenges

Despite the immense potential for enriching virtual team-building and social events, certain challenges demand attention. Ensuring active participation from all employees, especially those susceptible to distractions while working remotely, calls for inventive strategies. Technical aspects are equally vital; ensuring seamless connectivity, device compatibility, and minimal latency is essential for delivering seamless experiences. Yet, the most intricate challenge lies in replicating the sense of camaraderie and genuine human connection characteristic of in-person interactions in the virtual realm. Striking the right balance between the novelty of technology and the authenticity of human bonds demands continuous refinement and innovation.

Use case 3 – remote customer support

In the ever-evolving landscape of customer service, forward-thinking organizations are embracing the power of immersive technologies, specifically VR and AR, to elevate customer support to new heights. In this paradigm shift, customer support agents operate from remote virtual call centers, leveraging AR and VR to provide not just assistance but an enriched, personalized customer experience.

The setup

In this groundbreaking setup, customer support representatives don VR headsets that transport them to virtual call centers that closely resemble physical offices. These virtual call centers are thoughtfully designed to foster a collaborative environment for support teams. Through the magic of AR, agents have access to a wealth of customer data displayed on AR screens within their virtual workspace, providing them with real-time insights to serve customers more effectively. As they put on their VR headsets, these agents seamlessly transition into their virtual roles, allowing them to interact with customers as if they were physically present, breaking down geographical barriers.

Interactivity

The interactivity in this setting transcends traditional customer support. Agents conduct real-time product demonstrations within virtual showrooms, creating a dynamic and engaging environment for customers. Complex product features come to life as agents employ immersive tools to showcase them. AI-driven chatbots stand ready to provide rapid issue resolution, offering customers quick and efficient solutions to their problems. What truly sets this experience apart is the high level of personalization. Customers can share their screens with agents, enabling the troubleshooting of issues in a visual, collaborative manner. Agents not only solve problems but actively engage with customers, forging a deeper connection that goes beyond the transactional.

Technical innovation

The heart of this transformative customer support setup lies in its technical innovation. Managing complex customer inquiries within virtual spaces necessitates cutting-edge AI and virtual communication tools. Advanced AI-driven sentiment analysis empowers agents to gauge customer emotions and tailor their responses accordingly. In these virtual environments, every interaction becomes an opportunity to enhance the customer's perception of the brand. With each customer conversation, AI algorithms continuously learn and adapt, ensuring that the support experience keeps improving.

Challenges

While the potential benefits are immense, this new frontier also presents significant challenges. Ensuring the utmost data security is paramount; customer information must be safeguarded with the highest standards. Maintaining a personalized and empathetic customer experience in remote interactions requires a delicate balance between the convenience of technology and the human touch. Addressing potential technical hiccups and ensuring a seamless experience for both agents and customers is an ongoing concern. Furthermore, organizations must invest in comprehensive training to equip support agents with the skills needed to thrive in this virtual customer support environment. In this new era of customer service, the fusion of technology and human expertise holds the key to delivering exceptional support that transcends boundaries.

Use case 4 – virtual sales presentations

In this dynamic landscape of sales, the fusion of immersive technology and personalized engagement has ushered in a new era of client interactions. Virtual sales presentations empower clients with a deeper understanding of products, create memorable experiences, and forge stronger client relationships, making them a transformative tool for businesses aiming to stand out in the competitive market.

The setup

Sales representatives, armed with the power of VR, have transcended the confines of traditional product presentations. Instead, they harness the potential of immersive 3D environments to craft captivating sales experiences. These environments take the form of virtual showrooms where products come to life in stunning detail. What sets this setup apart is its adaptability; presentations can be meticulously customized to cater to the unique needs and preferences of each client. No longer limited to static brochures or flat images, these presentations immerse clients in a dynamic and interactive product showcase.

Interactivity

The cornerstone of virtual sales presentations is interactivity. Clients, no longer passive observers, step into a virtual world where they can experience products firsthand. They have the opportunity to explore, interact, and engage with products on an entirely new level. Real-time questions find immediate answers, fostering a deeper understanding of products. Sales representatives become virtual tour guides, leading clients through these immersive experiences. They have the power to highlight key features, benefits, and nuances of products, all while adapting the presentation to the client's preferences.

Technical innovation

Cutting-edge technology fuels this transformation. Real-time analytics is at work during these virtual sales presentations, tracking customer engagement and behavior. Every interaction is logged, providing valuable insights into client preferences and interests. The application of AI-driven pricing and configuration tools empowers clients to customize product options on the fly, creating a sense of ownership and personalization. The once static sales pitch has evolved into a dynamic and adaptive conversation.

Challenges

While the potential for transformation is immense, embracing virtual sales presentations comes with its set of challenges. Ensuring accessibility to VR content for all clients is a priority. Addressing potential technical issues during presentations is crucial to maintaining a seamless experience. Keeping clients engaged throughout the virtual sales process, especially in longer presentations, requires thoughtful design and content curation. Additionally, organizations must strike a balance between virtual and traditional sales channels to cater to diverse client preferences. This integration poses a unique challenge in managing a seamless transition between physical and virtual sales interactions.

Use case 5 – virtual project management

In this era of dynamic and complex projects, virtual project management not only simplifies the management process but also elevates it to new heights of efficiency and collaboration. By leveraging AR/VR technologies, project teams gain a comprehensive, real-time view of their projects, enabling them to make data-driven decisions and optimize project performance for successful outcomes.

The setup

Project teams gain access to immersive virtual project rooms where project data, timelines, and tasks are visualized in dynamic and interactive ways. This setup not only simplifies project management but also enhances the efficiency of handling complex projects. Project managers wield the power to create interactive project boards, allowing team members to intuitively grasp project progress and milestones. The virtual environment offers a holistic view of the project's life cycle, streamlining the planning, execution, and monitoring phases.

Interactivity

The heart of virtual project management lies in its interactivity. Teams can actively engage within the virtual workspace, updating tasks, discussing project developments, and facilitating real-time communication. Collaborative tools enable team members to edit documents simultaneously and share vital project-related information seamlessly. This immersive environment fosters a sense of presence and collaboration, transcending geographical boundaries and time zones.

Technical innovation

The technical innovation behind virtual project management is multifaceted. AI-driven project analytics provide invaluable data-driven insights, empowering project teams to make informed decisions swiftly. These analytics not only provide historical data but also predict potential project risks and suggest mitigation strategies, optimizing project performance and outcomes. With predictive analytics, project managers can proactively address challenges before they escalate.

Challenges

While the potential for transformation is immense, integrating virtual project management with existing tools and workflows poses a challenge. Ensuring data security within virtual workspaces is paramount, especially when dealing with sensitive project information. Facilitating a smooth transition for project teams, some of whom may be new to AR/VR technologies, requires thoughtful planning and support. Organizations must invest in training programs to ensure that team members are proficient in virtual project management tools and can harness their full potential.

Negative implications of office and virtual work in the Metaverse

In an era where the boundaries of reality and the digital realm blur, the Metaverse has emerged as a transformative arena poised to redefine the landscape of office and virtual work. However, beneath the promises of innovation lies a series of challenges that demand our attention. From the isolation of virtual cubicles to the intrusive data gathering and the looming tech disparities, the Metaverse introduces a complex web of concerns. Here, we will uncover the major negative implications of redefining office and virtual work within this digital frontier. These challenges serve as a stark reminder that as we embrace the Metaverse, we must navigate a landscape fraught with potential pitfalls.

Isolation in virtual cubicles

Isolation in virtual cubicles impacts mental well-being. Extended periods in virtual office spaces may lead to feelings of isolation and detachment from real-world interactions, impacting mental well-being.

Intrusive data gathering

- **Extensive user data collection**: Extensive user data collection for personalization can raise concerns about unauthorized data access, data breaches, and privacy
- **Privacy breaches**: This could result in privacy breaches and undermine the trust of virtual workers in the Metaverse

Tech disparity

- **Unequal access**: Unequal access to Metaverse technology could deepen existing digital disparities in job opportunities and resources
- **Unequal opportunities**: This digital divide could result in unequal job opportunities and disparities in access to resources, potentially exacerbating existing socioeconomic inequalities

Virtual cybersecurity risks

- **Fertile ground**: The interconnected nature of virtual office spaces in the Metaverse creates fertile ground for cybersecurity risks.
- **Vulnerability**: Virtual meetings, communication platforms, and data storage systems can be vulnerable to hacking, data breaches, and other digital threats. Confidential company information and sensitive data may be at risk, leading to potential financial and reputational damage for businesses operating in the virtual workspace.

Job displacement in virtual work

- **Automation and AI**: As automation and AI become more prevalent in virtual office environments, there's a risk of job displacement in certain industries.

- **Unemployment even for virtual workers**: Tasks that were previously performed by human workers may become automated, potentially leading to unemployment for some virtual workers. Traditional office roles that rely heavily on physical presence may face challenges in adapting to this digital transformation.

Dependence on virtual collaboration tools

- **Skill erosion**: Overreliance on virtual collaboration tools within the Metaverse might erode individuals' physical teamwork and communication skills.

- **Nonvirtual incompetence**: When communication primarily occurs in virtual spaces, professionals may become less adept at face-to-face interactions and physical teamwork. This could potentially affect their ability to collaborate effectively in non-virtual settings, impacting the overall quality of interpersonal relationships and teamwork.

Virtual fatigue and mental health

- **Digital fatigue**: Extended exposure to virtual work environments can contribute to digital fatigue.

- **Burnout and other effects**: Virtual workers may experience burnout from prolonged screen time and a constant presence in virtual meetings and spaces. Additionally, the pressure to maintain a digital persona in the virtual workspace can lead to stress and mental health challenges, including anxiety and feelings of inauthenticity.

Identity and trust challenges

- **Artificial or not**: In the Metaverse, distinguishing authentic interactions from simulations can be challenging.

- **Trust and credibility issues**: Virtual workers may encounter difficulties in discerning genuine connections from simulated ones, leading to issues of trust and credibility. Authenticity and trustworthiness become crucial concerns, potentially impacting the quality of virtual work relationships and collaboration.

Regulatory uncertainties in virtual work

- **Too fast a pace**: Policymakers may struggle to keep pace with the rapid evolution of the Metaverse, leading to regulatory uncertainties in virtual work settings.

- **Legal and compliance miasma**: This includes concerns related to data privacy, intellectual property rights, taxation, and online governance. A lack of clear and standardized regulations can create legal and compliance challenges for businesses operating in virtual spaces.

Environmental impact of virtual work

- **Server and data center needs**: Supporting the servers and data centers required to sustain virtual work in the Metaverse can have a significant environmental footprint.

- **High energy consumption**: The energy consumption associated with these digital infrastructures contributes to environmental concerns, particularly in an era where sustainability is a pressing global issue. Addressing the environmental impact of virtual work will be crucial in promoting responsible and sustainable practices within the Metaverse.

In summary, the Metaverse offers great potential for the future of work, but it also presents significant challenges. To navigate this digital frontier successfully, we must prioritize mental well-being, bridge technological disparities, fortify cybersecurity, and maintain authenticity and trust in virtual interactions. By doing so, we can harness the benefits of the Metaverse while mitigating its negative effects, ensuring a future of work that is both innovative and responsible.

Walk-around work unleashed in the Metaverse

In the dynamic landscape of work, where technological innovation drives change, a new frontier is emerging. Welcome to the world of *walk-around jobs* transformed by the Metaverse, a concept that encompasses professions requiring physical presence and mobility. These are the jobs where professionals move, interact, and operate in real-world spaces, from retail and field operations to on-site diagnostics.

Here, we will take you on a journey through practical use cases, offering valuable insights into how the Metaverse is reshaping walk-around work. Picture a retail associate in a bustling virtual store, a field technician repairing equipment in a digital landscape, or a healthcare professional conducting diagnostics through augmented reality. Through these use cases, we'll provide real-world examples and information that highlight the transformative power of the Metaverse in walk-around jobs.

As we explore further, we'll uncover the synergy between AR and VR in transforming these professions. Real-world examples will illustrate how businesses successfully integrate the Metaverse into walk-around job environments. Through these examples, we'll discover best practices that empower both professionals and organizations to thrive in this new virtual paradigm.

Whether you're a business leader seeking innovative solutions for your walk-around job workforce or an individual eager to understand how your profession may evolve in this digital age, the insights within this chapter are your compass. They offer more than just a glimpse; they provide a comprehensive understanding of how the Metaverse is redefining walk-around jobs, offering exciting opportunities for growth and innovation. Welcome to a world where physical and digital seamlessly intertwine, ushering in a new era for walk-around work in the Metaverse.

Use case 1 – VR/AR-assisted construction inspections

In the dynamic and constantly evolving field of construction and engineering, the integration of VR and AR technologies is sparking a transformative shift in the way construction inspections are executed, ushering in a new era of unparalleled precision and efficiency within the boundless expanse of the Metaverse. This groundbreaking fusion of virtual and augmented reality not only streamlines traditional inspection processes but also redefines the very essence of how construction professionals interact with and assess projects, blurring the lines between physical and virtual worlds to bring about an unprecedented level of accuracy and effectiveness. As these technologies continue to advance, they are not merely tools but catalysts for innovation, reshaping the landscape of construction inspection methodologies and propelling the industry into uncharted realms of possibilities.

The setup

Construction inspectors don VR headsets, creating a seamless bridge between the physical construction site and the virtual world. Within the Metaverse, they gain access to virtual replicas of construction projects, allowing for detailed inspections and quality control without being physically present. These virtual construction sites are meticulously recreated, complete with real-time data feeds and blueprints, providing inspectors with a holistic view of the project.

Construction inspectors don VR headsets, creating a seamless bridge between the physical construction site and the virtual world. Within the Metaverse, they gain access to virtual replicas of construction projects, allowing for detailed inspections and quality control without being physically present. These virtual construction sites are meticulously recreated, complete with real-time data feeds, augmented AR overlays, and blueprints, providing inspectors with a holistic view of the project. AR annotations and overlays offer inspectors additional contextual information during their virtual inspections, enhancing their ability to identify and address potential issues.

Interactivity

The interactivity within this virtual construction inspection process is nothing short of transformative. Inspectors don't merely observe; they become active participants. With VR headsets, they navigate the virtual construction site, exploring every nook and cranny. AR annotations come into play here, allowing inspectors to highlight potential issues or areas of concern within the virtual environment. In real time, they can collaborate with on-site personnel, discuss findings, and provide precise guidance, all while being miles away from the actual construction site.

Technical innovation

At the core of this innovation is AI-driven defect detection. AI algorithms meticulously analyze construction elements for anomalies or deviations from the plan. This automated system not only streamlines the inspection process but also ensures a higher degree of accuracy. Furthermore, VR simulations enable inspectors to do something previously unthinkable – virtually walk through a completed construction project. This immersive experience provides valuable insights and allows for a final quality check before physical inspection.

Challenges

While VR/AR-assisted construction inspections hold immense potential, they do come with their set of challenges. One of the primary considerations is ensuring the accuracy of remote inspections. While AI algorithms can detect many issues, there may be nuances that only the human eye can catch. Connectivity issues, particularly in remote construction sites, may affect the seamless flow of data. Effective communication with on-site personnel is crucial to ensure that instructions are clear and actions are taken promptly. Lastly, data security is paramount, as construction projects often involve sensitive information that must be protected in the digital realm.

Use case 2 – virtual training for emergency responders

In the realm of emergency response, VR and AR technologies are orchestrating a transformation in the way emergency responders prepare for the unpredictable challenges of their field within the expansive domain of the Metaverse.

The setup

Within the boundless Metaverse, emergency responders are granted access to a groundbreaking training paradigm facilitated by VR simulations. These simulations faithfully replicate a myriad of high-stress scenarios, from towering infernos to complex search-and-rescue missions. Responders are equipped with AR-enhanced gear that seamlessly integrates with their VR training, creating a unified experience where real-world data overlays the virtual environment.

Interactivity

Responders find themselves immersed in hyper-realistic emergencies, making critical decisions, coordinating with team members, and executing life-saving procedures. The synergy between VR and AR allows for real-time information to be displayed in their field of view, providing vital situational awareness. This dynamic interactivity fosters the development of split-second decision-making skills and enhances their ability to respond effectively under immense pressure.

Technical innovation

At the heart of this revolution lies AI-driven scenario generation. Responders can fine-tune training scenarios to match their skill levels, progressively increasing the complexity as they advance. The fusion of AR enriches their training by overlaying essential information such as maps, vital signs, and procedural instructions. Realistic haptic feedback devices simulate the physical sensations of emergency situations, from the vibrations of a collapsing structure to the heat of a blazing fire.

Challenges

The effectiveness of transferring skills honed in the virtual realm to real-world emergencies is a paramount challenge. Additionally, addressing potential motion sickness during high-stress simulations necessitates ongoing adjustments and acclimatization. To maintain the program's relevancy and effectiveness, responders must stay attuned to the ever-evolving landscape of emergency response technologies, continuously adapting their training to harness the latest innovations.

Use case 3 – digital twin oil rig inspections

In the context of oil rig operations, the application of AR and VR technologies is facilitating a transformative shift in the way maintenance crews conduct inspections and training activities within the vast and immersive Metaverse.

The setup

Within this virtual realm, detailed virtual oil rigs are meticulously replicated, featuring AR-enhanced equipment and environments. These digital twins serve as the backdrop for a range of essential maintenance tasks and immersive training experiences.

Interactivity

Maintenance crews leverage AR technology to conduct remote inspections of oil rig components, transcending geographical barriers. Through AR-enhanced interfaces, inspectors can remotely access and assess critical equipment, identifying potential issues without the need for physical presence. Meanwhile, VR technology immerses workers in realistic training simulations where they can rehearse emergency procedures. These virtual training exercises provide a risk-free environment for honing skills and decision-making under high-pressure scenarios.

Technical innovation

The innovative use of AI-driven predictive maintenance algorithms further enhances the AR-assisted inspection process. By analyzing real-time data and historical performance metrics, these algorithms predict maintenance needs, allowing proactive measures to be taken to prevent potential issues. In the VR domain, training scenarios include highly realistic emergency situations, providing workers with practical experience and the ability to test their responses in a controlled virtual environment.

Challenges

While AR enables accurate remote inspections, addressing potential differences between physical and virtual equipment remains a challenge. Ensuring that the virtual representations of equipment are accurate and reflective of their real-world counterparts is critical to maintaining safety and operational standards. Additionally, in the VR training sphere, the challenge lies in maintaining safety protocols and ensuring that the skills acquired in virtual simulations effectively translate to real-world applications, making the training as practical and valuable as possible.

Use case 4 – AR/VR-powered firefighter training

In the realm of firefighter training, the fusion of AR and VR technologies is revolutionizing how firefighters prepare for challenging scenarios within the expansive Metaverse. Fire stations now feature AR-enhanced firefighting gear and immersive VR training simulators, offering trainees an unprecedented level of experiential learning. AR visors overlay critical data and real-time hazard assessments, while VR scenarios replicate real-world firefighting challenges. AI-driven fire simulations in AR create dynamic and responsive conditions. However, ensuring effective training, addressing differences between physical and virtual environments, and maintaining safety in intense VR scenarios pose challenges. This integration represents a significant advancement, enhancing skills and decision-making in the Metaverse.

The setup

At the core of this innovative approach lies deploying cutting-edge technology within fire stations. Firefighters are equipped with state-of-the-art AR-enhanced firefighting gear, including advanced visors that seamlessly integrate AR elements into their field of view. Furthermore, dedicated VR training simulators are installed within fire stations, serving as the cornerstone of immersive firefighter preparation.

Interactivity

The effectiveness of this training approach hinges on the interactive capabilities offered by AR and VR. AR visors allow firefighters to simulate real fire emergencies within controlled training environments. These visors overlay critical information and dynamic fire simulations onto the firefighter's view, enabling them to practice firefighting techniques in a virtual yet realistic setting. In parallel, VR training modules provide immersive scenarios that expose firefighters to high-stress situations. These simulations encompass a wide range of firefighting challenges, including structure fires, hazardous material incidents, and rescue operations. The interactivity within these virtual environments fosters rapid decision-making, skill development, and enhanced preparedness for real-world emergencies.

Technical innovation

The technical innovation inherent to this training methodology is two-fold. AR technology introduces AI-driven fire simulations that dynamically adapt to the training scenario. These simulations recreate the unpredictable nature of fires, including their spread, behavior, and intensity, allowing firefighters to hone their abilities in responding to evolving situations. In the VR training space, realism is paramount. VR modules incorporate advanced physics and fire dynamics, ensuring that the training scenarios faithfully replicate the challenges faced during real firefighting missions. The synergy of AI-driven AR and realistic VR simulations ensures that firefighters receive training that prepares them for the most demanding situations they may encounter.

Challenges

Despite the numerous advantages of AR/VR-powered firefighter training, several challenges must be overcome. Ensuring the effectiveness of firefighter training with AR and VR technology is paramount, as the training must closely mirror the complexity and intensity of real-world firefighting situations. Addressing potential disparities between physical and virtual fire conditions is crucial to ensure that the skills acquired in VR training readily transfer to practical firefighting. Moreover, simulating intense fire scenarios in VR demands the utmost attention to detail, including accurate replication of fire dynamics, smoke behavior, and environmental hazards. Meeting these challenges head-on is essential to harnessing the full potential of AR and VR in preparing firefighters for their critical roles.

Use case 5 – AR/VR-enhanced healthcare for paramedics

In the dynamic area of healthcare for paramedics, the fusion of AR and VR technologies stands as a transformative paradigm, reshaping how paramedics prepare for the most demanding scenarios within the expansive world of the Metaverse. This immersive integration of advanced technologies empowers paramedics with unprecedented access to critical medical data and the opportunity to hone their skills in a risk-free, yet hyper-realistic, training environment.

The setup

Within this innovative healthcare paradigm, virtual ambulances are equipped with cutting-edge AR medical devices, seamlessly interfacing with the Metaverse's vast medical data repositories. Additionally, paramedics are provided access to highly immersive VR training environments that replicate real-world medical scenarios with astonishing fidelity.

Interactivity

Paramedics, equipped with AR glasses, gain instantaneous access to a treasure trove of medical data, including patient histories, vital signs, and real-time medical instructions. This AR-enhanced data accessibility significantly enhances decision-making during critical situations. Simultaneously, paramedics engage in highly immersive VR training scenarios that precisely replicate a wide range of medical emergencies. These scenarios provide paramedics with invaluable experience in handling complex medical procedures and responding to life-threatening situations with the utmost confidence.

Technical innovation

The heart of this use case lies in the exceptional technical innovation. AR devices are powered by state-of-the-art AI-driven data analysis algorithms. These algorithms not only provide paramedics with rapid access to relevant patient information but also offer predictive insights, aiding in more accurate diagnoses and treatment plans. VR training modules, on the other hand, incorporate cutting-edge simulations that perfectly mimic real-world medical challenges. From intricate surgical procedures to high-pressure emergency situations, these simulations offer an unprecedented level of realism.

Challenges

Despite the incredible advantages offered by this AR/VR healthcare integration, several significant challenges must be addressed. Ensuring the swift and efficient use of AR devices during medical emergencies is of paramount importance. Paramedics must seamlessly integrate these technologies into their workflows to maximize their effectiveness. Additionally, reconciling potential disparities between physical and virtual healthcare environments is essential. The transition from the virtual world to the physical must be smooth, ensuring that paramedics can apply their training effectively in real-world scenarios. Furthermore, the management of high-pressure situations during VR training is an ongoing challenge, as these scenarios push paramedics to their limits, demanding quick thinking and decisive action.

Negative implications for walk-around work in the Metaverse

As we venture into the expansive Metaverse, where the boundaries between physical reality and the digital realm blur into a seamless tapestry of experiences, we find ourselves at a pivotal juncture in the evolution of "walk-around work." This immersive and interactive form of employment not only promises to redefine how we engage with the world but also poses a range of intricate challenges that demand our attention.

Walk-around work within the Metaverse presents us with a transformative exploration of the convergence between physical and virtual domains. It beckons us to reimagine professions that traditionally required a physical presence, spanning diverse sectors such as retail, field operations, and tasks involving on-site engagement and mobility. Yet, as we embrace this paradigm shift, we must confront the multifaceted landscape of opportunities and obstacles that it unfolds.

Here, our focus centers on the negative implications of the Metaverse for walk-around work. We embark on a journey into a virtual realm where digital avatars step in for physical presence, where real-world spaces are mirrored in lines of code, and where our sensory experiences are redefined. From the challenges of physical strain and safety hazards to concerns about technological disparities and privacy breaches, we traverse the intricate terrain of this transformation.

Physical strain and health risks

- Prolonged use of VR headsets and AR devices can lead to physical strain, discomfort, and potential health risks.
- Risks include motion sickness, eye strain, and ergonomic issues associated with long-term use.

Mental fatigue

- Extended periods of physical movement and interaction in virtual and augmented spaces can contribute to mental fatigue.
- Users may experience cognitive overload from processing both virtual or augmented and real-world stimuli simultaneously.

Spatial awareness challenges

- Navigating virtual and augmented spaces while physically moving in the real world can lead to spatial awareness challenges.
- Users may accidentally bump into objects or individuals in the real world while immersed in the virtual or augmented one, posing safety risks.

Virtual disconnection

- VR and AR work can create a sense of disconnection from the physical world.
- This disconnection can impact real-world relationships and hinder face-to-face interactions.

Equipment and mobility barriers

- Specialized VR headsets, AR devices, and a spacious physical environment are often required for immersive work.
- Limited access to these resources can result in disparities in job opportunities and access to virtual or augmented work environments.

Virtual training challenges

- While immersive work offers opportunities for virtual training and simulations, they may not fully replicate real-world conditions.
- This could lead to challenges in transitioning skills from the virtual or augmented environment to the physical workplace.

Privacy and data security

- Immersive work may require the collection of user movement and interaction data for tracking and analysis.
- Privacy and data security concerns arise as sensitive location, behavior, and movement data may be at risk of unauthorized access or misuse.

Technological dependencies

- Dependence on VR and AR technology for immersive work may lead to workflow disruptions when equipment malfunctions or technical issues arise.

Economic disparities

- Economic disparities can emerge as access to VR and AR equipment and high-speed internet becomes a prerequisite for immersive work.
- This can exacerbate existing socioeconomic inequalities.

Digital addiction

- Immersive work's immersive nature can contribute to digital addiction as users may find it challenging to disconnect from virtual or augmented environments.

- This can impact work-life balance and overall well-being.

As we conclude our exploration of the Metaverse's impact on walk-around work, we find ourselves at the intersection of innovation and complexity. The digital realm offers a new frontier for immersive experiences and interactive employment, but it is not without its intricacies and challenges.

Walk-around work within the Metaverse has allowed us to rethink the boundaries of physical engagement and redefine how we approach various sectors. However, it has also presented us with issues in terms of physical strain, safety hazards, disparities in technology access, and privacy concerns.

In our journey through the Metaverse, we've sought to understand these challenges and identify potential solutions. It's important to acknowledge that while these complexities exist, they are not insurmountable. However, with careful consideration and responsible adoption of technology, we can harness the Metaverse's potential while addressing its negative implications.

The AR and VR work revolution in factories

In the domain of industry and manufacturing, a transformative wave is sweeping through, propelled by cutting-edge digital technologies. Welcome to the future, where factory work, training, and diagnostics are undergoing profound changes. Here, we will explore practical use cases, illustrating how advanced AR and VR technologies are reshaping these areas and providing invaluable insights into the future of manufacturing.

Imagine a factory floor where human workers seamlessly collaborate with intelligent robots, guided by the immersive capabilities of AR and VR. This is just one glimpse of how the digital age is revolutionizing factory work. In this section, we will present a series of use cases that showcase the practical applications and transformative potential of the technologies in this domain.

As we explore further, we'll uncover how AR and VR technologies synergize to enhance training and diagnostics within the factory setting. These use cases aren't just theoretical; they provide real-world examples and practical information, serving as a guide for businesses seeking to harness these capabilities for improved efficiency and innovation.

Whether you're a manufacturing leader in search of cutting-edge solutions or an individual looking to stay ahead in the world of factory work, training, and diagnostics, the insights included here are your compass. Welcome to an era where the factory floor extends into the digital frontier, and where practical use cases illuminate the path to mastery in factory operations.

Use case 1 – AR/VR-powered workplace safety audits

The integration of AR and VR technologies in workplace safety audits represents a significant leap forward in ensuring the safety of factory workers within the vast landscape of the Metaverse. This immersive fusion of advanced technologies empowers auditors with real-time safety inspection capabilities and invaluable training experiences, enabling them to navigate the ever-evolving challenges of workplace safety effectively.

The setup

Factory sites serve as the frontier of innovation in workplace safety. Here, AR safety inspection tools take center stage, offering auditors a suite of advanced resources for real-time safety inspections and assessments. In parallel, dedicated VR safety training environments provide auditors with immersive and hands-on safety training experiences, enhancing their preparedness.

Interactivity

Safety auditors wield AR tools that are empowered by AI-driven safety audit assistance and hazard detection capabilities. These tools enable real-time safety audits, hazard identification, and compliance checks, allowing auditors to assess workplace safety with unparalleled precision. VR training complements this by immersing auditors in realistic factory safety scenarios. Here, auditors engage in immersive safety simulations, mastering emergency response protocols through lifelike drills.

Technical innovation

This use case thrives on cutting-edge technical innovation. AR tools, guided by AI, redefine safety audits by offering real-time assistance and hazard detection. The AI-driven features ensure that auditors can swiftly and accurately identify potential safety hazards and non-compliance issues, enhancing overall workplace safety. VR training modules are equally groundbreaking, offering auditors realistic factory safety scenarios, including dynamic environments with evolving safety challenges. These scenarios ensure that auditors are not just proficient in established safety procedures but are also equipped to adapt and respond effectively to unforeseen safety challenges.

Challenges

While the benefits of AR and VR in workplace safety audits are substantial, challenges persist. The seamless integration of AR safety inspection tools into audit processes is essential. Auditors must become adept at utilizing these tools to perform comprehensive safety inspections. Addressing potential disparities between virtual and physical safety environments is a continuous consideration. Auditors must be able to transition seamlessly between the virtual and physical worlds to apply their insights effectively. Furthermore, VR training presents its own set of challenges, requiring auditors to maintain focus and adaptability within complex simulations.

Use case 2 – AR-enhanced maintenance and repairs

AR is emerging as a game-changer in equipment maintenance and repairs, offering maintenance technicians advanced tools for efficient and precise operations. The fusion of AR tablets, remote AR assistant systems, smart tools, and AI-driven predictive maintenance ushers in an era of unparalleled precision and efficiency. This technological marvel empowers maintenance technicians to navigate the complexities of equipment upkeep with confidence, ensuring that the Metaverse's vast array of machinery runs at peak performance.

The setup

In the heart of the Metaverse, virtual maintenance bays stand as the epitome of innovation. These meticulously designed spaces mirror real-world equipment but with a twist of AR enhancement. Here, technicians have access to state-of-the-art AR tablets and smart tools that seamlessly bridge the gap between the digital and physical realms. These AR-enhanced features provide technicians with real-time data streams, schematics, and invaluable repair guidance.

Interactivity

Armed with AR tablets, maintenance technicians initiate a new era of equipment maintenance and repair. Their tablets serve as gateways to a wealth of diagnostic information, allowing them to scan equipment efficiently. The AR overlay on the tablets provides contextual data, guiding technicians through step-by-step repair procedures. Simultaneously, smart tools equipped with real-time data feeds ensure that technicians have access to the latest information on equipment conditions, performance, and potential issues.

Technical innovation

The hallmark of this use case is its technical innovation. AI-driven predictive maintenance becomes the linchpin of equipment longevity, alerting technicians to potential issues before they escalate. AR tablets add another layer of innovation, seamlessly overlaying contextual data onto the equipment being serviced. This feature enhances repair efficiency by providing technicians with real-time insights and instructions, ultimately reducing downtime and optimizing equipment performance.

Challenges

While AR offers substantial advantages, it brings its own set of challenges. Ensuring the seamless integration of AR with physical equipment is paramount, requiring meticulous calibration and synchronization to prevent discrepancies. Addressing potential connectivity issues, which can disrupt the flow of real-time data to AR devices, is a persistent consideration. Additionally, effectively training technicians to utilize AR to its full potential remains an essential aspect of this transformative technology.

Use case 3 – virtual supply chain management and optimization

In the expansive Metaverse, the fusion of VR and cutting-edge analytics heralds a new era for supply chain management and optimization. Supply chain professionals are equipped not just with tools, but with immersive experiences that empower them to navigate the complexities of modern logistics with unparalleled precision and insight. Through AI-driven guidance and VR-enabled visualization, this use case represents a paradigm shift in the art of supply chain orchestration, promising heightened efficiency and competitiveness in the global marketplace.

The setup

Anchored in the virtual expanse, supply chain professionals find themselves amid a digital network of interconnected nodes. Here, entire supply chain networks come to life, intricately woven with virtual threads that bind factories, warehouses, logistics centers, and distribution channels. These virtual supply chain hubs replicate the entire logistical ecosystem, providing a comprehensive canvas for analysis and improvement.

Interactivity

As they traverse this intricate web of virtual supply chain nodes, managers assume the role of orchestrators, navigating and fine-tuning supply chain processes. They monitor inventory levels, simulate real-time demand fluctuations, and engage in the art of dynamic decision-making. The immersive capabilities of VR introduce a new dimension to data visualization and analysis, enabling managers to perceive complex supply chain dynamics in three dimensions.

Technical innovation

The backbone of this transformative journey is a symphony of technical innovation. AI-driven predictive analytics and optimization algorithms act as guiding lights, illuminating potential bottlenecks and routes to efficiency. Within the virtual domain, VR provides a 3D panoramic view of the entire supply chain, facilitating not just understanding, but true mastery of its intricacies.

Challenges

As with any transformative technology, challenges abound. Ensuring that the virtual representation of supply chain dynamics faithfully mirrors the real world remains a cornerstone concern. Addressing potential information overload in this immersive environment necessitates judicious design and user-friendly interfaces. Integrating seamlessly with existing supply chain management systems presents its own set of intricacies, requiring careful consideration and technical finesse.

Use case 4 – AR-powered maintenance drones

In the relentless quest for efficiency and precision, maintenance technicians harness the power of AR-equipped drones, unleashing them onto the factory floor for remote inspections and repairs, all within the transformative embrace of the Metaverse.

The setup

Imagine a squadron of drones – not ordinary drones, but AR-enhanced marvels of technology. These maintenance drones are equipped with cameras and sensors that grant them unparalleled vision. They patrol the factory floor like vigilant sentinels, capturing live video feeds of machinery and equipment, ready to dive into action at a moment's notice.

Interactivity

Here's where the magic unfolds. Technicians, donned in AR headsets, become the maestros of this aerial orchestra. With the stroke of a virtual interface, they control these drones remotely, directing their flight and commanding their cameras. Live video feeds from the drones provide a real-time window into the world of factory equipment. But it doesn't stop there. AR overlays offer more than just observation – they provide guidance, instructions, and insights, turning technicians into the conductors of a symphony of precision.

Technical innovation

The orchestra of maintenance drones dances to the tune of AI. Autonomous and adaptive, these drones navigate the factory floor with grace, effortlessly avoiding obstacles and selecting optimal paths. It's not just remote inspection; it's remote equipment maintenance with visual guidance, where AI and AR collaborate to create a ballet of efficiency.

Challenges

In this high-flying endeavor, reliability is the cornerstone. Ensuring that remote drone operations run like a well-choreographed performance is a task that demands constant attention. The potential for network latency, that unpredictable stutter in the symphony, is a concern that requires mitigation. And for those who wield these technological marvels, comprehensive training is not just desirable; it's imperative.

Use case 5 – augmented reality training for collaborative robots

In manufacturing, where automation takes center stage, a revolution unfolds: workers embarking on a journey to master the art of collaboration with their metallic counterparts. This revolution is powered by AR training, a digital realm where human workers and collaborative robots (cobots) unite to redefine the future of production.

The setup

Imagine donning AR glasses that seamlessly blend the physical and virtual worlds. In this state-of-the-art training environment, workers step into a virtual factory where cobots stand as digital allies. These AR glasses become their windows to a new era, offering real-time information and guidance, their indispensable companions in this digital frontier.

Interactivity

As the cobots come to life in the virtual factory, workers wield their AR interfaces as if they were conducting a symphony. With precise gestures, interactions, and voice commands, they orchestrate collaborative tasks, harmonizing their movements with their mechanical partners. Safety is paramount, and workers learn the delicate dance of sharing their workspace with cobots, guided by digital tutors and augmented cues.

Technical innovation

Underneath this digital ballet lies a sophisticated AI-driven choreographer. Cobots adapt to every nuance of worker interaction, their responses finely tuned to the human touch. AR weaves its magic, enhancing worker-cobot collaboration to new heights while safeguarding the workplace with digital layers of protection.

Challenges

In this groundbreaking endeavor, challenges abound. Confidence must grow in every worker's heart as they interact with cobots. The intricacies of cobot programming, though digital, demand mastery. Comprehensive AR training, the bridge between humans and machines, must be meticulously crafted to ensure workers are prepared for this brave new world.

Negative implications for factory work, training, and diagnostics in the Metaverse

As we step into the Metaverse, a realm where digital innovation intertwines with reality, we encounter a landscape rich with potential and complexities. Within this chapter, our focus is on the intricate web of challenges and concerns that emerge within the Metaverse, particularly as they pertain to factory work, training, and diagnostics.

While the Metaverse promises groundbreaking advancements in these areas, it simultaneously presents us with a set of formidable challenges. The transformation of factory processes, the evolution of training methodologies, and the integration of diagnostics into this digital domain come with a host of complexities. From workforce adaptation to job displacement, and data privacy to the resilience of technology, these challenges require our keen attention and proactive solutions.

Skills gap and training challenges

- **Growing skills gap**: As factory work becomes increasingly automated and digital, there may be a growing skills gap among workers.

- **Significant training needed**: Traditional workers may require significant training to adapt to new technologies and processes in the Metaverse.

Job displacement and economic shifts

- **Automation effects**: The automation of factory tasks in the Metaverse can lead to job displacement for some workers.

- **Support**: This shift may require governments and industries to address unemployment and provide opportunities for upskilling.

Technical failures and downtime

- **Dependence on technology and digital systems**: Factory operations in the Metaverse rely heavily on technology and digital systems.

- **Downtime**: Technical failures, software glitches, or cyberattacks can result in downtime and production interruptions.

Dependency on data accuracy

- **Real-time accuracy**: Factory processes in the Metaverse often depend on real-time data accuracy.

- **Errors and quality issues**: Data errors or inaccuracies can lead to production errors and quality issues.

Privacy and security concerns

- **Data needs**: Collecting and transmitting data for diagnostics and factory operations may raise privacy and security concerns.

- **Sensitivity**: Unauthorized access or data breaches can compromise sensitive production and operational data.

Environmental impact

- **Operational issue**: The energy requirements for sustaining virtual factory operations in the Metaverse can have a significant environmental footprint.

- **Balance**: Balancing productivity with sustainability will be a key challenge.

Worker well-being

- **Increased use**: Extended periods of digital factory work may have implications for worker well-being.
- **Balance**: Striking a balance between virtual and real-world responsibilities is important for mental and physical health.

Dependency on connectivity

- **Dependency**: Factory operations in the Metaverse require robust and reliable internet connectivity.
- **Work disruption**: Poor or unstable connections can disrupt manufacturing processes.

Cost of technology adoption

- **Expense**: Implementing and maintaining the necessary technology for virtual factory work can be costly.
- **Financial barriers to entry**: Smaller manufacturers may face financial barriers to entry.

Quality control challenges

- **Quality control**: Maintaining quality control standards in the virtual factory environment can be challenging.
- **Prioritization**: Ensuring product quality and safety remains a top priority.

Regulatory and compliance issues

- **Adaptation**: Policymakers and regulatory bodies may face challenges in adapting existing regulations to virtual factory environments.
- **Oversight**: Ensuring safety, quality, and compliance will require careful consideration.

Dependence on digital twins

- **Dependence**: Virtual factory operations often rely on digital twins or simulations of physical processes.
- **Inaccuracies**: Inaccuracies in digital twins can lead to real-world production problems.

Remote work challenges

- **Geographic dissonance**: Remote factory work in the Metaverse may require employees to work from different geographic locations.
- **Coordination**: Coordinating remote teams and ensuring effective collaboration can be complex.

Lack of physical feedback

- **Lack of cues**: Virtual factory work may lack physical feedback and sensory cues.
- **Lack of tactile sense**: Workers may not have a tactile sense of the materials they are working with, which can impact precision.

Worker isolation

- **Social isolation**: Virtual factory work can lead to social isolation among workers.
- **Team dynamic hit**: Reduced face-to-face interactions with colleagues may affect team dynamics.

In our exploration of the Metaverse's impact on factory work, training, and diagnostics, we've encountered both opportunities and challenges. The digital realm presents a promising future, but it also demands our careful consideration.

As we address the challenges of workforce adaptation, data privacy, and job displacement, let's remember that the Metaverse is a canvas for innovation. To navigate its complexities, we must strike a balance between progress and responsibility.

In closing, we should approach this digital frontier with a commitment to harness its potential while safeguarding against its negative implications. With purpose and vigilance, we can shape a Metaverse that benefits us all, advancing into the future with confidence and ethical clarity.

Best business practices

In the dynamic landscape of the Metaverse, businesses face the exciting challenge of harnessing the potential of AR and VR technologies to redefine how work is conducted. To navigate this digital frontier effectively, a set of best business practices emerges as the guiding principles for success. These practices span diverse domains, from comprehensive training programs that empower employees with AR/VR skills to the integration of real and virtual worlds, promoting seamless collaboration. Data security, continuous improvement, and environmental responsibility take center stage as organizations strive for excellence in this digital realm. Inclusivity and accessibility ensure that no one is left behind, while safety protocols, regulatory compliance, and cross-functional collaboration form the foundation of responsible and innovative Metaverse operations. Supporting the mental well-being of virtual workers, upholding quality standards, and fostering a sense of community round out this comprehensive guide to thriving in the Metaverse's dynamic landscape.

Comprehensive training programs

- **AR/VR training excellence**: Develop thorough training programs to familiarize employees with AR and VR technologies relevant to their roles.
- **Ongoing skill development**: Ensure ongoing training to keep employees updated with the latest developments in the Metaverse.

Integration of real and virtual worlds

- **Seamless real-virtual blend**: Strive for seamless integration between physical and virtual work environments to ensure a smooth transition for employees.

- **Augmenting physical workspace**: Use AR and VR to augment the physical workspace, enhancing collaboration and productivity.

Data security and privacy

- **Robust data protection**: Implement robust data security measures to protect sensitive information within the Metaverse.

- **Compliance and audits**: Ensure compliance with data protection regulations and regularly audit data security practices.

Remote monitoring and supervision

- **Real-time oversight with AR/VR**: Utilize AR and VR for remote monitoring and supervision, enabling real-time oversight of tasks and employee safety.

- **Accessible critical data**: Implement systems that provide supervisors with immediate access to critical data and visuals.

Continuous improvement and feedback

- **Feedback-driven enhancements**: Establish mechanisms for employees to provide feedback on AR and VR tools, processes, and workflows.

- **Continuous workflow improvement**: Use this feedback to drive continuous improvement in the use of these technologies.

Environmental responsibility

- **Sustainable Metaverse practices**: Consider the environmental impact of virtual workspaces and virtual factory operations.

- **Eco-friendly virtual operations**: Strive for energy efficiency and sustainable practices within the Metaverse.

Inclusivity and accessibility

- **Accessibility for All**: Ensure that AR and VR experiences are accessible to all employees, including those with disabilities.

- **Inclusive features**: Implement features that support inclusivity and provide necessary accommodations.

Safety protocols and hazard awareness

- **Comprehensive safety measures**: Develop comprehensive safety protocols for all roles within the Metaverse, especially for walk-around jobs.

- **Hazard awareness training**: Provide training to raise awareness of potential hazards and safe practices.

Regulatory compliance

- **Legal compliance assurance**: Stay informed about regulatory changes related to virtual work and factory operations.

- **Regulatory alignment**: Ensure that all Metaverse practices align with legal requirements and industry standards.

Cross-functional collaboration

- **Interdepartmental collaboration**: Promote collaboration between different departments, such as IT, operations, and safety, to ensure the effective integration of AR and VR technologies into workflows.

- **Innovative cross-functional culture**: Encourage a culture of cross-functional problem-solving and innovation.

Mental health and well-being support

- **Well-being resources**: Offer resources and support for employee mental health and well-being, particularly in virtual work environments where isolation may be a concern.

- **Virtual community building**: Provide a sense of community and social interaction within the Metaverse.

Quality assurance and diagnostics

- **Product quality assurance**: Implement quality assurance processes in virtual factory work and diagnostics to maintain product quality and safety standards.

- **Enhanced diagnostics with AR/VR**: Utilize AR and VR for remote diagnostics, expedited troubleshooting, and enhanced maintenance procedures.

In conclusion, as businesses embark on their journey into the Metaverse, these best business practices, encompassing both AR and VR, serve as the compass to navigate this transformative digital realm. By prioritizing comprehensive training, seamless integration of real and virtual worlds, robust data security, and ongoing feedback, organizations can thrive in the Metaverse while safeguarding their employees' well-being and privacy.

Environmental responsibility and inclusivity underscore a commitment to a sustainable and accessible digital future, while safety protocols and regulatory compliance ensure responsible operations. Cross-functional collaboration fosters innovation, and support for mental well-being reinforces a sense of community within the Metaverse.

Quality assurance and diagnostics, alongside a focus on continuous improvement, solidify the commitment to delivering excellence in virtual work environments. As businesses strive to excel in the Metaverse, these best practices provide a roadmap to success, allowing them to harness the full potential of AR and VR while shaping the future of work.

Through adherence to these guiding principles, organizations not only embrace the opportunities offered by the Metaverse but also contribute to its responsible and inclusive evolution, ensuring a brighter digital future for all.

Summary

In this chapter, we explored how AR and VR technologies are reshaping workplaces across various industries. These technologies facilitate immersive virtual office spaces, enhancing team collaboration with customizable avatars and spatial audio. Among other things, they revolutionize training and onboarding through realistic simulations and AI-driven virtual mentors, while also empowering customer support with remote product demonstrations and personalized assistance. Sales presentations benefit from 3D product demos and AI-driven analytics, while project management thrives on real-time updates and data-driven insights. Multilingual meetings become seamless with AR glasses and AI-driven translation algorithms. Virtual recruitment processes are streamlined with AI-driven matching algorithms, and AR and VR-based expos offer immersive product exploration and recommendations. These technologies find applications in employee wellness programs, mentorship, construction inspections, and emergency responder training, driven by innovations such as AI-driven analytics, real-time data feeds, and predictive algorithms. Challenges include addressing data privacy, hardware limitations, and training requirements.

Overall, AR and VR are driving innovation, transforming the way organizations operate within virtual environments.

Our next chapter, *Chapter 12, 3D and 2D Forms and Creation*, explores the vibrant landscape of the Metaverse through use cases that encompass 3D and 2D games, streaming entertainment, and art.

12
3D and 2D Content Forms and Creation

In this chapter, we will explore the multifaceted domains of gaming, streaming entertainment, and art within the Metaverse. As 3D games, immersive streaming content, and creative expression take center stage, we will examine the distinctions between AR and VR in their diverse applications. From the fusion of 3D and 2D content to the democratizing influence of generative AI, we will uncover how accessibility and contribution to this digital realm extend to everyone. Through a series of use cases and real-world examples, we will reveal best practices that enhance business efficiency and effectiveness, equipping you with valuable skills to navigate this dynamic virtual frontier.

In this chapter, we're going to cover the following main topics:

- How video games, streaming entertainment, and art will be featured in the Metaverse
- How 3D and 2D content forms and creation benefit the Metaverse and businesses within it
- How 3D and 2D content forms and creation in the Metaverse differ when using AR versus VR
- Best business practices that are exemplified in each use case in this chapter

Step into the Metaverse, a realm brimming with boundless possibilities and unparalleled creativity, where 3D games, streaming entertainment, and art converge to shape the future of digital experiences. In this chapter, we will embark on an extraordinary journey through the multifaceted dimensions of this virtual universe, exploring the distinct ways AR and VR gaming transform the landscape.

Gaming, a cornerstone of the Metaverse, serves as a testament to human ingenuity and technological advancement. Within this domain, we will delve into the profound differences that set AR and VR gaming apart, unveiling a rich tapestry of use cases that cater to diverse audiences and preferences. From immersive adventures to collaborative simulations, gaming in the Metaverse transcends traditional boundaries, offering a glimpse into the limitless potential of these technologies.

Streaming entertainment takes center stage as we navigate the ever-evolving landscape of digital content. Here, the fusion of 3D and 2D streams paves the way for a new era of entertainment. Witness the transformative power of generative AI, a groundbreaking technology that empowers individuals to craft engaging 2D (and soon, 3D) videos with remarkable ease, simply by inputting text prompts. The Metaverse becomes a global stage for storytellers and creators, where content knows no limits.

Art, in its most innovative and immersive form, finds a home in the Metaverse. Generative AI emerges as a game-changer, allowing anyone with a spark of creativity to sculpt breathtaking 3D art pieces. The democratization of art creation transcends expertise, inviting all to embark on a journey of digital expression and exploration.

Throughout our exploration, we won't merely dwell in the theoretical realm – we will breathe life into these concepts through a rich tapestry of real-world examples and immersive use cases. By revealing the best practices that enhance business efficiency and effectiveness within the Metaverse, we will equip you with invaluable skills to navigate this dynamic digital frontier.

With this chapter, you will not only possess a deep understanding of the various forms and creation methods of 3D and 2D content within the Metaverse but also acquire a profound appreciation for the limitless opportunities it offers. Welcome to the Metaverse, where creativity knows no bounds, and the future of digital expression unfolds before your eyes.

Gaming in the Metaverse

In the Metaverse's expansive gaming landscape, several compelling use cases emerge. Gamers become creators and modifiers, democratizing game development, with quality control as a challenge. Cross-platform gaming integration fosters an inclusive gaming community, while blockchain-backed virtual merchandise and collectibles introduce new opportunities with authenticity and copyright concerns. Virtual esports tournaments become global events, requiring stringent security measures. In-game advertising and product placement offer marketing potential, but striking a balance with player experience is vital. These use cases exemplify the diverse facets of gaming in the Metaverse, highlighting innovation and challenges in the pursuit of immersive digital gaming experiences.

Let's take a closer look at some use cases.

Use case 1 – game creation and modification

This use case exemplifies how the Metaverse empowers gamers to become active contributors to the gaming industry, shaping its future through their creativity and innovation. It highlights the democratization of game development and the dynamic synergy between technology, interactivity, and the challenges that come with it in this evolving digital realm.

The setup

Within the expansive and thriving Metaverse gaming landscape, a remarkable facet emerges where 3D and 2D virtual gamers are not just players but empowered creators and modifiers of games themselves. The Metaverse offers a vast canvas, brimming with opportunities for individuals and teams to craft unique gaming experiences that cater to a global audience.

Interactivity

In this immersive gaming domain, players transition into creators as they engage with innovative game creation and modification tools which include the use of generative AI. These tools empower users to design levels, characters, and gameplay mechanics, breathing life into their imaginative concepts. Collaborative platforms within the Metaverse foster teamwork, allowing multiple creators to combine their skills and ideas seamlessly.

Technical innovation

The Metaverse's technical innovation shines through in the form of user-friendly game development platforms that bridge the gap between novice creators and experienced developers. These platforms offer intuitive interfaces, drag-and-drop functionality, and pre-built assets, making game design accessible to a wide range of enthusiasts. AI-driven game design assistance provides suggestions and optimizations, reducing the learning curve for newcomers. And with generative AI, soon whole 3D, as well as 2D, games could be fully developed.

Challenges

While the Metaverse fuels creativity and democratizes game development, several challenges emerge on this vibrant frontier. Balancing the influx of user-generated content with quality control becomes pivotal. Moderation systems must ensure that games meet basic quality standards and are free from malicious or inappropriate content. Additionally, striking a harmonious balance between open creativity and maintaining fair play in modified games poses an ongoing challenge. Ensuring that user-created content doesn't disrupt the gaming experience for others is a priority. Continuous development and refinement of moderation and quality control mechanisms are essential to maintain a thriving and enjoyable gaming ecosystem within the Metaverse.

Use case 2 – cross-platform gaming integration

This use case illustrates how the Metaverse transcends the limitations of individual gaming platforms, fostering a more inclusive and interconnected gaming community. Cross-platform gaming integration enhances the social and competitive aspects of gaming, enabling players to unite in a shared virtual gaming universe. As the Metaverse continues to evolve, it reshapes the way we perceive and engage in gaming, offering a glimpse into the future of interactive entertainment.

The setup

Within the expansive Metaverse gaming landscape, cross-platform gaming integration becomes a prominent feature. This innovation allows players from various gaming platforms and devices to seamlessly interact and play together, breaking down traditional gaming silos.

Interactivity

In this interconnected Metaverse, players can engage in cross-platform gaming experiences with friends and gamers from around the world. Whether you're on a PC, console, VR headset, or mobile device, you can join the same virtual gaming universe. Gamers can form diverse teams and alliances, fostering a sense of community that transcends hardware preferences. This integration offers unprecedented opportunities for collaboration and competition.

Technical innovation

The technical innovation driving this use case is the development of cross-platform compatibility protocols and infrastructure. These innovations bridge the gaps between different gaming ecosystems, allowing for cross-device gameplay. Advanced matchmaking algorithms ensure that players of similar skill levels can enjoy fair and balanced gaming experiences, regardless of their chosen platform. This technical integration transforms the Metaverse into a truly inclusive gaming space.

Challenges

While cross-platform gaming integration is a remarkable achievement, it comes with its own set of challenges. Ensuring a level playing field for all players, regardless of their platform, requires ongoing fine-tuning of matchmaking algorithms. Addressing potential disparities in hardware capabilities, such as graphics processing power, can be complex. Additionally, maintaining a secure gaming environment across diverse platforms is essential to prevent cheating, unauthorized access, and other security concerns.

Use case 3 – game-related merchandise and collectibles

This use case showcases how the Metaverse transforms the concept of gaming merchandise and collectibles, offering a virtual marketplace where gamers can not only enhance their in-game experiences but also indulge in their passion for collecting virtual treasures. The integration of blockchain technology adds a layer of trust and scarcity to these digital possessions, creating a virtual economy that mirrors the real-world collectibles market.

The setup

Within the Metaverse, a vibrant and bustling marketplace dedicated to gaming-related merchandise and collectibles emerges. This dynamic digital marketplace transforms the concept of gaming memorabilia, offering a diverse range of 3D and 2D virtual goods that hold significant value for gamers and collectors alike. It's a virtual bazaar where gamers can immerse themselves in the culture of their favorite games beyond the confines of traditional gameplay.

Interactivity

In this immersive Metaverse marketplace, players gain the opportunity to personalize their avatars with a rich array of virtual gaming apparel and accessories. Gamers can browse an extensive catalog of virtual merchandise, including iconic character costumes, in-game items, and exclusive skins. This personalized customization allows players to showcase their gaming identity and immerse themselves even deeper into their favorite game worlds.

Technical innovation

At the heart of this use case lies the groundbreaking implementation of blockchain technology. This innovation plays a pivotal role in securing virtual collectibles, offering gamers a sense of rarity and ownership verification akin to physical collectibles. Each virtual item is tokenized on the blockchain, ensuring its uniqueness and provenance. Gamers can confidently buy, sell, and trade virtual merchandise, knowing that their digital possessions are genuine and scarce.

In terms of the companies that offer game-related merchandise and collectibles, generative AI provides an inexpensive, fast, and easy way to create assets.

Challenges

While this Metaverse marketplace promises exciting opportunities, it also presents unique challenges. Ensuring the authenticity of virtual merchandise is paramount. The presence of counterfeit or unauthorized virtual items could undermine the trust and value within the marketplace. Additionally, addressing potential copyright issues related to virtual merchandise is a central concern. Striking a balance between allowing creative expression and protecting intellectual property rights is essential to maintaining a thriving and ethical marketplace.

Use case 4 – virtual esports tournaments

This use case exemplifies how the Metaverse reimagines esports tournaments, offering an immersive and accessible platform for both participants and spectators. The integration of advanced technologies enhances the overall esports experience, making it a major attraction within the Metaverse. However, addressing challenges related to fair play, security, and scalability is essential to sustain the integrity and success of virtual esports tournaments.

The setup

Within the expansive Metaverse, esports tournaments evolve into major virtual events of global significance. This transformation redefines the landscape of competitive gaming, providing gamers with unparalleled opportunities to showcase their skills and engage with a vast audience. In this virtual arena, esports enthusiasts can immerse themselves in the thrill of competitive gaming like never before.

Interactivity

In this dynamic Metaverse esports ecosystem, gamers can participate in or spectate competitive esports matches, creating an electrifying atmosphere reminiscent of real-world sports events. Virtual esports arenas replicate the grandeur of traditional stadiums, and players can take on the roles of their favorite esports athletes. Whether you're an aspiring professional gamer or a passionate spectator, the Metaverse offers an inclusive space to experience the excitement of esports firsthand.

Technical innovation

The technical innovation at the core of this use case lies in the creation of advanced virtual tournament platforms. These platforms leverage cutting-edge technologies to deliver immersive esports experiences. High-quality graphics, seamless live streaming, and real-time data integration elevate the virtual esports experience to new heights. Gamers can engage in thrilling competitions, while spectators can enjoy broadcasts that rival the production values of traditional sports events.

Challenges

As esports tournaments within the Metaverse reach unprecedented levels of popularity, they also present unique logistical and security challenges. Ensuring fair competition in virtual environments is essential. The absence of physical oversight makes it crucial to implement robust anti-cheating measures and maintain the integrity of competitive gameplay. Additionally, managing large virtual audiences poses logistical challenges related to server capacity and user experience. Security measures to protect against potential cyber threats and data breaches become paramount to the success of virtual esports events.

Use case 5 – in-game advertising and product placement

This use case illustrates how the Metaverse transcends the limitations of individual gaming platforms, fostering a more inclusive and interconnected gaming community. Cross-platform gaming integration enhances the social and competitive aspects of gaming, enabling players to unite in a shared virtual gaming universe. As the Metaverse continues to evolve, it reshapes the way we perceive and engage in gaming, offering a glimpse into the future of interactive entertainment.

The setup

Within the Metaverse, advertisers and brands explore new opportunities for in-game advertising and product placement to reach a global audience.

Interactivity

Players encounter 3D, as well as 2D, virtual billboards, branded items, and interactive advertisements seamlessly integrated into the gaming experience. Brands collaborate with game developers to create immersive marketing campaigns within virtual worlds.

Technical innovation

Advanced ad targeting and analytics enable precise tracking of player engagement and ad performance. Brands leverage AR and VR technologies for interactive ad experiences, allowing players to interact with products in virtual space.

Challenges

Balancing the integration of advertising with the overall gaming experience while avoiding intrusive or disruptive ads is a design challenge. Ensuring that advertisements do not compromise gameplay fairness or introduce pay-to-win elements is a critical concern. Striking a balance between monetization and player experience is key to successful in-game advertising.

Negative implications of gaming in the Metaverse

Gaming in the Metaverse, while promising incredible innovation and immersive experiences, also carries negative implications that span technological, social, and ethical dimensions. These potential drawbacks must be considered alongside the benefits to ensure a balanced perspective on this digital frontier.

Technological implications

- **Dependency on technology**: As gaming in the Metaverse becomes increasingly sophisticated, there is a risk of individuals becoming overly dependent on technology for their entertainment and social interactions. This dependence may lead to issues related to screen time, addiction, and reduced physical activity.

- **Technical glitches**: The reliance on advanced technology for immersive gaming experiences introduces the possibility of technical glitches, server outages, or compatibility issues. These disruptions can frustrate players and disrupt their gaming experiences.

- **Privacy concerns**: The collection and utilization of user data within the Metaverse for targeted advertising and analytics can raise privacy concerns. Users may feel uncomfortable with the extent to which their online activities are monitored and analyzed.

Social implications

- **Social isolation**: Immersive gaming experiences in the Metaverse could lead to social isolation as individuals spend more time in virtual environments and less time in physical social interactions. Loneliness and a lack of real-world social skills can result from excessive immersion.

- **Economic disparities**: Access to the Metaverse and its premium gaming experiences may be limited by socioeconomic factors. Those with greater financial resources may enjoy a significant advantage, potentially creating digital divides and exclusivity.

- **Loss of physical interaction**: The allure of the Metaverse may lead to a reduction in face-to-face social interactions, which are crucial for human well-being. The diminished importance of real-world connections could have adverse effects on mental health and relationships.

Ethical implications

- **Exploitative monetization**: In-game purchases and microtransactions within the Metaverse can sometimes exploit players, particularly younger individuals who may not fully understand the financial implications. This raises ethical questions about the gaming industry's practices.

- **Digital addiction**: The highly immersive nature of gaming in the Metaverse may contribute to digital addiction, where individuals struggle to disengage from virtual experiences and prioritize them over real-world responsibilities.

- **Content regulation**: Balancing freedom of expression and maintaining a safe and inclusive gaming environment can be challenging. The Metaverse may struggle with regulating hate speech, inappropriate content, and cyberbullying.

Psychological implications

- **Escapism**: While gaming can be a form of entertainment, excessive escapism into the Metaverse may indicate underlying psychological issues or a desire to avoid real-world problems.

- **Impact on mental health**: Long hours spent in virtual gaming worlds may lead to mental health issues such as anxiety, depression, and a distorted sense of reality.

- **Cognitive overload**: The complexity of immersive gaming experiences within the Metaverse can lead to cognitive overload, especially in younger players, potentially impacting their academic performance and cognitive development.

Environmental implications

- **Energy consumption**: The infrastructure required to support the Metaverse's immersive experiences and multiplayer environments can consume significant amounts of energy, contributing to environmental concerns.

- **Electronic waste**: As technology evolves rapidly, older gaming equipment and hardware can quickly become obsolete, leading to electronic waste disposal challenges.

In conclusion, while gaming in the Metaverse offers exciting possibilities, it is essential to address these negative implications to ensure that the digital frontier is both enjoyable and responsible. Mitigating these challenges will be crucial in shaping a balanced and ethical future for gaming in the Metaverse.

Streaming entertainment in the Metaverse

In the captivating domain of streaming entertainment within the Metaverse, we will embark on a journey through immersive use cases that redefine our content consumption experience. Our first exploration, *immersive live concert experiences*, unveils the fusion of AR and VR technologies, revolutionizing how music artists connect with a global audience. These experiences transcend geographical boundaries, enabling fans to attend concerts virtually, engage with their favorite artists, and relish live performances from the comfort of their chosen surroundings. We have included technical innovations that power these virtual concerts while also shedding light on the challenges that are encountered in this dynamic landscape.

Use case 1 – immersive live concert experiences

In the dynamic Metaverse, immersive live concert experiences showcase the fusion of AR and VR technologies, redefining how music artists engage with their global audience. This use case explores how fans can virtually attend concerts, interact with artists, and enjoy live performances regardless of their location. We will explore the technical innovations that make these concerts possible and the challenges faced in delivering these immersive musical experiences. Join us as we enter the world of immersive live concert experiences in the Metaverse, where music transcends borders, and fans connect with artists like never before.

The setup

Within the Metaverse's vibrant entertainment landscape, artists and event organizers host immersive live concerts. These events are accessible to a global audience through AR-capable devices for AR experiences and VR headsets for VR experiences.

Interactivity

Fans attending these concerts via AR can bring the concert to their physical surroundings. They use AR glasses or smartphone apps to overlay virtual stages and holographic performers onto their environment. They can invite friends to virtual watch parties and enjoy synchronized live music experiences together. VR attendees, on the other hand, are fully immersed in virtual concert venues, where they can interact with other avatars, dance, and even have virtual meet-and-greets with the artists.

Technical innovation

Cutting-edge AR technology augments the real world with lifelike holograms of artists and dynamic virtual stages. VR technology transports users into meticulously crafted virtual concert venues with realistic visuals and spatial audio. Real-time streaming ensures that fans across the globe can enjoy live performances simultaneously, creating a sense of unity among the audience.

Challenges

Challenges include optimizing AR experiences for various AR-capable devices, ensuring server stability to accommodate large VR audiences during peak concert times, and addressing privacy concerns related to AR data collection. Additionally, preventing unauthorized streaming or recording of concerts and capturing the energy of live shows in a virtual environment are ongoing technical and creative challenges.

Use case 2 – AR sports enhancement

In AR sports enhancement, sports enthusiasts in the Metaverse leverage AR to enrich their live sports experiences. AR enhances both physical sports events and tabletop setups at home or sports bars, bridging the gap between fans and their favorite games. Whether attending live events or recreating matches at home, fans can access real-time statistics, player profiles, and interactive features through AR devices like smart glasses or smartphones. This innovation seamlessly integrates AR technology to create immersive sports experiences. However, it comes with challenges such as real-time synchronization, privacy concerns, and ensuring user-friendliness for all fans, regardless of their tech-savviness, to promote widespread AR adoption in sports within the Metaverse.

The setup

In the Metaverse, sports enthusiasts have the opportunity to integrate AR into their live sports experiences. This can occur either at physical sports events or within tabletop setups at home or sports bars. AR enhances the connection between fans and the games they love.

Interactivity

For fans attending live sports events, AR overlays real-time statistics, player profiles, and interactive elements onto their view. They can use AR-enabled devices such as smart glasses or smartphones to access additional information, such as player stats and instant replays, or even participate in interactive challenges. In tabletop AR sports viewing, fans can recreate entire sports matches on their tables, with digital players and elements enhancing the gameplay.

Technical innovation

AR technology is the key innovation, seamlessly blending digital and real-world elements. At live games, AR provides fans with an immersive layer of information, enhancing their understanding of the game. In tabletop scenarios, AR apps or devices project digital players and gaming elements onto the physical table, creating an engaging and interactive experience.

Challenges

While AR sports enhancement within the Metaverse is exciting, it does come with its set of challenges. Ensuring that AR experiences are synchronized with the live action in real time can be technically challenging. Privacy concerns related to the use of AR, especially in public venues, need to be addressed. Additionally, making AR sports enhancements accessible and user-friendly for a broad range of fans, including those less familiar with technology, is an ongoing challenge to ensure widespread adoption and enjoyment of AR in sports contexts within the Metaverse.

Use case 3 – Metaverse talk shows and podcasts

Metaverse talk shows and podcasts, hosts, and creators embrace the Metaverse's potential by establishing virtual studios with cutting-edge technology that mirrors their show's style. This shift revolutionizes audience engagement, allowing virtual attendees to participate as avatars, fostering a sense of community and co-creation. Some shows even offer virtual meet-and-greet sessions, deepening fan connections. Technical innovations ensure high-quality broadcasts, with spatial audio technology enhancing immersion. Challenges include managing audience participation respectfully, sustaining engagement in a distracting Metaverse, and addressing technical issues for a seamless experience.

The setup

In this use case, talk show hosts and podcasters embrace the potential of the Metaverse by establishing virtual studios and recording spaces within its expansive landscape. These virtual studios are equipped with cutting-edge technology and customizable settings to create unique and captivating environments that reflect the show's themes and style.

Interactivity

The Metaverse revolutionizes the way audiences engage with talk shows and podcasts. Virtual attendees can join live recordings as avatars, transcending geographical boundaries to become part of the studio audience. They have the opportunity to ask questions, provide feedback, and actively participate in discussions, fostering a sense of community and co-creation.

Some talk shows and podcasts take interactivity a step further by offering virtual meet-and-greet sessions with hosts and guests. This intimate engagement allows fans to connect with their favorite personalities on a personal level, deepening their connection with the content.

Technical innovation

The Metaverse empowers talk show hosts and podcasters with advanced streaming and recording equipment, which ensures high-quality broadcasts. Crystal-clear audio and high-definition video contribute to a professional and immersive viewer experience.

Spatial audio technology is a standout innovation in this use case. It enables a three-dimensional soundscape where voices and sounds emanate from specific locations within the virtual studio. This spatial audio creates a sense of presence, making listeners feel as though they are physically present in the studio, enhancing the immersion factor.

Challenges

While the Metaverse offers exciting opportunities for talk shows and podcasts, several challenges come to the forefront. Managing audience participation and ensuring a balanced and respectful exchange of ideas in a virtual setting can be complex. Hosts must navigate the nuances of moderating discussions and maintaining a welcoming atmosphere.

Sustaining audience engagement in the Metaverse, where distractions are abundant, requires innovative approaches. Hosts must continuously adapt to new trends and technologies to keep their content fresh and appealing.

Technical issues, such as connectivity problems or software glitches, are also considerations for hosts. Ensuring a seamless and glitch-free experience for both the virtual audience and in-studio participants is essential for the success of Metaverse talk shows and podcasts.

Use case 4 – interactive Metaverse storytelling

Within the Metaverse, a realm distinguished by its infinite creativity and ceaseless innovation, the art of storytelling undergoes a profound metamorphosis through the medium of live streaming. This use case delves into the realm of interactive Metaverse storytelling, where narratives cease to be static and solitary but instead become dynamic, responsive, and co-authored by the audience. Here, the conventional boundaries that separate storytellers from their audience dissolve as viewers step into the role of active participants, capable of shaping the course of the story's journey. These narratives are not set in stone but are malleable and adaptive, evolving in real time based on the collective decisions and choices made by the engaged audience. It's a collaborative odyssey through storytelling, where the final destination of the narrative emerges from the shared creativity and imagination of those who inhabit this expansive and limitless digital domain.

The setup

Within the Metaverse, content creators and master storytellers embark on a journey to craft live-streamed storytelling adventures, making them accessible to a global audience through Metaverse streaming platforms. These platforms serve as the dynamic stage where these narrative experiences come to life, offering an interactive canvas for storytellers to engage with their viewers.

Interactivity

What distinguishes this storytelling experience is the profound level of interactivity it offers. Viewers transcend the role of mere spectators, becoming integral participants in the unfolding narrative. During live broadcasts, viewers are invited to make decisions that directly influence the story's course. These decisions trigger real-time branching narratives, where each choice unlocks a unique path and outcome. Additionally, viewers can engage in polls and chat interactions, allowing them to voice their preferences, collaborate with fellow viewers, and collectively shape the story's direction. This heightened level of engagement fosters a sense of co-creation and community among viewers.

Technical innovation

At the core of interactive Metaverse storytelling lies streaming technology that seamlessly integrates narrative elements and viewer choices in real time. This innovative approach empowers storytellers to adapt to viewer decisions on the fly, ensuring the narrative remains fluid and responsive. The result is a highly personalized and interactive storytelling experience that blurs the lines between creator and audience, offering an unprecedented level of immersion and agency not seen in traditional storytelling.

Challenges

While the potential of interactive Metaverse storytelling is boundless, it does present its set of challenges. Crafting live narratives that can captivate a diverse audience with varying tastes and preferences demands a delicate balance of creativity and adaptability. Managing viewer choices and maintaining a coherent and engaging narrative during live broadcasts can be intricate, requiring meticulous planning and the ability to improvise. Storytellers must also contend with the unpredictability of viewer decisions, making each live broadcast a unique and dynamic experience. Nonetheless, these challenges are also opportunities for innovation, pushing the boundaries of storytelling and viewer engagement within the Metaverse.

Use case 5 – blended 3D and 2D streaming entertainment

In the dynamic Metaverse, entertainment events have evolved into mesmerizing spectacles that seamlessly blend 3D and 2D streaming technologies, creating immersive and unforgettable experiences for audiences worldwide. This use case presents blended 3D and 2D streaming entertainment, where traditional forms of live entertainment coexist harmoniously with cutting-edge virtual experiences.

The setup

Entertainment organizers and content creators within the Metaverse have harnessed the power of both AR and VR streaming to craft virtual venues that host these extraordinary blended streaming events. These virtual spaces are designed to be versatile, accommodating a wide range of entertainment forms, from live concerts and talk shows to gaming tournaments and interactive storytelling sessions.

Interactivity

What sets these events apart is the dynamic interplay between 3D and 2D elements facilitated by AR and VR technologies. During 2D segments, audiences can enjoy live performances, engage in panel discussions, or watch interviews on virtual stages or screens, replicating the experience of attending a traditional event using AR. As the event seamlessly transitions into 3D segments, attendees wearing MR or VR headsets find themselves immersed in a parallel universe of interactive wonders. Here, they can explore virtual theme park rides, participate in immersive movie screenings, or even embark on epic quests alongside their favorite gaming personalities.

Technical innovation

The heart of this transformative entertainment experience lies in state-of-the-art AR and VR streaming technology. This technology not only ensures the seamless integration of 3D and 2D content but also guarantees a coherent and engaging narrative throughout the event. Real-time interaction capabilities empower attendees to effortlessly switch between dimensions, creating a personalized and fluid entertainment journey.

Challenges

As organizers navigate this new frontier of blended AR and VR streaming entertainment, they face unique challenges. Coordinating smooth transitions between 3D and 2D segments to maintain audience engagement is paramount. Additionally, managing the technical aspects of blended streaming, such as optimizing bandwidth and ensuring consistent audiovisual quality in both AR and VR, presents its own set of considerations. However, it's these very challenges that drive innovation in this exciting realm of entertainment, where the Metaverse blurs the lines between physical and virtual experiences like never before.

Negative implications of streaming entertainment in the Metaverse

In the Metaverse, where streaming entertainment is becoming an integral part of digital experiences, it's essential to explore the potential negative implications that accompany this technological leap. Streaming entertainment offers a world of possibilities, from immersive gaming streams to captivating live events, but it also brings forth a set of challenges and concerns across various dimensions.

In this section, we will dissect the negative implications of streaming entertainment in the Metaverse, categorizing them into distinct domains for a comprehensive understanding. We will scrutinize the technical, social, ethical, economic, psychological, and legal aspects, shedding light on the complexities and risks that emerge in this digitally connected realm.

As the Metaverse continues to unfold, it's crucial to navigate its complexities with a keen awareness of both its promises and pitfalls, ensuring a balanced and informed journey through this ever-expanding virtual frontier.

Technical implications

- **Server overload**: High demand for streaming events can lead to server overloads, causing interruptions and frustration for viewers.
- **Technical glitches**: Technical issues such as lag, buffering, and audio/video desynchronization can disrupt the viewing experience.
- **Privacy concerns**: Streaming platforms may collect user data, raising privacy concerns regarding data security and usage.

Social implications

- **Isolation**: Excessive immersion in streaming entertainment could lead to social isolation, as users spend less time in the physical world.

- **Decreased physical activity**: Sedentary streaming activities may contribute to a decline in physical activity, impacting health.

- **Social comparison**: Constant exposure to curated online personas can lead to social comparison and feelings of inadequacy.

Ethical implications

- **Exploitative content**: Some streaming content may exploit vulnerable individuals or perpetuate harmful stereotypes.

- **Invasive advertising**: Aggressive advertising and product placement within streams can feel intrusive and manipulative.

- **Disinformation**: Misinformation and disinformation can spread quickly through live streaming, affecting public perceptions.

Economic implications

- **Monetization pressure**: Content creators may face pressure to monetize their streams, potentially sacrificing content quality for profit.

- **Market saturation**: An oversaturated market of streamers may lead to intense competition and challenges in building a sustainable audience.

Psychological implications

- **Addiction**: Excessive streaming can lead to addiction-like behaviors, negatively affecting mental health.

- **Depersonalization**: Constant online presence can lead to depersonalization and detachment from one's physical self.

- **Attention span**: Prolonged exposure to short-form content may affect users' attention spans and ability to focus.

Legal implications

- **Copyright infringement**: Streaming copyrighted content without proper authorization can lead to legal issues and penalties.

- **Regulatory challenges**: The Metaverse's global nature may pose regulatory challenges, with varying laws and standards.

In summary, as streaming entertainment takes center stage in the Metaverse, it brings with it a multitude of potential negative implications that span various facets of our digital lives. From technical hurdles such as server overloads and privacy concerns to social impacts such as isolation and social comparison, the Metaverse poses complex challenges. Ethical considerations encompass exploitative content and invasive advertising, while economic pressures may compromise content quality and market sustainability. The psychological toll includes addiction-like behaviors, depersonalization, and reduced attention spans, while legal issues such as copyright infringement and regulatory disparities loom large. Navigating these treacherous waters demands vigilance, responsibility, and a commitment to ethical and equitable digital experiences to ensure that the Metaverse remains a space of enrichment rather than harm.

Art in the Metaverse

In the boundless expanse of the Metaverse, where the digital realm intertwines with the world of art, creativity takes on an entirely new dimension. Within this dynamic digital landscape, artists, enthusiasts, and educators are reshaping the way we perceive, create, and interact with art. Through the fusion of cutting-edge technologies, including AR, VR, and generative AR, the Metaverse has become a canvas for innovation and imagination. In this section, we will embark on a captivating journey through five distinct use cases, each a testament to the transformative potential of art in this ever-evolving virtual frontier. From immersive virtual art galleries that transcend geographical boundaries to collaborative art creation that defies the constraints of physical space, these use cases not only redefine how we experience art but also push the boundaries of what's possible in the Metaverse.

Use case 1 – immersive virtual art galleries

In the Metaverse, artists and galleries embrace the digital realm by creating immersive virtual art spaces that revolutionize how art is explored, appreciated, and acquired. These virtual galleries seamlessly integrate 3D and 2D artworks, providing art enthusiasts with an unprecedented art experience.

The setup

Visionary artists and forward-thinking galleries construct meticulously detailed virtual art galleries within the Metaverse. These galleries are accessible through a variety of immersive technologies, including AR, VR, and MR headgear. This broad accessibility ensures that art lovers can engage with art using the devices they prefer, making art more inclusive and accessible.

Interactivity

Visitors to these virtual galleries are granted the freedom to navigate through digital exhibition halls, where they can view artworks from various angles, delve into the minutiae of each piece, and access comprehensive information about the artists and their creations. Some galleries offer guided tours, led either by advanced AI guides with deep art knowledge or real-life art experts who provide insight into the artworks' historical and cultural significance. Additionally, users have the unique opportunity to acquire digital or physical copies of the artworks directly from the virtual gallery, facilitating a seamless transition from digital exploration to tangible art ownership.

Technical innovation

The key to the transformative art experience within these virtual galleries lies in the cutting-edge AR, VR, and MR technologies that are employed. These technologies meticulously recreate the textures, colors, and dimensions of physical artworks, bringing them to life within the digital realm. Secure transactions, often backed by blockchain technology, assure users of the authenticity and provenance of the digital and physical art pieces, fostering trust and confidence in this evolving art ecosystem.

Challenges

While the possibilities in virtual art galleries are boundless, several challenges arise. One significant consideration is striking a balance between digital and physical art, ensuring that the immersive experience doesn't overshadow the traditional gallery ambiance but instead complements it. Copyright protection remains a paramount concern, with the need for robust digital rights management systems to safeguard artists' creations. Lastly, replicating the ambiance and aura of traditional galleries, which often play a crucial role in the art experience, poses an intriguing challenge in the digital realm. Nevertheless, these challenges propel the industry to innovate and redefine the boundaries of art presentation and appreciation.

Use case 2 – virtual art auctions

In a pioneering move, traditional art auctions seamlessly transition into the Metaverse, where galleries and auction houses host captivating virtual events featuring a diverse array of 3D and 2D artworks. This use case explores the transformative impact of virtual art auctions, redefining how art is showcased, appreciated, and acquired.

The setup

Forward-thinking galleries and established auction houses embrace the digital realm by establishing virtual venues within the Metaverse that are dedicated to hosting art auctions. These virtual auction spaces serve as platforms where art aficionados, collectors, and enthusiasts can converge to explore, bid on, and acquire artworks created by a wide range of artists. By transcending geographical constraints, these virtual auctions democratize access to art, fostering a global art community.

Interactivity

Attendees at virtual art auctions experience a dynamic and immersive environment. They can actively participate in real-time auctions, placing bids on coveted artworks that pique their interest. The integration of AR technology adds an extra layer of engagement, enabling users to visualize how a particular artwork might complement their living spaces. This feature enhances the decision-making process, making art acquisition a more personalized and informed experience. For attendees equipped with VR headsets, the auction experience transcends traditional boundaries, allowing them to immerse themselves in the auction as if they were physically present within a gallery, creating an unparalleled sense of presence.

Technical innovation

The heart of virtual art auctions lies in the innovative use of VR and AR technologies. VR delivers an immersive viewing and bidding experience, blurring the lines between the digital and physical worlds. Meanwhile, AR empowers users to envision artworks within their environments, fostering a deeper connection between art and collector. To ensure the security and authenticity of high-value art transactions, robust and trustworthy payment systems, often incorporating blockchain technology, are implemented.

Challenges

Virtual art auctions introduce unique challenges. One primary consideration is the security of high-value art transactions conducted within the digital realm. Ensuring the protection of both buyers and sellers against potential fraud and unauthorized access is paramount. Additionally, replicating the aura and ambiance of a physical auction, which can significantly influence bidding dynamics, poses an intriguing challenge in the digital landscape. However, these challenges underscore the need for innovation and adaptation within the art industry, pushing the boundaries of what is possible in the Metaverse.

Use case 3 – collaborative art creation

A groundbreaking paradigm shift unfolds within the Metaverse as artists hailing from diverse corners of the globe converge to embark on collaborative art projects that defy geographical constraints. This use case delves into the realm of collaborative art creation, where artists unite in virtual studios to paint, sketch, sculpt, and experiment with both 3D and 2D art forms, all facilitated by state-of-the-art AR, VR, and generative AI tools.

The setup

Within the vast expanse of the Metaverse, virtual studios equipped with cutting-edge AR, VR, and generative AI tools provide artists with an immersive playground for creative expression. These digital ateliers transcend the limitations of physical space and conventional mediums, enabling artists to sketch, paint, and sculpt together seamlessly. These studios serve as incubators for both traditional and digitally inspired art, bridging the gap between 3D and 2D art creation.

Interactivity

Collaboration takes on a new dimension as artists immerse themselves in real-time artistic symphonies. Through the magic of the Metaverse, they can view each other's work, offer instant feedback, and actively participate in co-creating unique and breathtaking art pieces. This process is not cloaked in secrecy; it's a shared experience that invites users and art enthusiasts to spectate and gain profound insights into the artistic journey. As they watch these sessions unfold, they witness the intricate dance of creativity, where individual visions blend harmoniously into collaborative masterpieces.

Technical innovation

The driving force behind collaborative art creation is the transformative power of VR and AR platforms. These technologies grant artists the ability to collaborate seamlessly, transcending the limitations of physical distance. Advanced tools within the Metaverse empower artists to create with unparalleled precision while fostering fluid communication among collaborators. Generative AI, a revolutionary addition to the toolkit, provides artists with a wellspring of inspiration and innovation. However, this innovation brings forth its own set of considerations, especially concerning copyright and ownership in the realm of AI-generated art.

Challenges

The realm of collaborative art creation within the Metaverse presents intriguing challenges. Coordinating the efforts of multiple artists in real time, particularly when integrating 3D and 2D elements, can be technically intricate. Achieving a seamless blend between different art forms and styles requires meticulous planning and execution. In the context of generative AI, the issue of copyrights and attribution becomes paramount, as artists navigate the fine line between creative collaboration and intellectual property rights. These challenges, while complex, drive innovation and underscore the Metaverse's potential to revolutionize the art world.

Use case 4 – AR street art

A captivating fusion of artistry and technology unfolds in the Metaverse as street artists harness the transformative potential of AR to elevate their physical creations. This use case delves into the world of AR street art, where urban landscapes become the canvas for interactive and immersive experiences, blurring the boundaries between the physical and the digital.

The setup

Within the dynamic Metaverse, visionary street artists embark on a creative journey that seamlessly melds the tangible with the virtual. Armed with AR tools, these artists add a layer of digital enchantment to their physical murals, breathing new life into public spaces. Users, equipped with AR glasses or AR-enabled smartphone apps, gain access to these augmented masterpieces as they wander through their urban landscapes. The Metaverse becomes an expansive gallery, with every city block offering a canvas for artistic expression.

Interactivity

AR street art unfolds like a hidden treasure trove waiting to be discovered. Viewers, when they engage with AR, are treated to a symphony of animated elements, concealed messages, and interactive games superimposed onto the physical mural. It's an art form that transcends the static constraints of traditional street art, inviting passersby to actively participate in the creative process. Beyond mere observation, users have the power to leave their mark by sharing comments and reactions, forging a dynamic connection between artists and their audience. It's a dialogue that blurs the boundaries of time and space, as viewers from around the world converge to experience the magic of urban artistry.

Technical innovation

At the heart of AR street art lies the groundbreaking AR technology, a bridge that effortlessly connects the real and the virtual. This transformative technology melds digital elements with the physical world, turning every street corner into an augmented canvas. Real-time updates further enhance the experience, granting artists the power to modify and evolve their virtual augmentations in response to shifting artistic visions or current events. The result is an ever-evolving urban art experience, where the cityscape serves as both canvas and stage for an ever-unfolding narrative.

Challenges

The realm of AR street art is not without its creative and technical challenges. Artists must embark on a journey of adaptation, mastering the tools and techniques required to seamlessly integrate AR into their artistic arsenal. Preserving the integrity of physical street art while infusing it with virtual components requires a delicate balancing act, as artists tread the fine line between creative expression and artistic preservation. The fusion of the traditional and the digital presents a unique set of challenges and opportunities, beckoning artists to explore new dimensions of creativity within the Metaverse.

Use case 5 – virtual art classes and workshops

The Metaverse opens a gateway to artistic exploration and learning, as artists and educators converge to offer immersive art classes and workshops. This use case dives deep into the realm of virtual art classes and workshops, where the fusion of 3D and 2D techniques enriches the learning experience and empowers aspiring artists.

The setup

Within the dynamic Metaverse, visionary art instructors transform traditional classrooms into virtual havens of creativity. These virtual studios and classrooms, accessible through a spectrum of AR, VR, or MR headgear, serve as the canvas for an array of art forms. From the strokes of a brush on canvas to the intricate digital sculpting of 3D masterpieces, these spaces cater to the diverse artistic appetites of learners.

Interactivity

In this digital realm of artistic education, students find themselves at the heart of an interactive and collaborative learning experience. They have the freedom to choose between attending live classes or accessing pre-recorded sessions, each offering unique advantages. Whether it's interacting with experienced instructors, seeking guidance from peers in real-time, or immersing themselves in hands-on creative activities, learners are spoilt for choice. Some classes even incorporate cutting-edge design tools, including the remarkable generative AI, which empowers students to conjure art within a shared virtual space. The Metaverse transforms from a passive learning environment into a vibrant arena where creativity knows no bounds.

Technical innovation

Virtual art classes and workshops in the Metaverse are underpinned by cutting-edge technology. AR and VR seamlessly facilitate real-time art instruction and collaboration, transcending the limitations of physical classrooms. Within these virtual spaces, 3D and 2D design tools reign supreme, unleashing a wave of creative potential. Students can sculpt intricate 3D models, paint vivid digital canvases, and experiment with a palette of digital hues. The Metaverse becomes a playground for innovation, where artistic expression finds new dimensions.

Challenges

As art educators embark on this digital journey, they encounter unique challenges that demand innovative solutions. Ensuring equal access to art materials for students from diverse backgrounds becomes a priority. Adapting teaching methodologies to cater to different learning styles within virtual environments presents an ongoing consideration. Striking a balance between traditional and digital art forms while harnessing the power of generative AI requires a delicate dance of creativity and technical proficiency. These challenges, however, are catalysts for innovation, propelling art education into a new era of accessibility and creative exploration.

Negative implications of art in the Metaverse

The emergence of art in the Metaverse represents an exciting fusion of creativity and technology, but it also brings forth a complex array of implications. As artists and audiences navigate this digital realm, they encounter technical challenges such as glitches and digital preservation concerns, while the potential exclusivity of high-end equipment raises social questions about accessibility and the loss of tangibility in art experiences. Ethical considerations revolve around copyright and the rise of AI-generated art, and economic pressures may push artists to compromise their integrity for profit. Additionally, the immersive nature of Metaverse art experiences raises psychological concerns related to digital addiction and detachment from reality, and the ease of copying and sharing digital art gives rise to legal issues, including copyright infringement and regulatory challenges.

Technical implications

- **Technical glitches**: Art in the Metaverse relies heavily on digital technologies, which can be prone to technical issues such as lag, glitches, or compatibility problems, potentially disrupting the art experience.

- **Digital preservation**: Ensuring the long-term preservation of digital art pieces poses challenges, as formats and platforms may evolve, making it difficult to access or appreciate historical digital artworks.

- **Access barriers**: Not all individuals have access to the necessary AR, VR, or MR equipment, limiting their ability to fully engage with Metaverse art experiences.

Social implications

- **Exclusivity**: The Metaverse art scene could become exclusive, with access limited to those who can afford high-end equipment, potentially excluding marginalized communities.

- **Loss of tangibility**: The shift to digital art may diminish the tangible and sensory aspects of traditional art experiences, impacting our connection to physical artworks.

- **Artificial influence**: The Metaverse may introduce AI-generated art that challenges the authenticity of human creativity, raising questions about the role of the artist.

Ethical implications

- **Copyright and ownership**: Determining ownership and copyright of digital artworks, especially those created collaboratively or with AI assistance, can be legally complex and may lead to disputes.

- **Exploitative practices**: The Metaverse may be vulnerable to exploitative art practices, including plagiarism, unauthorized distribution, or unethical use of AI-generated content.

- **Artificial aesthetics**: The influence of AI on art may homogenize artistic styles, potentially leading to a loss of diversity in artistic expression.

Economic implications

- **Monetization pressure**: Artists may face pressure to commercialize their work in the Metaverse, potentially compromising artistic integrity for profit.

- **Market oversaturation**: An influx of artists into the Metaverse could saturate the art market, making it challenging for individual artists to gain recognition or income.

Psychological implications

- **Digital addiction**: The immersive nature of Metaverse art experiences may contribute to digital addiction, potentially affecting mental health and well-being.

- **Detachment from reality**: Overindulgence in digital art experiences could lead to detachment from physical reality and social isolation.

Legal implications

- **Copyright infringement**: The ease of copying and sharing digital art in the Metaverse may lead to copyright infringement issues and legal disputes.
- **Regulatory challenges**: The global nature of the Metaverse may pose challenges for art-related regulations, as laws and standards may vary between regions.

In conclusion, the advent of art in the Metaverse is undeniably a captivating fusion of human creativity and cutting-edge technology, offering boundless possibilities. However, it simultaneously presents a complex tapestry of concerns that necessitate careful navigation. From the technical glitches that can disrupt the art experience to the social implications of potential exclusivity and the loss of tangibility in art, the Metaverse challenges us to address accessibility and authenticity. Ethical considerations, including copyright disputes and the impact of AI on art, underscore the need for a balanced approach. Economic pressures may tempt artists to prioritize profit over their artistic integrity, while the immersive nature of Metaverse art experiences can raise psychological concerns. Legal issues such as copyright infringement and regulatory variances add another layer of complexity. As we venture deeper into the Metaverse, addressing these negative implications becomes crucial, allowing us to build a prosperous digital art ecosystem where creativity thrives while respecting ethical, social, and psychological boundaries.

Best business practices

The Metaverse has ignited a revolution in the art world, providing a dynamic platform for artists and creators to push the boundaries of their craft. To navigate this digital frontier effectively, it is essential to establish and uphold best business practices that uphold authenticity, innovation, and accessibility within the Metaverse's art ecosystem. In this section, we will explore key business practices, including digital art authentication, copyright, transparency, digital rights management, monetization models, collaborative partnerships, education, sustainability, quality control, legal compliance, customer support, market research, and long-term vision. Collectively, these practices define the landscape of art in the Metaverse, ensuring that this new era of creativity remains vibrant and inclusive, catering to a diverse audience of art enthusiasts and creators.

Digital art authentication

- **Ensure the legitimacy of digital art in the Metaverse**: Implement blockchain technology or secure methods for verifying the authenticity of digital artworks.
- **Prevent counterfeiting and maintain the integrity of digital artworks**: Use unique digital signatures or certificates of authenticity to protect against forgery.

Copyright and licensing

- **Establish legal frameworks**: Clearly define ownership and copyright terms for digital artworks, including collaborative or AI-generated pieces.
- **Safeguard**: Develop licensing agreements to protect artists' rights and outline fair usage terms.

Transparency

- **Maintain transparent pricing and sales processes**: Build trust through openness and comprehensive information.

Digital rights management (DRM)

- **Safeguard art from piracy while balancing accessibility**: Implement effective DRM solutions to protect digital art from unauthorized distribution and reproduction.
- **Balance protection with accessibility**: Protect art without hindering legitimate use.

Diverse monetization models

- **Explore various revenue streams**: Diversify income sources for artists and creators.

Collaborative partnerships

- **Drive innovation**: Foster collaborations between artists, galleries, and tech companies.

Education and outreach

- **Promote knowledge**: Offer educational resources and workshops.

Accessible platforms

- **Reach a broader audience with inclusive technology**: Ensure compatibility with a wide range of AR, VR, and MR devices.

Community engagement

- **Belonging and collaboration**: Build and nurture a supportive and engaged community.

Sustainability

- **Promote eco-friendly practices**: Consider the environmental impact of Metaverse art creation and sales.
- **Encourage environmentally responsible choices**: Seek eco-friendly solutions and contribute to sustainable practices in the digital art space.

Quality control

- **Ensure consistent excellence in art**: Maintain high standards of quality in the creation, curation, and presentation of digital artworks.

Legal compliance

- **Up to date**: Stay up-to-date with evolving legal frameworks and comply with legal requirements and standards.
- **Uphold legal responsibilities**: Ensure full compliance with regulations and tax obligations.

Customer support

- **Be responsive**: Offer excellent service to users and customers.

Market research

- **Stay informed and adaptable**: Continuously monitor trends, market dynamics, and emerging technologies.
- **Evolve with the dynamic art market**: Adapt and innovate based on insights and changing market conditions.

Long-term vision

- **Develop a sustainable long-term strategy**: Plan for the future in a rapidly changing environment.
- **Foster innovation and adaptation**: Embrace change and growth in the art world.

In conclusion, embracing best business practices in the Metaverse art industry is vital for its growth and integrity. Implementing blockchain authentication and clear copyright frameworks ensures authenticity and artist rights. Transparent pricing, effective DRM solutions, and diverse monetization models foster trust, accessibility, and revenue. Collaboration, education, and community engagement drive innovation and inclusivity. Sustainability efforts and maintaining quality uphold ethical standards. Legal compliance and customer support are essential, as is market research for adaptability. A forward-looking vision that embraces change and growth in the art world ensures long-term success, collectively shaping a robust future for Metaverse art.

Summary

In the expansive realm of the Metaverse, our deep exploration of 3D and 2D content forms and their creation unveils a captivating tapestry of creative expression and technological prowess. This multifaceted journey immerses us in the dynamic world of virtual content, where the boundaries of reality blur, and creative possibilities abound. From the very inception of digital art, 2D content has been a steadfast beacon of expression, offering a canvas where artists, illustrators, and designers craft intricate narratives, breathtaking landscapes, and emotionally resonant visuals. The pixel's power remains undiminished as 2D artists weave timeless tales with elegance and precision, captivating audiences with the allure of visual storytelling.

Yet, it is in the dimensionality of 3D content that we find ourselves truly transported into a realm of limitless potential. Within this space, creators sculpt virtual landscapes, infuse life into lifelike characters, and construct immersive environments that defy the confines of the physical world. The Metaverse becomes a stage where storytelling transcends mere observation, enabling participants to actively engage with narratives, explore fantastical realms, and forge their own unique experiences. The collaborative spirit flourishes, with artists, developers, and visionaries uniting their talents to breathe life into the digital ether.

Throughout this chapter, innovation served as the guiding light that propels creators to push the boundaries of what's achievable. Emerging technologies, from AR and VR to blockchain and AI, become not just tools but conduits for boundless imagination. Experimentation becomes a cornerstone, with creators fearlessly venturing into uncharted territories, fusing the artistry of the past with the technology of the future. Adaptability remains paramount as the ever-evolving digital landscape demands continual evolution to remain relevant and engaging.

In this odyssey through the realms of digital artistry and innovation, we have witnessed the transformative power of creativity within the Metaverse. This exploration has celebrated the fusion of technology and art, igniting the imagination and inspiring artists and innovators to envision new horizons. As we traverse deeper into the Metaverse, we'll carry forward the lessons learned, embracing the limitless potential it offers as a canvas for boundless creativity and innovation. In a world where possibilities know no bounds, the Metaverse beckons as a realm of endless artistic exploration, where the convergence of 3D and 2D content forms continues to shape the future of digital expression.

Our next chapter, *Chapter 13*, *Retail Experiences*, explores the transformative power of AR and VR technologies in shaping the future of retail within the Metaverse. From clothes, footwear, and cosmetics shopping to food and furniture purchasing, the next chapter elaborates on how these technologies facilitate personalized and efficient retail experiences. Users can virtually try on products, view furniture in their own homes, and even interact with 3D images of food items when selecting a restaurant, all contributing to a seamless and enriched consumer journey.

13
Retail Experiences

Shopping using AR and VR in the Metaverse is one of its most anticipated features for consumers. The Metaverse can deliver on-demand online retail experiences from query to fulfillment, which includes getting personalized matches and unsolicited (but opted-in) recommendations based on data from AI digital assistants, digitally trying on goods such as clothes, footwear, and cosmetics, viewing furniture and home goods as they would appear in a room, and pulling up 3D images and videos of food dishes when choosing restaurants and also once seated in one. Much of this could be replicated when a person who is wearing glasses is outside, be it walking or driving around. Use cases that use AR and VR to exemplify particular Metaverse retail scenarios are included in this chapter.

In this chapter, we're going to cover the following main topics:

- How retail experiences benefit the Metaverse and businesses within it
- How retail experiences operate in the Metaverse
- How retail experiences in the Metaverse differ when using AR versus VR
- Best business practices that are exemplified in each use case provided

In the ever-evolving landscape of the Metaverse, one of the most highly anticipated developments is the transformation of online retail experiences. This evolution promises to revolutionize the way consumers shop, offering a holistic and immersive journey from the initial query to the final fulfillment of a purchase. At the heart of this metamorphosis are AI-powered digital assistants, which will guide users through their shopping adventures, providing personalized product recommendations that align with individual preferences and needs.

Imagine a world where shopping becomes a seamless and engaging experience, where you can virtually try on clothing, shoes, and cosmetics with the simple touch of a button. You'll have the power to visualize how furniture and home goods would fit perfectly into your living space, all within the confines of the Metaverse. When you're in a physical retail store, AR overlays that indicate store navigation and product information will be fully available. When deciding on a dining experience, whether it's ordering in or dining out, you'll be able to access vibrant 3D images and videos of restaurant dishes, helping you make the perfect choice.

But what makes this transformation truly groundbreaking is its integration into the physical world. For those equipped with AR glasses, the boundary between the digital and physical realms blurs. You can effortlessly incorporate these virtual shopping experiences into your outdoor activities, whether you're strolling through a park or driving around town. The Metaverse seamlessly extends beyond screens, becoming an integral part of your everyday life.

This chapter serves as your gateway into the immersive world of Metaverse retail. Here, we will offer you a glimpse into these innovative shopping scenarios, providing practical use cases that vividly illustrate their transformative potential. As you explore the chapters ahead, you'll gain insights into how these retail experiences enhance business efficiency and effectiveness, operating seamlessly within the Metaverse's ever-expanding digital ecosystem.

Clothes, footwear, and cosmetic shopping

In the Metaverse, innovative use cases are redefining the retail experience. Virtual fashion shows and pop-up stores bring exclusive fashion events and curated collections to users in an immersive blend of AR and VR technologies. Virtual shopping assistants offer AI-driven guidance, whether in virtual or physical stores, enhancing the shopping journey. Augmented reality mirrors revolutionize in-store try-ons, allowing users to virtually experiment with outfits and cosmetics. Cosmetic augmented testing labs empower users to create personalized beauty products through AR simulations. Fashion sustainability audits provide shoppers with AR-enabled insights into clothing brands' sustainability practices. These use cases showcase how the Metaverse is shaping the future of retail, offering convenience, personalization, and sustainability awareness.

Use case 1 – virtual fashion shows and pop-up stores

This use case reimagines the fashion industry within the Metaverse, where brands host exclusive virtual fashion shows and pop-up stores in AR and VR. It offers users the chance to immerse themselves in virtual fashion extravaganzas, explore exclusive collections, and make purchases in an environment that combines the best elements of both physical and digital retail experiences.

Here are the further details of this use case.

The setup

Within the Metaverse's dynamic fashion landscape, brands embrace a revolutionary concept: hosting virtual fashion shows and pop-up stores. These digital showcases are meticulously designed to replicate the charm and exclusivity of real-world fashion events.

Interactivity

Users have the opportunity to attend virtual fashion shows that rival their physical counterparts in terms of glamour and innovation. These immersive experiences transcend geographical boundaries, enabling users to witness the latest trends and limited-time collections from the comfort of their Metaverse-connected spaces.

Moreover, virtual pop-up stores beckon users to explore curated collections that are available for a limited time. The interactive nature of the Metaverse allows users to engage with clothing, accessories, and cosmetics in a virtual environment that mirrors the charm of traditional brick-and-mortar pop-up stores.

Technical innovation

The technical innovation driving this use case lies in the seamless integration of AR and VR technologies to create immersive fashion experiences. Users are transported into virtual venues where they can enjoy high-definition fashion shows and explore meticulously rendered virtual pop-up stores.

Fashion brands leverage 3D modeling and rendering techniques to ensure that every garment and accessory is represented with precision and attention to detail. Lighting, textures, and animations enhance the realism of these virtual spaces, offering users an unparalleled fashion encounter.

Challenges

While virtual fashion shows and pop-up stores hold immense potential within the Metaverse, they are not without challenges. Ensuring a seamless and visually stunning user experience requires ongoing investment in technological advancements. Brands must continually refine their 3D modeling and rendering techniques to maintain a high level of realism.

Use case 2 – virtual shopping assistants

This use case spotlights AI-driven virtual shopping assistants in the Metaverse, providing guidance and personalized recommendations. AI-driven VR shopping assistants guide users through virtual stores, while AI-driven AR shopping assistants assist users in physical stores. They help users find products, answer questions, and provide real-time assistance, enhancing the shopping experience.

Here are the further details of this use case.

The setup

In the immersive Metaverse shopping ecosystem, virtual shopping assistants take center stage. These AI-driven companions are designed to enhance both virtual and physical shopping journeys. Whether users are navigating virtual stores or exploring real-world retail spaces, they have access to these helpful digital assistants.

Interactivity

Users can interact with AI-driven VR shopping assistants when exploring virtual stores within the Metaverse. These virtual companions guide users through the digital aisles, helping them locate products, make informed decisions, and navigate the expansive virtual shopping landscape.

In physical stores, AI-driven AR shopping assistants come to the forefront. Equipped with AR technology, these assistants provide users with real-time information and guidance. They can identify products, answer questions about features and pricing, and offer personalized recommendations, creating a seamless shopping experience.

Technical innovation

The technical innovation powering this use case revolves around AI and AR/VR technologies. AI algorithms drive the intelligence of these virtual shopping assistants, allowing them to understand user preferences and offer tailored product suggestions. In the virtual realm, VR technology creates immersive environments for users to explore, while AR enhances real-world shopping experiences.

These assistants leverage real-time data and inventory information to provide accurate and up-to-date product recommendations. Whether users are in the Metaverse or physical stores, they can rely on these digital companions for assistance and guidance.

Challenges

While virtual shopping assistants promise to enhance the shopping experience, challenges persist. Ensuring the accuracy and effectiveness of AI-driven recommendations remains a critical consideration. Fine-tuning AI algorithms to provide personalized suggestions without compromising user privacy and data security is an ongoing endeavor.

In the physical realm, AR shopping assistants must seamlessly integrate with real-world store environments. Addressing technical challenges related to AR tracking and real-time data synchronization is crucial for a smooth user experience. Additionally, ensuring that users are comfortable, trusting of, and receptive to the presence of AR assistants in physical stores requires strategic design and user education efforts.

Use case 3 – augmented reality mirror

This use case explores the introduction of AR mirrors in physical stores within the Metaverse. These AR mirrors enable customers to virtually try on outfits and experiment with makeup without the need for physical fitting rooms. This innovative approach enhances the shopping experience and empowers customers with personalized styling options.

Here are the further details of this use case

The setup

In the evolving landscape of Metaverse-enhanced shopping, physical stores embrace AR mirrors as a transformative feature. These mirrors are strategically placed within the store environment, offering customers an exciting and interactive shopping experience.

Interactivity

Customers can engage with AR mirrors to explore virtual outfits and makeup options. By simply standing in front of these mirrors, they trigger the AR technology to superimpose clothing and cosmetics onto their reflection. This allows customers to see themselves in various styles, colors, and makeup looks in real time.

AR mirrors are equipped with customization features, enabling customers to mix and match clothing items and experiment with makeup products. This interactivity extends to virtual try-ons for different occasions, whether it's casual wear, formal attire, or themed makeup.

Technical innovation

The technical innovation driving this use case centers on the integration of AR technology with physical mirrors. These AR mirrors are equipped with cameras and sensors that capture the customer's image in real time. Advanced AR algorithms overlay virtual clothing and makeup onto the customer's reflection with remarkable accuracy and realism.

Additionally, the customization features are powered by AI-driven recommendations. Based on user preferences, previous purchases, and style choices, the AR mirror suggests clothing combinations and makeup looks that align with the customer's taste, providing a personalized shopping experience.

Challenges

While AR mirrors offer a compelling shopping experience, they come with unique challenges. Ensuring that the AR overlay accurately represents the fit, texture, and appearance of clothing items is essential. Brands must invest in high-quality 3D modeling and rendering to maintain a high level of realism in virtual try-ons.

Privacy considerations also come into play as the use of AR mirrors involves capturing customer images. Ensuring robust data protection measures and user consent mechanisms is crucial to address privacy concerns. Additionally, accommodating diverse body types and skin tones in virtual try-ons is an ongoing challenge that requires continuous refinement.

Use case 4 – cosmetic augmented testing labs

This use case introduces cosmetic augmented testing labs within the Metaverse, leveraging AR to simulate virtual laboratories. Users can experiment with mixing and matching cosmetic products to create personalized formulations, fostering creativity in cosmetics and empowering users to tailor their beauty products to their preferences.

Here are the further details of this use case.

The setup

In the Metaverse's ever-evolving beauty industry, cosmetic augmented testing labs emerge as a groundbreaking feature. These laboratories are designed to replicate real-world cosmetic formulation environments, offering users an immersive and interactive cosmetic creation experience.

Interactivity

Users can immerse themselves in these virtual laboratories, where they have the opportunity to experiment with various cosmetic products. Whether it's creating the perfect shade of lipstick, customizing skincare serums, or developing unique makeup palettes, the interactivity allows users to mix and match ingredients and formulations in real time.

AR technology enables users to visualize the effects of their cosmetic creations on themselves. This hands-on approach to cosmetic customization fosters creativity and empowers users to tailor beauty products to their unique preferences.

Technical innovation

The technical innovation driving this use case lies in the sophisticated AR technology used to create realistic virtual laboratories. Users interact with a range of cosmetic ingredients, textures, and colors, all rendered with precision and accuracy.

AI-driven algorithms offer guidance and suggestions, ensuring that users create safe and effective cosmetic formulations. Realistic simulations of skin types and tones enable users to see how their creations would look and feel on different individuals.

Challenges

While cosmetic augmented testing labs hold tremendous potential, they come with their own set of challenges. Ensuring that the VR environment accurately replicates the effects of real cosmetics on diverse skin types and tones is a continuous endeavor. Brands must invest in research and development to maintain a high level of realism.

Privacy and safety considerations are paramount as users interact with cosmetic ingredients. Ensuring that users have access to accurate information about ingredient safety and product usage is essential to prevent adverse reactions.

Use case 5 – fashion sustainability audits

This use case concerns fashion sustainability audits within the Metaverse, harnessing AR to provide shoppers with vital information on the sustainability practices of clothing brands. It includes details on materials, ethical sourcing, and carbon footprint, empowering consumers to make eco-conscious shopping decisions.

Here are the further details of this use case.

The setup

In the ever-evolving landscape of sustainable fashion, the Metaverse introduces fashion sustainability audits as a transformative feature. Shoppers can access augmented information while exploring clothing brands within the Metaverse, providing transparency and insights into the sustainability practices of various fashion labels.

Interactivity

Shoppers can engage with AR-enhanced displays and information overlays when exploring clothing brands. By scanning clothing items or brand labels with AR-enabled devices, users access a wealth of sustainability-related data. This interactivity allows them to learn about materials, ethical sourcing, and carbon footprint with a simple scan, fostering eco-conscious shopping habits.

AR overlays offer an immersive experience, presenting sustainability information in engaging and informative ways, ensuring that users have access to relevant data to inform their purchasing decisions.

Technical innovation

The technical innovation driving this use case revolves around AR technology and information integration. Brands and platforms collaborate to provide real-time sustainability data, which is seamlessly integrated into the Metaverse shopping experience.

AI-driven algorithms ensure that the information presented is up-to-date and relevant, reflecting the latest sustainability efforts of clothing brands. The AR interface offers an intuitive and user-friendly way for shoppers to access and digest this information, enhancing the overall shopping experience.

Challenges

While fashion sustainability audits offer valuable transparency, they come with challenges. Ensuring the accuracy and reliability of sustainability data is crucial as it directly impacts shoppers' purchasing decisions. Brands must commit to providing comprehensive and truthful information to maintain consumer trust.

Technical challenges related to data integration and real-time updates must be addressed to ensure that shoppers receive the latest sustainability insights. Additionally, user education plays a vital role in encouraging shoppers to engage with AR overlays and make informed decisions based on sustainability criteria.

Negative implications of clothes, footwear, and cosmetic shopping in the Metaverse

Clothes, footwear, and cosmetic shopping in the Metaverse have profound implications across various domains. These implications can be categorized into several key areas.

Technological implications

- **Data privacy**: As users engage with AR and VR technologies, there are concerns regarding data privacy. Personal data related to appearance and preferences could be captured, raising privacy issues.

- **Digital exhaustion**: The extended use of Metaverse technology for shopping may lead to digital exhaustion, with users experiencing fatigue and potential disconnection from the real world.

Social implications

- **Loss of physical interaction**: The convenience of virtual shopping may reduce physical interactions in traditional stores, impacting local businesses and the joy of in-person shopping.

- **Social isolation**: Over-reliance on virtual shopping could contribute to social isolation as people choose virtual experiences over physical outings.

Ethical implications

- **Overconsumption**: The ease of shopping in the Metaverse may encourage overconsumption, potentially contributing to unsustainable consumer behavior and environmental concerns.

- **Digital addiction**: The immersive nature of the Metaverse could lead to digital addiction, where individuals spend excessive time shopping in the virtual world, neglecting other aspects of life.

Psychological implications

- **Dependency on technology**: Excessive use of AR and VR for shopping may lead to a dependency on technology for everyday activities, potentially disconnecting individuals from the tangible world.

- **Impersonal shopping**: Despite attempts to personalize shopping, the Metaverse may still lack the personal touch and assistance provided by physical stores.

Environmental implications

- **Reduced sustainability**: Despite sustainability audits, the Metaverse may promote a fast fashion culture, potentially exacerbating sustainability issues.

- **Economic disruption**: A significant shift toward virtual shopping could disrupt traditional retail industries, potentially leading to job losses and economic challenges in the physical retail sector.

In conclusion, clothes, footwear, and cosmetic shopping in the Metaverse offer convenience and innovation, but they also bring along various challenges and implications that need to be carefully addressed for a balanced and responsible transition to this emerging shopping paradigm.

Restaurants and food shopping

In the dynamic world of the Metaverse, culinary enthusiasts and food shoppers are embarking on immersive and dynamic journeys that redefine their dining and shopping experiences. This collection of use cases explores how the Metaverse seamlessly integrates with the world of food, offering a diverse range of virtual culinary adventures, from grocery shopping in meticulously designed virtual stores to immersive cooking classes guided by expert instructors. These experiences leverage cutting-edge technology, including AI algorithms and 3D modeling (including the use of generative AI) and real-time integration with real-world systems, to create realistic and engaging encounters. As users navigate this culinary Metaverse, they are presented with a rich tapestry of opportunities and innovations that cater to their dining and food shopping needs, all within the confines of the digital space.

Use case 1 – virtual grocery stores

In this use case, grocery stores in the Metaverse offer users an immersive and convenient way to shop for everyday food items. This use case leverages technical innovation to provide realistic representations of products and personalized recommendations while addressing the challenges of accuracy and technical stability in a virtual shopping environment.

Here are the further details of this use case.

The setup

Within the Metaverse, virtual grocery stores are meticulously designed to replicate the layout and aesthetics of real-world supermarkets. These virtual stores feature a wide range of products, from pantry staples to fresh fruits and vegetables, ensuring that users have access to a comprehensive selection of items.

Interactivity

The interactive aspect of virtual grocery shopping in the Metaverse is a key attraction. Users can navigate through virtual aisles, mimicking the experience of walking through a physical supermarket. They can visually explore products, examine labels, and read detailed descriptions. When users find items they want to purchase, they can simply select them, adding them to their digital shopping cart with a virtual click or tap.

Technical innovation

The technical innovation in this use case revolves around providing users with realistic and engaging virtual grocery shopping experiences. To achieve this, advanced 3D modeling and rendering technologies are used to create lifelike representations of food products. Every item is meticulously designed to look and feel as close to its real-world counterpart as possible, down to the texture and color.

Challenges

While virtual grocery shopping in the Metaverse offers numerous benefits, it comes with its own set of challenges. Ensuring the accurate representation of food products is a top priority, as users rely on visual cues when making selections. Any discrepancies in the appearance or description of items can lead to confusion and mistrust.

Technical glitches and server issues can disrupt the shopping experience, causing frustration for users. Maintaining a seamless and glitch-free virtual grocery store environment requires ongoing maintenance and monitoring.

Additionally, ensuring that the virtual grocery store's product selection remains up-to-date and reflective of real-world inventory can be challenging. Coordinating with real-world grocery retailers to synchronize virtual and physical stock can be a logistical hurdle.

Use case 2 – recipe visualization and ingredient shopping

This use case leverages the Metaverse to offer users an immersive and personalized culinary journey. Through the integration of AI algorithms, users can explore recipes, visualize ingredients, and manage their shopping lists conveniently. However, addressing challenges related to ingredient accuracy, real-time availability, and data security is essential to provide a seamless and trustworthy culinary experience within the Metaverse.

Here are the further details of this use case.

The setup

Virtual recipe platforms within the Metaverse serve as culinary hubs, offering users access to an extensive collection of detailed recipes. These platforms provide not only the recipes themselves but also comprehensive ingredient lists and step-by-step cooking instructions. Users can explore a diverse range of recipes, from international cuisines to dietary-specific dishes, all within the virtual culinary world.

Interactivity

The Metaverse empowers users to interact virtually with recipe ingredients, taking culinary exploration to new heights. When users explore a recipe, they are immersed in a 3D environment where each ingredient is rendered with remarkable realism. Users can closely examine ingredients from various angles, allowing them to identify quality and freshness.

Moreover, interactivity extends to the convenience of ingredient management. As users browse through recipes, they can seamlessly add the required ingredients to a digital shopping list. This dynamic shopping list keeps track of quantities and ensures users never miss an item needed for their culinary adventures. Users can also customize their shopping lists based on dietary preferences or ingredient availability.

Technical innovation

The technical innovation behind this use case revolves around AI algorithms that enhance the recipe experience in several significant ways. These algorithms provide intelligent recommendations for ingredient substitutions, accommodating various dietary preferences such as dairy-free or gluten-free options. They also consider users' dietary choices when suggesting recipes, ensuring a personalized culinary journey. Additionally, the AI algorithms suggest complementary dishes, enhancing the overall dining experience. For example, if a user selects a pasta recipe, the system might recommend suitable appetizers, side dishes, or desserts.

Challenges

While this use case offers an immersive culinary experience, it also poses unique challenges. Accurately representing ingredients in the virtual environment is crucial, as users rely on visual cues for selection. Real-time ingredient availability, especially for seasonal or fluctuating items, is complex to manage. Additionally, protecting users' data, preferences, and culinary choices is essential for privacy and data security.

Use case 3 – culinary classes and ingredient shopping

This culinary use case immerses users in the art of cooking within the Metaverse. Attendees can participate in virtual cooking classes, shop for ingredients in real time, and have those ingredients delivered to their doorstep. Overcoming challenges related to ingredient availability, delivery logistics, and instructor expertise is essential to providing a seamless and enriching culinary education experience within the Metaverse.

Here are the further details of this use case.

The setup

The Metaverse plays host to a diverse array of culinary experiences, including interactive cooking classes and culinary workshops conducted by skilled chefs and instructors. These virtual gatherings bring together passionate food enthusiasts, aspiring home cooks, and culinary experts in a shared digital space.

The virtual cooking classrooms are designed to replicate the ambiance of real-world kitchens, complete with state-of-the-art appliances, utensils, and cookware. Users can choose from a wide range of culinary classes, covering everything from basic cooking techniques to specialized cuisines. Expert instructors guide participants through each step, sharing their culinary wisdom and techniques in real time.

Interactivity

Interactivity lies at the heart of this culinary use case within the Metaverse. Users can actively engage with culinary classes by following along with cooking demonstrations and practical cooking sessions. Whether it's learning to prepare a gourmet meal, bake artisanal bread, or master the art of sushi-making, users have the opportunity to acquire culinary skills from the comfort of their virtual kitchen.

One of the standout features of this use case is the seamless integration of ingredient shopping. While attending a cooking class, users can access virtual stores filled with 3D goods created with generative AI, specifically curated for the class they're participating in. These virtual stores offer all the necessary ingredients, allowing users to add items to their digital shopping cart as they progress through the class.

Technical innovation

The technical innovation within this use case centers on real-time ingredient sourcing and delivery options. As users select ingredients during the cooking class, the Metaverse leverages advanced technology to source these ingredients from virtual or real-world suppliers. Users can choose between virtual ingredients, which are digitally replicated for cooking practice, or opt for real-world ingredients for a more immersive experience.

Furthermore, ingredient delivery logistics are seamlessly integrated into the culinary classes. Users can select delivery preferences, including delivery time and location, ensuring that the ingredients arrive precisely when needed. This level of real-time coordination enhances the authenticity of the culinary learning experience.

Challenges

While this culinary use case offers an extraordinary opportunity for culinary exploration, it comes with its own set of challenges. Coordinating ingredient availability involves ensuring that the required ingredients are consistently accessible in both virtual and real-world stores. This logistical challenge demands a seamless experience where users can easily obtain the items needed for their culinary classes. Addressing these delivery logistics is essential. This encompasses coordinating timely and accurate deliveries of ingredients to users' physical locations. It involves considerations such as delivery preferences, fees, and the crucial need to maintain the freshness of perishable items.

In addition, the success of this use case relies heavily on the expertise of instructors. To provide high-quality virtual classes, it's vital to maintain a roster of skilled chefs and culinary instructors. Ensuring that these instructors possess not only culinary expertise but also the necessary technical skills for effective virtual instruction is paramount.

Use case 4 – restaurant exploration and reservations

This culinary use case within the Metaverse offers users a multifaceted dining experience, from exploring virtual restaurant environments to making reservations for various dining options. The seamless integration with real-world restaurant systems ensures accuracy and convenience. Overcoming challenges related to real-time synchronization and user expectations is essential for providing an exceptional dining adventure within the Metaverse.

Here are the further details of this use case.

The setup

Within the Metaverse, users can embark on a culinary journey that transcends the boundaries of the physical world. Virtual restaurant platforms provide users with the unique opportunity to step inside digital dining establishments created using generative AI. These virtual restaurants meticulously replicate the ambiance, decor, and atmosphere of their real-world counterparts. Users can explore restaurant interiors, soak in the ambiance, and interact with staff avatars who provide a warm and welcoming dining experience.

Interactivity

Interactivity is at the core of this culinary exploration use case. Users can seamlessly browse through virtual restaurant menus, each offering an array of tantalizing dishes and culinary delights. This immersive menu exploration goes beyond static text; users can view vivid 3D representations of dishes, complete with mouthwatering visuals that help them make informed dining choices.

Furthermore, users have the freedom to interact with the menu, placing orders for dine-in experiences or virtual dining adventures. Whether they're reserving a table for an intimate dinner date or ordering a virtual feast to enjoy in the comfort of their virtual abode, the Metaverse offers a diverse range of dining experiences to suit every palate.

Technical innovation

The technical innovation within this use case is centered on the seamless integration of virtual and real-world restaurant systems. The Metaverse has established robust connections with real-world restaurant booking systems, ensuring the accuracy of reservations made within the virtual environment. This integration offers users the convenience of booking a table at their favorite real-world restaurants directly from the Metaverse.

In addition to reservation accuracy, the Metaverse leverages advanced technology to provide users with real-time availability information. This means that users can check restaurant availability, review wait times, and select reservation slots that align with their dining preferences – all from within the Metaverse.

Challenges

While this culinary exploration use case offers a remarkable dining experience, it also presents its share of challenges.

Ensuring the real-time synchronization of restaurant availability between virtual and real-world systems is a complex task. Users expect that the availability displayed within the Metaverse aligns with the actual reservations made in real-world restaurants. Maintaining this synchronization requires ongoing technical coordination.

Additionally, the performance of staff avatars or generative AI chatbots, including responsiveness, professionalism, and user engagement, is critical to the success of this use case.

Use case 5 – food allergen and dietary information

In the dynamic world of the Metaverse, ensuring user safety and catering to diverse dietary needs is paramount. This use case focuses on providing users with an extensive database of information regarding food allergens and dietary restrictions within the Metaverse's virtual dining establishments.

Here are the further details of this use case.

The setup

Virtual food labels and product descriptions have transformed into comprehensive sources of information. These digital labels provide users with detailed insights into the ingredients and nutritional content of virtual food items. Moreover, they include thorough information about common food allergens, such as peanuts, gluten, dairy, and more. The Metaverse has also integrated real-time updates from food manufacturers, ensuring the accuracy of allergen and dietary data.

Interactivity

Users can now scan virtual food, as well as physical labels, using their AR or VR devices. This scanning capability triggers an instant display of allergen and dietary alerts. For instance, if a user has a peanut allergy, scanning a food item will immediately inform them if it contains peanuts or was produced in a facility that processes peanuts. Additionally, users can customize their dietary preferences within the Metaverse, enabling tailored alerts for their specific restrictions, whether they follow a vegan, gluten-free, or nut-free diet.

Technical innovation

The backbone of this use case is advanced AI-powered technology. AI algorithms analyze virtual food item compositions, cross-reference them with dietary preferences and allergen information, and generate instant alerts. These algorithms are not only capable of detecting explicit allergens but can also identify potential cross-contamination risks. Furthermore, they provide dietary recommendations based on health goals, ensuring a holistic approach to dining in the Metaverse.

Challenges

As with any ambitious endeavor, challenges are present. One of the primary concerns is the continuous accuracy of allergen information. Keeping virtual food databases up to date and synchronized with real-world ingredient changes and regulations requires ongoing effort and coordination with food manufacturers. Addressing potential cross-contamination risks also poses a challenge as these can vary across different virtual dining establishments. Ensuring that the AI algorithms provide reliable and context-aware alerts and recommendations is an ongoing process that involves refining machine learning models and constantly improving data sources.

Negative implications of restaurants and food shopping in the Metaverse

While the culinary Metaverse offers an enticing menu of experiences, it also presents a smorgasbord of technological, social, ethical, psychological, and environmental concerns. As we savor the possibilities, it is imperative to navigate these challenges with care, ensuring that our journey into this digital culinary realm is both delectable and responsible.

Technological implications

- **Realism and accuracy**: Ensuring lifelike representations in the virtual environment is crucial. Discrepancies can lead to inefficiencies and user frustration.
- **Technical stability**: The immersive nature of the Metaverse relies on glitch-free experiences. Disruptions can hinder user progress, demanding continuous maintenance.

Social implications

- **Disconnection from the physical world**: Overreliance on virtual food shopping may lead to reduced physical interactions.
- **Possible reduction in employment opportunities**: Automation and AI may reduce employment opportunities in traditional restaurants and grocery stores, impacting livelihoods.
- **Nutritional choices**: Overreliance on AI recommendations may influence users to make unhealthy food choices.
- **Privacy concerns**: Protecting users' data and preferences is essential for privacy and trust.

Ethical implications

- **Inequality**: Unequal access to advanced technology may create disparities in culinary experience.
- **Responsibility**: Handling user data and dietary information responsibly is a moral obligation.

Psychological implications

- **Addiction potential**: Excessive time in the culinary Metaverse may impact mental well-being.

- **Social isolation**: Overindulgence may lead to reduced social interactions in physical spaces.

Environmental implications

- **Increased resource use**: VR and AR technologies demand energy and resources, contributing to environmental concerns.

- **Delivery impact**: Frequent ingredient deliveries can lead to environmental consequences.

Furniture and home goods shopping

The Metaverse, with its fusion of AR and VR, has already started to revolutionize the way we shop for furniture and home goods. These use cases dive deep into a world where users can seamlessly navigate virtual showrooms, collaborate with AI interior designers, and envision their dream spaces. From personalized home styling assistance to interactive virtual furniture exploration, the Metaverse promises to provide an immersive and efficient shopping experience like never before. However, with this technological marvel come challenges of privacy, data security, and real-world integration. Join us on a journey through these use cases to discover the future of furnishing our homes in the Metaverse.

Use case 1 – personalized home styling assistance (mixed reality (MR))

In this use case, the Metaverse offers personalized home styling assistance through a seamless combination of AR and VR technologies. AI-driven interior designers become invaluable partners in helping users create cohesive and stylish living spaces.

Here are the further details of this use case.

The setup

Users enter mixed reality design studio apps within the Metaverse, creating a bridge between their physical homes and the digital realm. These studios serve as collaborative spaces where users interact with AI-powered interior designers. Once users are satisfied with their virtual designs, using MR, they overlay the proposed design elements onto their physical living spaces, ensuring a seamless transition between the virtual and real worlds. This mixed reality step enables users to assess the practicality of design changes and make adjustments as needed.

Interactivity

The interaction begins with users utilizing MR tools to scan their existing living spaces, capturing details of their current decor, furniture arrangements, and color schemes. This mixed reality component provides a comprehensive understanding of the user's starting point.

Technical innovation

The technical innovation lies in the seamless integration of AR and VR technologies within the MR experience. MR tools capture the user's physical environment while providing an immersive platform for design exploration. Advanced AI algorithms then analyze this MR-generated data, considering user style preferences and the unique characteristics of their living spaces. The AI interior designers leverage this information to generate initial style recommendations. Integration with eCommerce platforms allows users to seamlessly purchase recommended furniture and decor items directly from virtual stores, providing a practical and streamlined shopping experience.

Challenges

While this personalized home styling assistance offers immense value, it comes with a set of challenges. Accuracy is paramount in design recommendations, and fine-tuning AI algorithms to consistently deliver spot-on style suggestions is an ongoing endeavor. Ensuring the privacy of user data throughout the design process is essential, with a strong emphasis on data security and compliance. Moreover, the Metaverse must strike a delicate balance between the virtual and physical worlds, ensuring that the design concepts generated within the Metaverse can be feasibly realized in users' physical homes. This coordination between the digital and real realms presents an ongoing challenge, as the Metaverse aims to bridge the gap between the two seamlessly.

Use case 2 – virtual antique and collectibles auctions

In this use case, the Metaverse unfolds a realm where users can immerse themselves in the world of antique and collectibles auctions. Here, rare and unique home decor pieces take center stage.

Here are the further details of this use case.

The setup

Virtual auction houses within the Metaverse transform into hubs of excitement, hosting antique and collectibles auctions that feature a captivating array of rare and valuable items. As users enter these digital auction houses, they step into a virtual world steeped in history and craftsmanship.

Interactivity

The heartbeat of this use case lies in real-time interactivity. Users become active participants in these virtual auctions, allowing them to place bids on coveted items, engage in strategic bidding wars with fellow collectors, and experience the thrill of securing their desired pieces. The Metaverse enhances this experience by providing users with comprehensive item histories and provenance information, enriching the narratives behind each collectible.

Technical innovation

The technical innovation is multi-faceted. First and foremost, the Metaverse leverages cutting-edge technologies such as generative AI and Gaussian splats to create lifelike 3D representations of collectibles. These representations capture intricate details, textures, and aesthetics, ensuring that users can appreciate the beauty and craftsmanship of each item as if they were physically present. Moreover, secure virtual bidding systems are seamlessly integrated, guaranteeing the authenticity and transparency of the auction process. AI-driven valuations, based on historical data and market trends, empower users to make informed bidding decisions, enhancing their overall auction experience.

Challenges

While virtual antique and collectibles auctions offer an unparalleled experience, several challenges require ongoing attention. Ensuring the security of virtual transactions is paramount, as users engage in high-value transactions within the Metaverse. Addressing auction-related disputes and discrepancies is another challenge that demands swift and fair resolution mechanisms. Moreover, maintaining the integrity of item provenance data is essential to preserving the trust and authenticity of collectibles in this digital realm. These challenges serve as reminders that even within the immersive Metaverse, real-world concerns and safeguards remain critical.

Use case 3 – custom furniture design studios

This intricate use case explores custom furniture design studios in the Metaverse, offering users a creative journey to collaboratively design and personalize furniture pieces to suit their unique preferences and needs.

Here are the further details of this use case.

The setup

Virtual design studios located within the Metaverse serve as the canvas for users to engage in the art of designing and customizing furniture items. As users enter these virtual studios, they step into an environment brimming with endless possibilities for creativity and self-expression.

Interactivity

The heart of this use case beats with real-time interactivity. Users become active participants in design sessions where they can meticulously customize furniture dimensions, select materials, experiment with colors, and explore various styles. What sets this apart is the opportunity for users to collaborate with virtual furniture artisans who lend their expertise to bring bespoke creations to life. It's a seamless blend of user vision and skilled craftsmanship.

Technical innovation

The technical innovation in this scenario is multifaceted. First and foremost, advanced design tools and 3D modeling capabilities (including generative AI) empower users to translate their creative visions into tangible furniture designs. This process is not only visually immersive but also practically informative, as AI-driven cost estimations and material recommendations facilitate informed decisions throughout the design journey. Users can explore a wide spectrum of design options, fine-tune their creations, and see the impact of their choices in real time.

Challenges

While the custom furniture design studios offer a realm of creative exploration, several challenges are integral to this experience. Coordinating collaborative design sessions in real time can be complex as it involves ensuring seamless communication between users and virtual artisans. Additionally, guaranteeing that virtual representations accurately reflect the customized items is crucial to maintaining user trust and satisfaction. Beyond the design phase, addressing manufacturing and delivery logistics presents a significant challenge as the Metaverse transitions from creative ideation to real-world implementation. These challenges underscore the intricate dance between digital creativity and the practicalities of physical production and delivery.

Use case 4 – virtual home organization workshops with AR and VR Integration

Within the Metaverse, users have the opportunity to participate in virtual home organization workshops, harnessing the power of AR and VR to declutter and optimize their living spaces.

Here are the further details of this use case.

The setup

Virtual workshops that seamlessly blend AR and VR technologies are central to this use case within the Metaverse. Users step into these immersive digital spaces, where expert organizers, represented as realistic avatars, guide them through the intricacies of optimizing home organization strategies. The setup mirrors real-world workshops, offering users a platform to transform their living spaces with the aid of technology.

Interactivity

The heartbeat of this use case is interactivity, fueled by the fusion of AR and VR. Users actively engage in decluttering and organization tasks, implementing expert tips and recommendations in their physical spaces. AR overlays superimpose digital guidance onto users' real-world environments, providing step-by-step instructions for effective organization. VR enables users to explore alternative arrangements and layouts within their living spaces, facilitating the visualization of optimized organization solutions. Additionally, integration with virtual home organization stores within the

Metaverse allows users to browse and select suitable storage solutions, seamlessly transitioning from digital guidance to practical implementation.

Technical innovation

AR and VR technologies are at the forefront of technical innovation in this use case. Realistic 3D simulations, powered by VR, enable users to visualize their living spaces transformed into organized and clutter-free havens. AI-driven organization recommendations, tailored to users' specific needs and spaces, leverage both AR and VR to provide personalized guidance. AR superimposes labels and digital markers onto physical items, aiding in identification and organization. The integration of these technologies ensures that users not only receive expert advice but also have the tools to execute efficient home organization.

By combining AR and VR, this use case offers a comprehensive home organization experience within the Metaverse, from expert guidance to real-time implementation, ultimately leading to clutter-free and optimized living spaces.

Challenges

In the context of virtual home organization workshops that seamlessly incorporate AR and VR technologies, several challenges arise. One core challenge revolves around the necessity for realism and precision. Maintaining a high degree of accuracy and lifelike representations within the virtual environment proves to be crucial. Users depend on AR overlays and VR simulations to effectively declutter and optimize their living spaces. Any discrepancies or misrepresentations can lead to inefficiencies in the organization and user frustration. Achieving a seamless and realistic experience that aligns with real-world spaces remains an ongoing challenge.

Another vital aspect pertains to the technical stability of AR and VR integration. The immersive nature of these technologies is highly reliant on a glitch-free and stable experience. Technical disruptions or interruptions can disturb the immersive flow of home organization tasks and hinder user progress. Ensuring a smooth and uninterrupted experience in the virtual environment demands continuous monitoring and maintenance.

Use case 5 – AR and VR furniture restoration workshops

Within the Metaverse, AR and VR technology come together to offer users a transformative experience in the world of furniture restoration. These virtual workshops immerse participants in the art of refurbishing and reviving home decor pieces. This integrated AR and VR furniture restoration workshop in the Metaverse brings together the best of both technologies, allowing users to gain practical skills, restore their furniture with confidence, and collaborate with others in an immersive and educational environment.

Here are the further details of this use case.

The setup

Virtual workshops in the Metaverse are accessible through AR glasses and VR headsets. Users enter a collaborative digital workspace designed to replicate a real-world furniture restoration studio, complete with virtual tools and materials.

Interactivity

In virtual furniture restoration workshops, participants begin by using an MR headset, AR glasses, headgear, or a smartphone to scan their physical furniture items, creating virtual replicas in the Metaverse. Inside the VR restoration studio, users can choose a virtual piece of furniture to work on and employ VR controllers that mimic real restoration tools. AI-driven avatars, representing instructors, provide step-by-step tutorials on furniture restoration techniques tailored to the user's skill level. Users have the option to collaborate with others in real time, working together on their respective projects and seeking advice from instructors and peers. If they're using an MR headset, they can seamlessly switch to AR passthrough mode to view their real-world furniture items alongside their virtual counterparts, facilitating progress visualization and accuracy checks.

Technical innovation

The integration of AR and VR in this use case is groundbreaking. AR accurately captures the dimensions and details of physical furniture items, creating virtual replicas with high fidelity. VR provides a realistic and immersive environment where users can apply restoration techniques using virtual tools and materials.

AI-driven instructors adapt to each user's pace and skill level, offering guidance and feedback in real time. The simulation accurately replicates the physics and behavior of materials, ensuring a true-to-life restoration experience.

Challenges

Maintaining a high level of realism in virtual restoration, including lifelike textures, materials, and the physicality of tools, is an ongoing challenge. Additionally, coordinating the collaborative efforts of users across different locations and ensuring seamless communication with instructors and peers can be complex.

Negative implications of furniture and home goods shopping in the Metaverse

In the world of furniture and home goods shopping within the Metaverse, there's a tantalizing fusion of technological innovation and creative possibilities. This virtual realm offers users the opportunity to explore unique and imaginative designs, collaborate with AI-powered interior designers, and envision their dream living spaces like never before. From personalized home styling assistance to virtual antique auctions that transport users into the past, the Metaverse redefines convenience and engagement in the shopping experience. Custom furniture design studios allow for boundless creativity and collaboration with virtual artisans, while virtual home organization workshops harness the power of AR and VR to optimize living spaces. Additionally, AR and VR furniture restoration workshops provide users with hands-on skills in a captivating digital environment.

Technological implications

- **Design limitations**: The virtual environment may impose constraints on furniture designs due to business models optimizing profitability, limiting creativity and customizability.

- **Technical stability**: Its immersive nature relies on glitch-free experiences. Disruptions can hinder user progress, demanding continuous maintenance.

Social implications

- **Reduced social interaction**: Overindulgence in virtual shopping can contribute to reduced physical social interactions, potentially leading to loneliness.

- **Isolation from local businesses**: Increased reliance on virtual furniture shopping may result in decreased support for local brick-and-mortar furniture stores, impacting local economies.

- **Reduced interior design skills**: Reliance on AI and pre-made designs in the Metaverse may erode users' creativity and interior design skills, making them less capable of decorating their real-world homes.

- **Isolation from local artisans**: Virtual shopping may neglect local artisans and craftsmen, diminishing the appreciation of handcrafted, unique furniture.

Ethical implications

- **Consumer manipulation**: AI-driven recommendations may steer users toward specific brands or styles, potentially compromising genuine choice.

- **Inequality**: Unequal access to advanced technology may create disparities in furniture shopping experiences.

Psychological implications

- **Addiction potential**: Excessive time in the "furniture Metaverse" may impact mental well-being.

- **Decision fatigue**: The overwhelming array of choices in virtual furniture shopping may lead to decision fatigue, impacting users' cognitive well-being.

Environmental implications

- **Excessive packaging**: Virtual furniture purchases still require packaging and delivery, contributing to the environmental impact of packaging waste.

- **Lack of sustainable practices**: The Metaverse may not prioritize sustainable sourcing and manufacturing of furniture, contributing to environmental degradation.

Best business practices

As we journey deeper into the boundless possibilities of the Metaverse, the world of retail is undergoing a profound transformation. Metaverse retail is not merely an extension of traditional eCommerce; it's a dynamic fusion of AR, VR, and immersive technologies that promises unparalleled shopping experiences. In this section, we will explore the crucial landscape of best business practices for Metaverse retail, where businesses are crafting innovative strategies to engage with digitally empowered consumers. These practices are essential guideposts for navigating the ever-evolving landscape of Metaverse commerce.

AI-powered personalization

- **Utilize AI algorithms**: Utilize AI algorithms to personalize product recommendations.
- **Offer virtual shopping assistants**: Offer virtual shopping assistants with AI-driven recommendations and assistance.
- **Understand user preferences** and adapt product offerings accordingly.

Seamless real-world integration

- **Facilitate easy transitions**: Facilitate easy transitions from virtual shopping to real-world product purchases.
- **Partner with physical retailers**: Partner with physical retailers for seamless in-store pickup or delivery options.
- **Ensure consistency**: Ensure consistency between virtual and physical inventory.

Social shopping experiences

- **Foster social interactions**: Foster social interactions within the Metaverse retail space.
- **Implement features**: Implement features such as virtual shopping parties, where friends can shop together.
- **Enable users to share**: Enable users to share their virtual shopping experiences on social media.

Security and trust

- **Prioritize**: Prioritize data security and user privacy.
- **Assure users**: Assure users of the safety of their virtual transactions and personal information.
- **Establish trust**: Establish trust through transparent business practices.

Sustainability and eco-consciousness

- **Promote**: Promote sustainable products and ethical sourcing within the virtual retail space.
- **Highlight options and practices**: Highlight environmentally friendly options and practices.
- **Consider the impact**: Consider the environmental impact of Metaverse retail operations.

Virtual product launches and events

- **Host**: Host virtual product launches and exclusive events.
- **Create excitement**: Create excitement and anticipation for new product releases in the Metaverse.
- **Offer**: Offer limited-time promotions and discounts during events.

Customer support and assistance

- **Provide support**: Provide real-time customer support within the virtual retail environment.
- **Offer assistance**: Offer live chat and virtual assistants to assist users with inquiries.
- **Ensure resolution**: Ensure prompt issue resolution for a seamless shopping experience.

Ethical pricing and transparency

- **Maintain**: Maintain fair and transparent pricing practices.
- **Avoid**: Avoid price manipulation and hidden fees.
- **Clearly communicate**: Clearly communicate pricing, including taxes and shipping costs.

Inclusivity and accessibility

- **Ensure**: Ensure that virtual retail experiences are accessible to users with disabilities.
- **Prioritize**: Prioritize inclusive design and consider diverse user needs.
- **Offer options**: Offer hardware and software options for a wide range of users.

Legal and compliance considerations

- **Stay informed**: Stay informed about legal regulations specific to Metaverse retail.
- **Comply**: Comply with virtual commerce laws, taxation, and consumer protection regulations.
- **Establish**: Establish clear policies and procedures for legal compliance.

Virtual event return on investment (ROI) analysis

- **Assess**: Assess the ROI of virtual events and product launches.
- **Analyze**: Analyze user engagement sales conversion rates and customer retention.
- **Optimize**: Optimize future virtual event strategies based on data insights.

Emerging technology adoption

- **Stay at the forefront**: Stay at the forefront of emerging AR and VR technologies.
- **Consider cutting-edge features**: Consider the integration of cutting-edge features like haptic feedback and advanced visual effects.
- **Embrace innovation**: Embrace innovation to enhance the Metaverse retail experience.

Data-driven decision-making

- **Collect and analyze user data**: Collect and analyze user data to make informed business decisions.
- **Utilize data insights**: Utilize data insights to optimize product offerings and marketing strategies.
- **Implement data-driven personalization**: Implement data-driven personalization for a tailored shopping experience.

Global market expansion

- **Explore international markets**: Explore international markets and user demographics within the Metaverse.
- **Tailor virtual retail strategies**: Tailor virtual retail strategies to meet the preferences of diverse global audiences.
- **Address**: Address localization and language considerations for a global presence.

Collaborative product development

- **Collaborate**: Collaborate with users and virtual product creators.
- **Involve community**: Involve the Metaverse community in shaping product offerings.
- **Encourage content**: Encourage user-generated content and collaborative design.

Community engagement and support

- **Foster community**: Foster an active and engaged Metaverse retail community.
- **Offer**: Offer user forums, discussion boards, and community events.
- **Provide support**: Provide responsive customer support within virtual environments.

Ethical AI and algorithm use

- **Responsible and ethical use**: Use AI responsibly and ethically in virtual retail.
- **Avoid bias and discrimination**: Avoid bias and discrimination in AI-driven product recommendations.
- **Provide transparency**: Provide transparency in AI algorithms and decision-making.

User privacy education

- **Educate**: Educate users about data privacy and security in the Metaverse.
- **Provide resources and guidelines**: Provide resources and guidelines for user data protection.
- **Promote**: Promote user awareness and responsible data-sharing practices.

User empowerment and ownership

- **Empower users**: Empower users with ownership of virtual assets and purchases.
- **Ensure user control**: Ensure that users have control over their virtual possessions and transactions.
- **Promote user confidence**: Promote user confidence in the security and ownership of virtual products.

Adaptation to Metaverse evolution

- **Prepare for the future**: Prepare for the evolving landscape of the Metaverse.
- **Embrace emerging technologies**: Embrace emerging technologies and trends in virtual retail.
- **Continuously adapt and innovate**: Continuously adapt and innovate to stay competitive in the dynamic Metaverse environment.

In the dynamic world of Metaverse retail, these best business practices serve as essential guideposts for businesses as they navigate the fusion of AR, VR, and immersive technologies. These practices are instrumental in crafting innovative strategies that cater to digitally empowered consumers. Metaverse retail transcends conventional eCommerce, offering unprecedented shopping experiences. From AI-driven personalization to sustainability and ethics, from global market expansion to embracing emerging technologies, these practices illuminate the path forward in this ever-changing Metaverse environment.

Summary

In the vibrant world of retail experiences within the Metaverse, a captivating fusion of cutting-edge technology and boundless creative potential unfolds. This digital frontier offers users a diverse array of opportunities that transcend traditional boundaries. From personalized home styling assistance that seamlessly blends the virtual and physical worlds to virtual antique auctions that transport users through time, the Metaverse redefines convenience and engagement in the shopping experience. Custom furniture design studios empower individuals to unleash their creativity and collaborate with virtual artisans, while virtual home organization workshops harness the transformative power of AR and VR to optimize living spaces. Furthermore, AR and VR furniture restoration workshops provide users with hands-on skills in an immersive digital environment. As we embrace these innovative retail experiences, we must also navigate the challenges of privacy, data security, and sustainability, ensuring that our journey through the Metaverse enriches our lives while safeguarding our values and resources.

In an age defined by digital connectivity and immersive technologies, the next chapter, *Chapter 14, Benefits and Possible Dangers Reframed*, revisits and sharpens the multifaceted nature of possibilities and pitfalls within the Metaverse we have covered, where an array of benefits intersects with potential dangers, shaping our digital existence in profound ways. As we traversed vibrant experiences of social interaction, creative expression, virtual workspaces, entertainment, and futuristic shopping, we uncovered the remarkable advantages offered by this digital frontier. Yet, this exploration does not shy away from the shadows lurking beneath the surface – privacy concerns, the pull of technology addiction, and the temptation of laziness. With an unwavering commitment to understanding and optimizing our Metaverse experiences, we will explore the best business practices that illuminate the path toward maximizing the rewards and minimizing the risks.

Part 4:
Why Metaverse Redux?

In *Part 4*, we will explore the diverse benefits and potential concerns presented by the Metaverse. These advantages encompass various aspects of life within the Metaverse, including social interactions, networking, creativity, work, entertainment, and shopping. We will also consider its broader societal impacts. We will address potential pitfalls, such as privacy issues, technology dependency, and complacency.

This concluding chapter examines the evolving roles of AR and VR within the Metaverse. It envisions a future where 3D visuals, driven by AI-driven personalized assistants, become commonplace, facilitating instant knowledge acquisition. The Metaverse is poised to revolutionize work, altering urban landscapes as remote work becomes standard, and autonomous vehicles and advanced robotics seamlessly interface with this digital realm.

In essence, the Metaverse promises to transform our digital experiences, offering a dynamic landscape where information becomes knowledge, and work, entertainment, and daily life converge in innovative ways.

This part has the following chapters:

- *Chapter 14, Benefits and Possible Dangers Reframed*
- *Chapter 15, Future Vision*

14

Benefits and Possible Dangers Reframed

The Metaverse offers a multitude of advantages while also raising some notable concerns. Throughout this book, we've uncovered the diverse benefits it brings, encompassing social interactions, professional pursuits, entertainment, and innovative shopping experiences. However, this expansive digital realm isn't without its share of potential pitfalls. Issues related to privacy, the risk of technology addiction, and the temptation of idleness emerge as noteworthy considerations. Finding the right balance between the wealth of opportunities and these evolving challenges becomes paramount. Navigating this uncharted terrain is an exhilarating journey, and a comprehensive understanding of its potential and pitfalls is crucial for a well-rounded and enriching engagement.

In this chapter, we're going to cover the following main topics:

- The main benefits of the Metaverse
- The possible dangers of the Metaverse, focusing on privacy issues, technology addiction, and laziness
- The best business practices to accomplish the most benefit and halt possible danger in the Metaverse

In an age where the digital and physical realms converge, the Metaverse emerges as a compelling landscape, teeming with opportunities and uncertainties alike. Throughout our exploration in preceding chapters, we've ventured into this multifaceted domain, uncovering its vast array of advantages. From enriching social interactions that transcend borders to enabling innovative avenues for professional growth, entertainment, and immersive shopping experiences, the Metaverse offers a tantalizing tapestry of benefits. Yet, as we navigate this ever-evolving digital expanse, we must also grapple with the shadows it casts. The concerns surrounding privacy infringements, the allure of technology addiction, and the specter of idleness loom large, demanding our attention and vigilance. Striking the equilibrium between harnessing the potential and mitigating the risks becomes a pivotal challenge. Our journey into this uncharted terrain is nothing short of exhilarating, and it compels us to acquire a comprehensive understanding of its rewards and pitfalls. In this chapter, we will embark on a thorough exploration of the Metaverse's main benefits, scrutinize the possible dangers it poses, and illuminate the path toward maximizing the rewards while minimizing the risks. Through a discerning examination of these facets, we will equip ourselves with the knowledge needed for a well-rounded and enriching engagement with this digital frontier.

Metaverse of benefits

The Metaverse, a digital frontier, is redefining the rules of engagement for businesses across the globe. It tears down geographical barriers, giving rise to unprecedented global reach and market expansion. In this limitless digital realm, companies find not only boundless opportunities but also significant cost savings as overhead costs associated with physical locations fade into obscurity.

But it's not just about efficiency; the Metaverse ushers in a new era of collaboration and remote work. No longer confined by physical distances, teams can seamlessly interact and innovate, supporting diverse work preferences.

Perhaps most strikingly, the Metaverse revolutionizes customer engagement. Brands craft immersive experiences that forge profound emotional connections with their audience. Data-driven insights gleaned from this digital landscape empower businesses to adapt in real time, creating a personalized and captivating brand narrative.

Yet, it doesn't stop there. The Metaverse fuels innovative marketing and advertising strategies that nurture increased brand loyalty, while also transforming training and skill development into interactive and cost-effective endeavors. Moreover, the monetization of virtual assets opens doors to diverse income streams, further enhancing financial sustainability.

Early adoption and adept utilization of Metaverse technologies offer companies a competitive edge, enabling them to set trends and secure their position as influential industry leaders. As we journey through this transformative landscape, it becomes clear that the Metaverse is a dynamic force reshaping businesses and industries alike.

Also, the Metaverse is a digital realm that offers numerous societal advantages. It allows for diverse digital identities, fostering self-expression and creativity. Virtual communities provide a sense of belonging, positively impacting mental well-being.

Societal fusion in the Metaverse encourages cultural exchange and understanding, creating innovative norms. It also serves as a platform for meaningful societal critique and commentary, amplifying voices for change.

For artists and creators, the Metaverse offers a dynamic canvas for innovation. In education, it revolutionizes learning, making it more engaging and accessible.

Collaboration knows no geographical boundaries in the Metaverse, benefiting various fields. Inclusivity is a core principle, accommodating individuals with disabilities and promoting social inclusivity.

Digital economies thrive, enabling entrepreneurship and content creation, while environmental benefits include reduced commuting and sustainability initiatives.

In essence, the Metaverse is a transformative force, fostering inclusivity, innovation, and positive societal change.

Business benefits

The Metaverse represents a digital frontier ripe with opportunities for businesses across various industries. This immersive digital realm is reshaping how companies engage with customers, streamline operations, and foster innovation.

In this exploration, we will uncover the myriad advantages that the Metaverse offers to businesses. From global reach and efficiency gains to immersive branding and data-driven insights, it's a landscape where innovation knows no bounds.

Global reach and market expansion

In an era marked by the proliferation of digital technologies, the Metaverse stands as a beacon of unparalleled opportunity for businesses seeking to expand their horizons. One of its most striking attributes is the dissolution of traditional geographical constraints, ushering in an era of boundless possibilities for enterprises of all scales.

For small startups, the Metaverse functions as a transformative gateway, granting them access to a global customer base that was once a distant dream. This democratization of reach empowers fledgling businesses to compete on a level playing field with established counterparts. They can now showcase their unique offerings to a diverse and international audience, transcending the limitations of brick-and-mortar establishments and local markets. This newfound global presence not only amplifies brand visibility but also unlocks a myriad of growth prospects.

On the other end of the spectrum, multinational corporations find an unparalleled avenue for diversification. Beyond their traditional strongholds, these industry giants can seamlessly extend their influence into uncharted territories. This expansion isn't merely about geographic reach; it's about the profound transformation of customer demographics. By venturing into the Metaverse, corporations can engage with a customer base that defies conventional categorizations. The Metaverse is a melting pot of individuals from diverse cultural, social, and economic backgrounds. As a result, businesses can fine-tune their strategies to cater to the specific preferences, needs, and behaviors of this dynamic global audience.

Moreover, the Metaverse fosters innovation in market exploration. Traditional market research often involves significant investments of time and resources. In contrast, the Metaverse provides a unique testing ground. Businesses can experiment with new products, services, and marketing approaches in real time, gauging immediate responses and adapting swiftly. This agility enables them to make informed decisions about market expansion strategies, minimizing risks and maximizing returns.

In essence, the concept of global reach and market expansion within the Metaverse isn't merely about breaking geographical barriers. It's a paradigm shift that empowers businesses to redefine their relationships with customers and markets. It's about tapping into a vast, diverse, and ever-evolving digital landscape where the boundaries of growth are limited only by innovation and imagination.

Cost savings and efficiency

In the realm of business, the Metaverse unveils a profound transformation in the way enterprises operate, ushering in an era of unprecedented cost savings and efficiency. At the forefront of this transformation are virtual offices and storefronts that serve as digital counterparts to their physical counterparts. These virtual spaces revolutionize the traditional business model by significantly reducing the overhead costs associated with brick-and-mortar locations.

One of the most tangible benefits of the Metaverse is the financial relief it offers through reduced operational costs. Businesses, both large and small, find themselves unburdened by the heavy financial obligations tied to physical premises. Consider the substantial savings on rent, which is often one of the largest expenditures for businesses. By shifting operations into the Metaverse, companies can eliminate or dramatically reduce this recurring cost, redirecting those funds toward growth initiatives or enhancing the quality of their products and services.

Beyond rent, numerous other expenses diminish when businesses embrace the Metaverse. Utilities, maintenance, and property insurance are no longer part of the financial equation, further bolstering cost-effectiveness. Additionally, the need for employees to commute to physical offices is greatly diminished or eliminated. This not only saves employees valuable time but also reduces their transportation costs and contributes to a greener, more sustainable future.

The cumulative effect of these cost savings is the enhancement of profit margins, allowing businesses to allocate their budgets more strategically. With financial resources freed from the shackles of traditional overhead, enterprises can invest in areas that directly impact their competitiveness and growth. Whether it's intensifying research and development efforts, expanding marketing campaigns, or nurturing talent development, the Metaverse provides a financial boon that empowers businesses to thrive in a fast-paced and competitive landscape.

Furthermore, the efficiency afforded by the Metaverse extends beyond mere cost savings. Virtual offices and storefronts provide an agile platform for streamlined operations. They enable businesses to adapt quickly to changing market dynamics and customer preferences. This newfound agility is invaluable in today's rapidly evolving business environment, where the ability to pivot and innovate is often the key to success.

In summary, the concept of cost savings and efficiency within the Metaverse is a game-changer for businesses. It liberates them from the financial constraints of physical locations and empowers them to allocate resources strategically. Beyond the financial aspect, it fosters a culture of innovation and adaptability that positions businesses for long-term growth and competitiveness.

Enhanced collaboration and remote work

In the dynamic world of modern business, the Metaverse offers a transformative paradigm shift, particularly concerning enhanced collaboration and remote work. It breaks down the traditional barriers of geographical separation, opening new avenues for seamless teamwork and innovation.

One of the most compelling aspects of the Metaverse is its ability to transcend physical boundaries, enabling teams distributed across different regions to collaborate as if they were in the same room. Virtual workspaces serve as hubs where team members can convene, communicate, and collaborate in shared digital environments. This dissolution of spatial limitations facilitates uninterrupted collaboration, eliminating the constraints imposed by time zones and physical distances. Furthermore, the Metaverse offers the intriguing concept of digital identity, allowing individuals to seamlessly adopt multiple identities for different purposes. You can have an identity or character in your career, another one in politics, and yet another as an artist, all within the same immersive digital realm, fostering a multifaceted and dynamic approach to engagement and interaction.

The advantages of this enhanced collaboration are manifold. First and foremost, it cultivates a sense of interconnectedness among remote teams. Colleagues can gather in virtual meeting spaces, engaging in real-time discussions, brainstorming sessions, and collaborative projects. The immersive nature of the Metaverse enables more natural and intuitive communication, with user avatars representing team members and spatial audio replicating face-to-face conversations. This not only strengthens team bonds but also stimulates creativity and problem-solving.

The benefits extend beyond Improved teamwork. Businesses can access a global talent pool without the constraints of physical proximity. The Metaverse serves as a gateway to a vast and diverse workforce, enabling organizations to source top talents, regardless of their location. This globalization of talent acquisition not only ensures a broader range of skills but also fosters diversity and inclusivity in the workforce.

Moreover, the flexibility offered by virtual workspaces aligns with the evolving preferences of the contemporary workforce. Many professionals value the autonomy to work from locations of their choice, whether it's from the comfort of their homes or while traveling. The Metaverse accommodates these preferences by providing a platform where remote work is not only feasible but also highly productive. This flexibility, in turn, enhances job satisfaction, boosts employee retention, and elevates overall productivity.

In summary, enhanced collaboration and remote work in the Metaverse transcends traditional boundaries, facilitating seamless teamwork, access to global talent, and increased work flexibility. It positions businesses to thrive in an era where remote work is not just a necessity but a strategic advantage, and where the potential for innovation knows no geographical constraints.

Immersive branding and customer engagement

In an age where the digital and physical realms converge, the Metaverse emerges as a compelling landscape, teeming with opportunities and uncertainties alike. Throughout our exploration in preceding chapters, we've ventured into this multifaceted domain, uncovering its vast array of advantages. From enriching social interactions that transcend borders to enabling innovative avenues for professional growth, entertainment, and immersive shopping experiences, the Metaverse offers a tantalizing tapestry of benefits. Yet, as we navigate this ever-evolving digital expanse, we must also grapple with the shadows it casts. The concerns surrounding privacy infringements, the allure of technology addiction, and the specter of idleness loom large, demanding our attention and vigilance. Striking the equilibrium between harnessing the potential and mitigating the risks becomes a pivotal challenge. Our journey into this uncharted terrain is nothing short of exhilarating, and it compels us to acquire a comprehensive understanding of its rewards and pitfalls. In this chapter, we will embark on a thorough exploration of the Metaverse's main benefits, scrutinize the possible dangers it poses, and illuminate the path toward maximizing the rewards while minimizing the risks. Through a discerning examination of these facets, we will equip ourselves with the knowledge needed for a well-rounded and enriching engagement with this digital frontier.

In today's dynamic business landscape, the Metaverse presents an innovative avenue for companies to reimagine their customer engagement strategies. This segment explores how businesses are leveraging the Metaverse to create captivating and interactive experiences that leave a lasting impact on their audience.

The allure of the Metaverse lies in its ability to offer immersive brand encounters that surpass traditional marketing methods. Companies can host virtual events, construct interactive showcases, and design 3D environments that empower customers to engage with their brands in unprecedented ways. Whether it's a virtual product launch, a store unveiling, or an entertainment journey, the Metaverse provides the canvas for constructing immersive narratives that captivate and resonate with audiences.

These immersive interactions serve as powerful tools for reinforcing brand identity and nurturing emotional connections with customers. Unlike conventional advertising, which often relies on passive consumption, the Metaverse allows customers to actively participate in the brand story. They can explore virtual spaces, interact with products in a three-dimensional context, and even engage with AI-driven brand representatives.

The impact of immersive branding in the Metaverse extends to increased brand loyalty and advocacy. When customers are deeply engaged and emotionally connected to a brand, they are more likely to become devoted patrons and enthusiastic advocates. Positive experiences in virtual settings foster a sense of community around the brand, nurturing a loyal customer base that not only returns but also spreads the brand's message to others.

Additionally, the Metaverse offers a wealth of data and insights into customer behavior and preferences. Companies can harness this information to refine their branding strategies, creating tailored experiences that resonate with their target audience. This data-driven approach enhances the effectiveness of branding efforts and leads to improved returns on marketing investments.

In summary, the use of the Metaverse for immersive branding and customer engagement represents a paradigm shift in how businesses connect with their audience. It provides a dynamic platform for crafting immersive brand narratives, strengthening emotional connections, and cultivating brand loyalty and advocacy. As the Metaverse continues to evolve, it is poised to become an integral part of forward-thinking brand strategies, enhancing the relationship between businesses and their customers in innovative ways.

Data-driven insights

In the Metaverse, data takes on a whole new dimension. Metaverse platforms are treasure troves of information, capturing every nuance of user behavior, preferences, and interactions. This wealth of data opens up unprecedented opportunities for businesses to gain profound insights into customer trends and behavior.

The sheer volume and granularity of data available in the Metaverse is staggering. Businesses can track how users navigate virtual environments with eye tracking and gaze detection, to discover what products or services they engage with, how long they spend in immersive experiences, and even their emotional responses in certain scenarios. This level of detail provides a comprehensive view of customer interactions that was previously unattainable.

One of the key advantages of data-driven insights in the Metaverse is its real-time nature. Companies can access up-to-the-minute data on user activities, allowing them to adapt and respond swiftly to changing market dynamics. For instance, if a new trend emerges in user preferences, businesses can pivot their offerings or marketing strategies accordingly, ensuring they stay relevant and competitive.

These insights are not limited to customer behavior alone. They extend to product and service development. By analyzing user feedback and engagement patterns, businesses can refine their offerings to better align with customer desires. This iterative approach to product development leads to more customer-centric solutions and enhances overall product-market fit.

Marketing strategies also benefit immensely from data-driven insights in the Metaverse. Businesses can tailor their advertising efforts to specific user segments based on their virtual behaviors and preferences. This level of personalization increases the effectiveness of marketing campaigns, reducing ad spend wastage and improving ROI.

Furthermore, businesses can identify emerging trends and anticipate customer needs through predictive analytics. By analyzing historical data and identifying patterns, they can proactively address market shifts and customer demands. This forward-looking approach positions companies as industry leaders and innovators.

In essence, the Metaverse offers businesses not just data but the power of data-driven decision-making. It empowers companies to transform raw data into actionable insights, driving product innovation, enhancing marketing strategies, and ensuring they remain agile in a rapidly changing digital landscape. Embracing data-driven insights in the Metaverse is not merely a competitive advantage; it's a necessity for thriving in the modern business landscape.

Innovative marketing and advertising

In the Metaverse, the possibilities for marketing and advertising are as boundless as the digital landscapes themselves. This expansive digital realm serves as an inventive canvas for brands to unleash their creativity and engage users in unprecedented ways.

One of the standout features of marketing in the Metaverse is the ability to host immersive and interactive events. Brands can create virtual product launches that transcend the limitations of physical spaces. Imagine unveiling a new product in a fantastical virtual environment, complete with interactive elements, live demonstrations, and real-time Q&A sessions with experts. Such events not only captivate audiences but also generate a buzz that reverberates across social media platforms, amplifying brand reach and engagement.

Storytelling takes on a whole new dimension in the Metaverse. Brands can craft intricate narratives that users become a part of, immersing themselves in the brand's world. Whether it's a thrilling adventure or a heartwarming tale, these immersive storytelling experiences forge deeper connections between brands and consumers. Users don't just passively consume content; they actively participate in the narrative, creating a sense of belonging and emotional attachment.

Personalization is another hallmark of innovative marketing in the Metaverse. Brands can tailor experiences to individual preferences and behaviors. Users can explore virtual spaces that reflect their interests, receive personalized product recommendations, and even interact with AI-driven virtual brand ambassadors who understand their unique needs. This level of personalization not only enhances user engagement but also boosts conversion rates as users are presented with offerings that resonate with them.

The Metaverse also opens doors to gamified marketing strategies. Brands can create interactive games and challenges that not only entertain but also reward users with exclusive virtual items or discounts. These gamified experiences foster a sense of competition and achievement, driving user engagement and loyalty.

Moreover, the Metaverse encourages user-generated content. Brands can initiate challenges or campaigns that encourage users to create and share their virtual experiences featuring the brand's products. This user-generated content serves as an authentic endorsement, amplifying brand advocacy, and trust among peer groups.

In summary, innovative marketing and advertising in the Metaverse transcend the boundaries of traditional advertising. Brands have the opportunity to captivate audiences with immersive events, intricate storytelling, personalization, gamification, and user-generated content. The result is not just increased brand awareness but a profound and lasting connection with users in this dynamic digital landscape.

Efficient training and skill development

Efficient training and skill development in the Metaverse represents a transformative leap in corporate education and professional growth. Virtual training and onboarding programs redefine how employees acquire knowledge and hone their skills, offering a host of advantages over traditional training methods.

One of the defining features of Metaverse-based training is its engagement factor. Traditional training often involves lengthy, text-heavy manuals or passive video presentations, which can lead to disengagement and limited knowledge retention. In contrast, the Metaverse leverages immersive simulations and interactive scenarios that actively involve employees. These simulations recreate real-world situations, allowing employees to apply their knowledge in a practical context. Whether it's a medical practitioner performing virtual surgeries or a technician troubleshooting complex machinery, these experiences provide hands-on learning that is engaging and memorable.

The effectiveness of Metaverse training goes beyond engagement; it extends to accelerated learning. Employees can access training modules at their convenience, eliminating the need for rigid schedules. Moreover, the iterative nature of virtual simulations allows employees to practice until they master a skill or procedure. This accelerated learning not only reduces the time required for skill acquisition but also ensures a higher level of proficiency.

Cost efficiency is another compelling aspect of Metaverse-based training. Traditional training often entails expenses related to physical venues, travel, printed materials, and instructors' fees. In the Metaverse, these overhead costs are significantly reduced or even eliminated. Training modules can be accessed remotely, saving both time and resources. Additionally, the scalability of virtual training means that the same program can be deployed to a large number of employees without incurring additional costs.

Furthermore, Metaverse training fosters a culture of continuous learning. Employees can access training modules whenever they wish to upskill or cross-skill, promoting ongoing professional development. This adaptability is particularly valuable in industries where rapid skill adaptation is essential, such as technology or healthcare.

The Metaverse also excels in tracking and assessing employee progress. Detailed analytics capture every interaction, enabling organizations to gauge individual and group performance accurately. This data-driven approach empowers organizations to identify areas where employees may require additional support and tailor training programs accordingly.

In summary, efficient training and skill development in the Metaverse offers an engaging, accelerated, cost-effective, and flexible approach to corporate education. By leveraging immersive simulations, flexible access, and data-driven insights, organizations can equip their workforce with the skills and knowledge needed to excel in an ever-evolving business landscape.

Monetization of virtual assets

The Metaverse represents not only a creative playground but also a fertile ground for innovative monetization strategies. Businesses can expand their revenue streams by venturing beyond their traditional offerings and exploring a plethora of virtual asset opportunities. These diverse income sources not only enhance financial sustainability but also contribute to a dynamic and thriving digital economy.

One lucrative avenue within the Metaverse is the sale of virtual real estate. Virtual lands and properties within digital worlds hold intrinsic value, often driven by their location, accessibility, and potential for development. Businesses can acquire virtual real estate within popular Metaverse platforms and then lease or sell these digital properties to other users or corporations. The demand for prime virtual real estate has recently tanked; we believe that some form of virtual real estate will re-emerge in the Metaverse.

Branded merchandise is another compelling monetization avenue. Within the Metaverse, businesses can design and sell virtual merchandise that aligns with their brand identity. These digital products, whether they be clothing, accessories, or even digital collectibles, enable users to express their affinity for a brand within the virtual world. Limited-edition and exclusive virtual merchandise can become highly sought-after, creating a sense of scarcity and driving sales. In some cases, users may even convert their virtual purchases into physical products, further solidifying the bridge between the digital and physical realms.

Exclusive digital experiences represent a premium monetization opportunity. Businesses can offer access to immersive, members-only events or virtual spaces within the Metaverse. These experiences can range from exclusive concerts and art exhibitions to private meetings and workshops. By creating a sense of exclusivity and value, businesses can attract a dedicated user base willing to pay for these unique encounters. These experiences not only generate direct revenue but also enhance brand prestige and customer loyalty.

Moreover, the Metaverse fosters a thriving creator economy. Businesses can collaborate with content creators, artists, and influencers to develop branded virtual assets. These digital creations can include custom avatars, branded skins, or virtual pets that users can purchase and integrate into their Metaverse experiences. Such collaborations leverage the creativity and reach of content creators, extending a brand's presence and revenue potential.

In the realm of virtual asset monetization, the Metaverse's decentralized nature plays a pivotal role. Blockchain technology underpins many virtual asset ecosystems, ensuring transparency, provenance, and ownership. **Non-fungible tokens** (**NFTs**) have become a cornerstone of virtual asset ownership, enabling businesses to create unique and verifiable digital assets. This technology not only secures the authenticity of virtual assets but also opens up secondary markets where users can trade, sell, and collect these digital treasures.

In conclusion, the monetization of virtual assets within the Metaverse represents a multifaceted opportunity for businesses to diversify their income streams and foster financial sustainability. From virtual real estate and branded merchandise to exclusive digital experiences and collaboration with content creators, these avenues empower businesses to thrive in the dynamic and ever-evolving digital economy. The Metaverse, with its innovative and decentralized infrastructure, presents a fertile ground for pioneering monetization strategies that redefine how businesses engage with their audience and generate revenue.

Competitive edge and industry leadership

In the Metaverse, early adoption and adept utilization of its technologies hold the key to a significant competitive edge. Businesses that proactively establish a robust presence within the Metaverse find themselves not merely as participants but as industry trailblazers, shaping trends, and redefining customer expectations. This leadership role isn't just a momentary advantage; it has the potential to forge long-term market dominance.

The Metaverse isn't merely an extension of current digital landscapes; it represents a paradigm shift in how businesses engage with their audience and operate in the digital realm. By venturing into the Metaverse and embracing its possibilities, businesses can cultivate an environment of innovation, expansion, and heightened competitiveness.

One of the primary drivers of competitive advantage within the Metaverse is the ability to set trends. As businesses pioneer novel and immersive experiences, they set the benchmark for what customers expect in the digital sphere. By being at the forefront of innovation, these industry leaders shape the Metaverse's landscape, molding it into a space where their unique vision becomes the industry standard. This proactive approach allows businesses to not merely follow trends but to define them, granting them lasting influence over the evolving Metaverse.

Moreover, early adoption of Metaverse technologies provides businesses with a unique opportunity to build deep and meaningful connections with their customers. By offering engaging and interactive experiences, companies foster brand loyalty and cultivate a dedicated user base. This connection extends beyond transactions; it becomes a part of customers' digital lives. As a result, businesses that lead the way in the Metaverse can secure a loyal customer base that actively participates in their Metaverse endeavors.

Beyond setting trends and cultivating customer loyalty, industry leadership in the Metaverse has the potential to translate into substantial market share. As businesses establish themselves as pioneers in this digital frontier, they naturally attract a growing audience that's eager to explore and engage with their offerings. This expanded customer base can lead to increased revenue streams, further solidifying their market presence.

In essence, the Metaverse isn't just a technological advancement; it's a transformative landscape where businesses can expand their horizons, optimize their operations, and create profound and enduring connections with customers. The benefits highlighted here underscore the vast potential for growth, innovation, and enhanced competitiveness in the ever-evolving business terrain of the Metaverse.

The Metaverse offers businesses a world of unparalleled opportunities for growth, innovation, and engagement. As we journeyed through the various facets of this digital frontier, it has become evident that it transcends traditional boundaries, offering a limitless canvas for businesses to explore. From expanding global reach to redefining customer engagement, embracing data-driven insights to pioneering innovative marketing, the Metaverse holds the potential to revolutionize every aspect of the modern business landscape. Furthermore, the efficiencies it brings to training and skill development, along with the diverse monetization avenues it offers, underscore its role as a catalyst for financial sustainability and dynamic growth. Early adoption and adept utilization of Metaverse technologies position businesses as industry leaders, setting trends, and influencing customer expectations.

As the Metaverse continues to evolve, it will undoubtedly play an integral role in the ever-changing business landscape, offering opportunities that are both exciting and transformative.

Societal benefits

The Metaverse offers a multitude of societal advantages that transcend the boundaries of the physical world. Within this dynamic landscape, individuals find the freedom to craft and explore their digital identities, fostering self-expression and creativity while championing inclusivity and diverse cultural expressions. Virtual communities bridge geographical divides, providing a sense of belonging and emotional support, while societal fusion encourages cultural exchange and innovative norms. Additionally, the Metaverse serves as a platform for meaningful societal critique, amplifying voices for change and driving progress. It fuels digital artistry, revolutionizes education, and enables global collaboration, all while facilitating entrepreneurship and contributing to sustainability efforts by reducing commuting. In essence, the Metaverse is a transformative force that not only enriches individual lives but also fosters a more inclusive, innovative, and positively impactful digital society.

Moreover, these societal benefits within the Metaverse create a fertile ground for businesses to thrive. Embracing inclusivity, fostering creativity, and championing cultural exchange is not only socially responsible but also strategically sound. Companies that leverage the inclusive and collaborative nature of the Metaverse can tap into a global talent pool, enhance their brand reputation, and engage customers in more meaningful ways, ultimately translating these societal advantages into competitive advantages and business success.

Digital identity and expression

In the Metaverse, individuals have the freedom to craft and explore their digital identities. This allows people to express themselves in ways that may not be possible or comfortable in the physical world. Users can experiment with different personas, styles, and avatars, fostering a sense of empowerment and self-discovery. Moreover, the acceptance of diverse digital identities promotes a more inclusive and open-minded digital society.

Beyond personal expression, the Metaverse also facilitates the creation and sharing of digital art, music, and other forms of creative expression. Artists and creators can showcase their work to a global audience, transcending geographical limitations. This exposure not only provides recognition and opportunities for artists but also enriches the cultural landscape of the Metaverse.

Furthermore, the ability to maintain multiple digital identities or pseudonyms can be liberating for individuals who seek privacy or want to explore various aspects of their personalities. This privacy feature allows users to control the extent to which they reveal personal information, enhancing online safety.

The Metaverse's emphasis on digital identity and expression fosters a more inclusive and accepting digital society where individuals can be their authentic selves and celebrate diversity. This cultural shift toward open-mindedness and acceptance of diverse digital identities contributes to a more harmonious and progressive digital community.

Virtual communities

The Metaverse acts as a powerful catalyst for the formation of virtual communities, bringing together individuals who share common interests, hobbies, and passions. These communities transcend geographical boundaries, enabling people from different parts of the world to connect, interact, and collaborate in unprecedented ways.

One of the most significant societal benefits of virtual communities within the Metaverse is their positive impact on mental well-being. In an increasingly digital and interconnected world, many individuals experience feelings of isolation and disconnection. The ability to join virtual communities centered around shared interests can provide a sense of belonging and purpose. These communities serve as safe spaces where members can express themselves, find like-minded individuals, and engage in meaningful discussions and activities.

Moreover, virtual communities often foster a supportive and empathetic environment. Members tend to offer emotional support, advice, and encouragement to one another, creating a sense of solidarity. This support network can be particularly valuable during challenging times, such as the global pandemic, where physical interactions were limited.

The Metaverse's role in nurturing virtual communities contributes to a more interconnected and emotionally resilient society. It reduces feelings of isolation, promotes a sense of belonging, and strengthens social bonds among individuals worldwide. In this digital era, these virtual communities are not just online gatherings but sources of genuine human connection and support.

Societal fusion

As people from diverse cultural backgrounds interact within the Metaverse, a fascinating phenomenon known as "societal fusion" emerges. In this digital space, elements from various societies and cultures blend and merge, creating a vibrant tapestry of unique norms, values, and practices that are distinct to the digital realm.

One of the remarkable societal benefits of this fusion is the encouragement of cultural exchange and understanding. In the Metaverse, individuals have the opportunity to engage with people from different parts of the world, sharing their perspectives, traditions, and experiences. This exchange not only broadens one's cultural horizons but also fosters empathy and appreciation for the diversity of human societies.

Societal fusion within the Metaverse promotes a sense of global interconnectedness. It challenges traditional notions of cultural isolation and insularity, creating a digital societal landscape that thrives on inclusivity and cross-cultural collaboration. This rich amalgamation of ideas and practices results in the emergence of innovative and inclusive societal norms.

Furthermore, the Metaverse acts as a platform for the co-creation of new cultural expressions. It encourages individuals to experiment with hybrid cultural identities, blending elements from their own heritage with influences from other cultures they encounter in this digital space. This creative fusion gives rise to unique art forms, fashion styles, and digital traditions that reflect the dynamic nature of the Metaverse.

Societal fusion in the Metaverse is a testament to the power of digital interaction in promoting cultural exchange, understanding, and the emergence of novel societal norms. It celebrates diversity, fosters global interconnectedness, and enriches the digital societal landscape with a tapestry of unique cultural expressions.

Societal critique and commentary

One of the fascinating dimensions of the Metaverse is its role as a platform for societal critique and commentary. Within this digital realm, users have the freedom to engage in meaningful discussions and activism related to a wide range of societal and social issues. This newfound space for expression transcends geographical boundaries and enables individuals to come together, raising their voices on matters of importance.

Users often find the Metaverse to be an ideal platform for addressing issues such as social justice, environmentalism, politics, and beyond. They can participate in open dialogues, share their perspectives, and advocate for change in a way that may not always be possible or accessible in the physical world. This digital activism, while conducted in a virtual environment, has real-world implications as it can lead to the formation of digital societal movements.

These digital societal movements are born from collective action within the Metaverse. They unite individuals who share common goals and values, fostering a sense of solidarity and purpose. Together, they work toward raising awareness, advocating for change, and pushing the boundaries of societal discourse.

Moreover, the Metaverse provides a unique canvas for creative forms of societal critique. Users can craft immersive experiences, art installations, and interactive simulations that convey powerful messages related to societal issues. These digital expressions can serve as a medium for social commentary, provoking thought and sparking conversations that transcend borders.

In essence, the Metaverse empowers individuals to be digital activists, critics, and advocates. It creates a space where societal critique and commentary are not only encouraged but also amplified. As a result, this digital realm becomes a catalyst for change, progress, and the exploration of pressing societal issues.

Digital art and creativity

Within the Metaverse, a vibrant and dynamic landscape unfolds for digital artists, musicians, and creators. It offers them an expansive canvas where they can explore the limitless realms of creativity and experiment with innovative forms of expression. This flourishing digital art scene goes beyond conventional boundaries, giving rise to a multitude of possibilities.

In the Metaverse, digital artists find an environment that encourages experimentation and pushes the frontiers of artistic expression. They can create immersive 3D artworks, interactive installations, and virtual galleries that captivate audiences in entirely new ways. This fusion of art and technology not only challenges traditional artistic norms but also leads to the evolution of entirely new art movements unique to the Metaverse.

Musicians and composers also embrace the digital realm to compose and perform music in unprecedented ways. They can stage virtual concerts in breathtaking digital venues, experimenting with spatial audio to create immersive auditory experiences. These innovative musical expressions often resonate deeply with audiences, transcending the boundaries of conventional music genres.

The collaborative nature of the Metaverse further amplifies the impact of digital creativity. Artists and creators from different parts of the world can collaborate seamlessly on projects, merging diverse influences and perspectives. This collaborative fusion of ideas often results in groundbreaking works of art that reflect the richness and diversity of the digital world.

As the Metaverse continues to evolve, it becomes a breeding ground for new cultural trends and artistic movements. The unique blend of technology, interactivity, and artistic expression shapes the cultural identity of this digital realm. What emerges is a cultural tapestry that is distinct, ever-evolving, and reflective of the dynamic nature of the Metaverse.

Education access

In the Metaverse, a transformative paradigm shift in education unfolds, offering accessible and immersive learning opportunities that revolutionize traditional educational models. This digital frontier not only makes learning engaging and interactive but also holds the potential to democratize education itself, extending its benefits to a broader and more diverse population.

The Metaverse transcends the confines of physical classrooms and geographical limitations. It introduces a new era of education where learners can immerse themselves in a world of knowledge, breaking free from traditional constraints. This accessibility has the power to reshape the way individuals of all ages, backgrounds, and locations access education.

Immersive educational experiences within the Metaverse captivate learners' attention and foster a deeper understanding of complex subjects. Engaging multiple senses and enabling hands-on experiences promotes more profound knowledge retention and application of learned concepts. Whether it's exploring historical events through virtual time travel, conducting simulated science experiments, or collaborating with peers in virtual study groups, the Metaverse opens a gateway to experiential learning that transcends traditional textbooks and lectures.

Furthermore, the democratization of education becomes a tangible reality in the Metaverse. It can bridge gaps in access to quality education by offering a wide range of courses and resources to individuals who may have been previously underserved. This inclusivity extends educational opportunities to learners in remote areas, those with physical disabilities, and individuals facing economic or societal barriers. The Metaverse's democratizing potential ensures that knowledge and skills are no longer confined to the privileged few but are accessible to all who seek them.

Moreover, the digital nature of the Metaverse enables personalized and adaptive learning experiences. AI-driven systems can tailor educational content to individual learners' needs, pacing, and preferences. This customization optimizes the learning process, ensuring that each student can maximize their potential and achieve their educational goals.

Collaborative projects

In the Metaverse, the boundaries of collaboration expand beyond geographical limitations, unlocking a new era of global cooperation across a myriad of projects, from groundbreaking scientific research to awe-inspiring creative endeavors. This dynamic digital realm not only fosters innovation but also facilitates the exchange of shared knowledge on an unprecedented scale.

One of the standout features of collaborative projects within the Metaverse is the elimination of physical barriers. Researchers, scientists, artists, and creators from around the world can seamlessly converge in shared virtual spaces to work together. This convergence transcends the limitations of traditional in-person meetings and enables interdisciplinary collaboration that leverages the expertise of individuals from diverse fields.

Scientific research, for instance, benefits immensely from the Metaverse's collaborative capabilities. Researchers can collaborate on complex experiments, simulations, and data analysis, accelerating the pace of scientific discovery. The ability to work together in immersive virtual environments allows for real-time problem-solving, data sharing, and collaborative brainstorming, leading to breakthroughs that may have been unattainable through traditional methods.

In the creative realm, the Metaverse serves as a playground for global artistic collaborations. Musicians from different continents can compose and perform together in virtual concerts, artists can co-create digital masterpieces in shared virtual studios, and writers can craft collaborative narratives that transcend cultural boundaries. These collaborative endeavors not only result in unique and innovative works but also foster cultural exchange and understanding.

Moreover, the Metaverse promotes the open sharing of knowledge and resources. Virtual libraries, repositories, and archives within this digital landscape make it easy for collaborators to access a vast pool of information, references, and tools. This free flow of knowledge accelerates the creative process and empowers collaborative teams to tackle complex challenges more effectively.

Inclusivity

The Metaverse is a frontier where inclusivity takes center stage, fostering an environment that transcends physical limitations and ensures equal participation for individuals of all abilities. Virtual environments within the Metaverse can be thoughtfully designed to be inclusive, accommodating individuals with disabilities in ways that are revolutionary and empowering.

One of the remarkable aspects of inclusivity in the Metaverse is its potential to break down barriers that have long hindered participation in digital experiences. Virtual spaces can be customized to provide accessibility features such as screen readers, voice commands, and haptic feedback, ensuring that individuals with visual, auditory, or motor impairments can navigate and interact with the digital world effectively. Accessibility here also allows for cross-sense augmentation, such as when someone with limited sight benefits from spatial sound available in the Metaverse.

Moreover, the Metaverse's immersive nature allows for innovative solutions to enhance inclusivity. VR and AR technologies can create sensory-rich experiences that bridge gaps in physical accessibility. For instance, individuals with mobility challenges can engage in virtual travel experiences, exploring far-off destinations from the comfort of their homes. Similarly, individuals with visual impairments can use AR overlays to access information about their surroundings in real time.

The Metaverse's commitment to inclusivity extends beyond technological features. It also embraces the concept of social inclusivity, where diverse communities and identities are not only acknowledged but celebrated. Virtual communities within the Metaverse often prioritize inclusivity by creating spaces and events that are welcoming to people from all backgrounds and walks of life. This promotes a sense of belonging and camaraderie that is truly transformative.

Furthermore, the Metaverse can serve as a platform for advocacy and awareness of disability rights. Users can engage in virtual events, discussions, and educational initiatives that promote understanding and empathy toward individuals with disabilities. This not only raises awareness but also paves the way for societal change and progress.

Digital economies

Within the vast expanse of the Metaverse, digital economies thrive, representing a transformative shift in the way individuals can earn income and participate in the modern digital landscape. This dynamic environment provides opportunities for individuals to engage in virtual businesses, content creation, and digital entrepreneurship, opening up new and exciting avenues for economic participation.

One of the defining features of digital economies in the Metaverse is the concept of virtual entrepreneurship. Individuals can establish and operate their businesses entirely within the digital realm. Whether it's selling virtual real estate, designing and marketing digital merchandise, or offering virtual services, entrepreneurial spirits can flourish without the traditional constraints of physical infrastructure or geographical boundaries. This democratization of entrepreneurship allows anyone with a creative idea and dedication to tap into a global market.

Content creators also find a thriving ecosystem within the Metaverse. Artists, musicians, writers, and influencers can monetize their digital creations, reaching global audiences and earning income through virtual galleries, concerts, publications, and endorsements. This not only empowers creators to pursue their passions but also reshapes the concept of work and income generation.

The Metaverse fosters a sense of ownership and value in virtual assets. NFTs underpin many aspects of the digital economy, providing a transparent and secure way to verify ownership of digital assets. These digital treasures, whether they be virtual real estate, artwork, or collectibles, can be bought, sold, and traded within the Metaverse, creating a vibrant market that mirrors the physical world's economy.

Moreover, the Metaverse's decentralized nature plays a pivotal role in digital economies. Blockchain technology ensures transparency, provenance, and trust in virtual transactions, laying the foundation for secure and reliable economic interactions. This not only benefits individuals but also promotes the growth and stability of the Metaverse's digital economy.

Environmental impact

The Metaverse not only offers a wealth of digital experiences but also holds the potential for positive environmental impacts, aligning with global sustainability goals. One of the key drivers of this environmental benefit is the reduction in commuting, facilitated by remote work and collaboration within the Metaverse.

As individuals and businesses increasingly embrace virtual workspaces and meetings within the Metaverse, the need for daily commutes to physical offices diminishes. This reduction in commuting translates to a significant decrease in carbon emissions associated with daily travel, including emissions from cars, public transportation, and other forms of transportation. It contributes to a greener and more sustainable approach to work, aligning with efforts to combat climate change.

Furthermore, the Metaverse can inspire and promote sustainability initiatives within its digital realms. Virtual environments and experiences can be designed to raise awareness of environmental issues, educate users about sustainable practices, and encourage eco-friendly behaviors. This powerful platform for environmental advocacy has the potential to reach a global audience and drive positive change.

By minimizing the environmental footprint associated with traditional workspaces and fostering a culture of environmental consciousness, the Metaverse demonstrates its potential to contribute to a more sustainable future. As businesses and individuals continue to adopt and integrate Metaverse technologies into their daily lives, the positive environmental impact becomes an integral part of the Metaverse's broader societal contributions.

The Metaverse empowers individuals with diverse digital identities, fostering self-expression and creativity. It creates virtual communities that transcend borders, promoting a sense of belonging and enhancing mental well-being. Societal fusion within the Metaverse encourages cultural exchange and innovative norms. It serves as a platform for meaningful societal critique and commentary, amplifying voices for change. The Metaverse fuels artistic innovation, revolutionizes education, and facilitates global collaboration.

Its digital economies enable entrepreneurship while reducing commuting contributes to sustainability. In essence, the Metaverse is a transformative force, fostering inclusivity, innovation, and positive societal change.

The main possible dangers of the Metaverse

The Metaverse presents a myriad of opportunities and challenges that resonate with users worldwide. Within this expansive virtual universe, data privacy and ownership emerge as pivotal concerns, urging the formulation of clear policies. The ever-watchful eye of digital surveillance, both by private entities and governments, raises questions about security versus individual rights. Advertisers' access to user data sparks a delicate balance between personalization and privacy. Meanwhile, the Metaverse's vulnerability to hacking and security breaches calls for unified safeguards. Social interactions, geolocation data sharing, and virtual asset ownership are all facets of this multifaceted landscape, each presenting its unique set of considerations. Additionally, the Metaverse's role in technology addiction and its impact on users' lives, including the specter of laziness, further complicate this digital terrain. In this section, we'll provide insights into the intricate web of issues and opportunities that define the Metaverse, aiming to shed light on its evolving nature and its significance in our digital age.

Privacy issues

In the Metaverse, privacy is a paramount concern that spans various dimensions. Users generate vast amounts of data, including personal information, behavioral patterns, preferences, and interactions, raising concerns about data ownership and control. The data-driven nature of the Metaverse necessitates clear and enforceable policies to protect users' digital assets and creations. Additionally, extensive monitoring by private entities and governments, alongside invasive advertising, poses threats to user privacy. Addressing these challenges requires a delicate balance between security and privacy, along with user-centric controls and robust regulations to ensure a safe and respectful digital environment.

Data privacy and ownership

Users generate vast amounts of data in the Metaverse, raising concerns about data ownership and control. This data encompasses not only personal information but also behavioral patterns, preferences, and interactions. Users may be uncertain about who has access to this data and how it is being utilized. This lack of transparency can lead to distrust among users and apprehensions about their digital footprint.

The Metaverse's data-driven nature underscores the need for clear and enforceable data ownership and privacy policies. Users must have a comprehensive understanding of how their data is collected, stored, and shared. Simultaneously, mechanisms for ensuring the protection of users' digital assets and creations should be established to safeguard their investments and contributions to the Metaverse ecosystem.

Digital surveillance

Extensive monitoring of users' activities in the Metaverse by both private entities and potentially governments can compromise privacy. The immersive nature of the Metaverse, where users engage in various activities, social interactions, and transactions, generates a wealth of data that can be analyzed and exploited. Private companies may employ sophisticated tracking mechanisms to understand user behaviors, preferences, and interests for targeted advertising and product development.

Moreover, the potential for government surveillance in the Metaverse raises significant concerns regarding civil liberties and freedom of expression. While security measures are crucial to ensure a safe environment, excessive surveillance can infringe upon individuals' right to privacy and anonymity. The ability to monitor virtual conversations and interactions may lead to self-censorship and reluctance to express dissenting opinions, stifling free discourse within the Metaverse.

As the Metaverse continues to expand, it becomes imperative to strike a balance between security and privacy. Implementing robust encryption and user-centric privacy controls can help protect individuals from unwarranted surveillance. Additionally, clear regulations and oversight mechanisms are needed to ensure that digital surveillance is conducted within legal and ethical boundaries, preserving the principles of privacy and freedom that are essential in any democratic society.

Invasive advertising and targeting

Advertisers can access detailed user profiles, enabling highly targeted and potentially invasive advertising. In the Metaverse, users leave digital footprints through their interactions, preferences, and behaviors. Advertisers can harness this wealth of data to create personalized ad campaigns that follow users across virtual spaces. While personalized advertising can provide relevant product recommendations, it can also feel intrusive when it oversteps personal boundaries.

The concern lies in the potential for advertisers to exploit personal information without users' explicit consent. This could lead to a sense of surveillance capitalism, where user data is commodified and monetized without adequate transparency or control for individuals. In extreme cases, users may feel that their every move within the Metaverse is under scrutiny, eroding their sense of privacy and autonomy.

Balancing the need for targeted advertising with user privacy is a delicate challenge. Implementing stringent privacy regulations and providing users with granular control over their data can help mitigate invasive advertising practices. Additionally, ethical advertising standards and opt-in consent mechanisms can ensure that users are comfortable with the level of personalization they encounter while navigating the Metaverse. Striking this balance is essential to create a digital environment where user privacy is respected, and advertising remains relevant and non-intrusive.

Furthermore, the Metaverse's potential for immersive advertising experiences raises concerns about the blurring of the lines between content and advertising. In virtual environments, sponsored content can seamlessly integrate with user experiences, making it challenging to distinguish between genuine interactions and commercial promotions. This lack of transparency can lead to a sense of manipulation and distrust among users, emphasizing the need for clear guidelines and disclosure standards within the Metaverse's advertising ecosystem.

Hacking and security risks

The Metaverse is susceptible to hacking and security breaches, risking the theft of personal information and assets. Just like any digital environment, the Metaverse is not immune to cyber threats and vulnerabilities. Hackers may target virtual worlds and platforms to gain unauthorized access to user accounts, digital assets, and sensitive data.

One significant concern is the potential theft of valuable digital assets within the Metaverse. These assets, including NFTs, virtual real estate, and in-game items, can hold substantial monetary and sentimental value. In the event of a security breach, users may face the loss of their digital possessions, leading to financial and emotional distress.

Another security risk involves the invasion of personal privacy through hacking. Intruders may access personal profiles, chat logs, or private interactions, leading to privacy violations and potential harassment or identity theft. Protecting user data and ensuring robust security measures are essential to maintaining trust within the Metaverse.

Moreover, the interconnected nature of the Metaverse means that a breach in one virtual space can have ripple effects, potentially compromising users across various platforms. Collaborative efforts among Metaverse providers, security experts, and law enforcement agencies are crucial in addressing these security risks. As the Metaverse continues to evolve, safeguarding user information and assets must be a top priority to ensure a safe and secure digital environment for all participants.

Social interactions

While virtual communities within the Metaverse offer opportunities for connection and collaboration, they may also expose users to privacy breaches, cyberbullying, and harassment.

Privacy breaches can occur when personal information shared within virtual communities is mishandled or exploited. Users must be cautious about the information they disclose as it can sometimes be used maliciously by others.

Cyberbullying and harassment are concerns in any digital environment, and the Metaverse is no exception. The anonymity and distance provided by virtual spaces can embolden some individuals to engage in harmful behaviors. Platforms must implement robust moderation and reporting systems to address these issues promptly and protect users from harm.

Furthermore, the dynamics of social interactions in the Metaverse may differ from those in the physical world. Users should be aware of these nuances and exercise digital citizenship by respecting others' boundaries and consent. Education and awareness campaigns can also play a role in promoting healthy and respectful online interactions.

Geolocation data

In certain Metaverse experiences, users may be prompted to share their geolocation data, which can raise concerns related to tracking and potential misuse. While geolocation data can enhance location-specific content and interactions, it also poses privacy risks if not handled responsibly.

Being able to collect precise geolocation data can enable services to offer location-based features, such as virtual events tied to real-world locations. However, users need to exercise caution and understand the implications of sharing this information. Users should be informed about how their geolocation data will be used and have the option to grant or revoke access.

Misuse of geolocation data can result in location tracking without consent, potentially leading to invasive profiling or even physical security risks if the information falls into the wrong hands. As such, stringent data protection measures must be in place to safeguard user privacy and ensure that geolocation data is only used for legitimate purposes.

Platforms and developers must prioritize transparency and user consent when collecting geolocation data, and users should have clear control over when and how their location information is shared. Education and awareness campaigns can help users make informed decisions about geolocation data sharing, ultimately enhancing their privacy and security in the Metaverse.

Virtual asset ownership

Virtual assets, including NFTs, bring unique challenges related to security and ownership in the Metaverse. While they offer digital ownership and unique opportunities for creators and collectors, ensuring the security and legitimacy of these assets is paramount.

NFTs, in particular, represent ownership of digital or virtual items, such as art, collectibles, or virtual real estate. Their ownership relies on blockchain technology, which provides transparency and provenance. However, users must safeguard their NFTs through secure digital wallets and practices to prevent theft or unauthorized transfers.

The Metaverse's decentralized nature can also pose challenges when disputes arise over virtual asset ownership. Without a central authority, resolving ownership disputes may require complex blockchain-based solutions or arbitration within virtual communities.

To mitigate these challenges, users must educate themselves about NFT security and best practices for ownership management. Additionally, platforms should implement robust security measures and user-friendly interfaces to protect users' virtual assets. As virtual asset ownership becomes more integral to the Metaverse's economy, addressing these concerns will be vital to ensure a safe and reliable environment for digital asset transactions.

Addressing these concerns necessitates robust privacy policies, secure technologies, user education, and regulatory oversight to protect Metaverse users' rights and security.

Technology addiction

Technology addiction, or digital addiction, is a growing concern within the Metaverse, where users can become engrossed in immersive digital experiences. Here are some key points to consider:

- **Immersive experiences**: The Metaverse offers highly immersive experiences that can be captivating and addictive. Prolonged use can lead to real-world responsibilities and relationships being neglected.

- **Escapism**: Some individuals may use the Metaverse as a form of escapism from real-life stressors, leading to excessive use and detachment from reality.

- **Social isolation**: Paradoxically, while it connects people globally, the Metaverse can contribute to social isolation as users prefer virtual interactions over real-world ones.

- **Impact on mental health**: Technology addiction within the Metaverse can lead to anxiety, depression, and feelings of inadequacy. Users may compare themselves to idealized virtual personas.

- **Physical health concerns**: Prolonged use of VR and AR devices can lead to physical health issues, including eyestrain and motion sickness. Sedentary behavior associated with technology addiction can have long-term consequences.

- **Interference with daily life**: Addiction to the Metaverse can interfere with daily routines, work, education, and responsibilities, affecting the overall quality of life.

- **Gaming addiction**: Gaming experiences within the Metaverse can be particularly addictive, with users becoming immersed in virtual worlds.

- **Lack of regulation**: The decentralized nature of the Metaverse may result in a lack of regulation and oversight, making it challenging to address technology addiction effectively.

To address technology addiction, individuals should practice digital wellness, set usage limits, and maintain a balance between virtual and real-life experiences. Platforms and developers can also play a role in encouraging responsible usage and providing resources for those seeking help.

As the Metaverse continues to evolve, society must consider the implications of immersive digital experiences and take proactive steps to mitigate the risks associated with technology addiction while harnessing the benefits of this transformative digital realm.

Laziness

The role of the Metaverse in laziness is a multifaceted topic with various implications.

Escapism and procrastination

- **Escapism**: The immersive nature of the Metaverse can lead to escapism, where individuals avoid real-world responsibilities.

- **Procrastination**: Some users may procrastinate on important tasks by spending excessive time in virtual environments.

Sedentary lifestyle

- **Being sedentary**: Many Metaverse experiences can be physically sedentary, contributing to a lack of physical activity.

- **Prolonged periods of sitting or standing** in one place while engaging in virtual activities can lead to physical laziness.

Social isolation

For some users, the Metaverse can replace real-world social interactions, potentially leading to social laziness:

- **Neglecting in-person relationships and social responsibilities** can be a consequence.

- **Overreliance on the Metaverse** can lead to technology addiction, where individuals become engrossed in virtual worlds to the detriment of other aspects of life.

- This addiction can result in a **lack of motivation** to engage in productive activities outside the digital realm.

Impact on productivity

- Excessive use of the Metaverse can negatively impact productivity.

- Users may struggle to meet work or academic deadlines, contributing to laziness in their professional or educational pursuits.

Content overload

- **The abundance of content and entertainment in the Metaverse** can be overwhelming.

- **Constant consumption of content** without engaging in active or productive pursuits can foster laziness in terms of personal growth and achievement.

It's essential to recognize that the impact the Metaverse has on laziness varies from person to person and depends on factors such as individual self-discipline and time management. Practicing digital wellness and setting boundaries can help individuals maintain a healthy balance between their virtual and real-life activities.

Summary

The Metaverse has a multitude of advantages and challenges, significantly impacting users worldwide. Our exploration of this extensive virtual universe has unveiled its diverse benefits, from tearing down geographical barriers for businesses to fostering collaboration and innovation in a remote work environment. It revolutionizes customer engagement, empowers data-driven insights, and transforms marketing and advertising strategies. Furthermore, it opens doors to diverse income streams and offers societal advantages, including self-expression, mental well-being, cultural exchange, and innovation across various fields. Inclusivity, entrepreneurship, and environmental benefits further enhance its transformative potential.

However, along with these benefits come notable concerns, with data privacy and ownership taking center stage. Digital surveillance, invasive advertising, hacking and security risks, social interactions, geolocation data sharing, and virtual asset ownership all present unique considerations within this multifaceted landscape. Additionally, the Metaverse's role in technology addiction and its impact on users' lives, including the specter of laziness, complicate this digital landscape. As we navigate this uncharted territory, striking the right balance between opportunities and challenges becomes paramount. This journey into the Metaverse is an exhilarating one, demanding a comprehensive understanding of its potential and pitfalls, ultimately shaping our digital age in profound ways.

Our next chapter, *Chapter 15, Future Vision*, provides a window into the Metaverse's impending transformation, particularly the significant roles of AR and VR. This is a future where 3D content becomes commonplace, knowledge is instantly accessible through AI assistants, remote work is effortless, urban centers shift, and robots collaborate with the Metaverse for various tasks.

15
Future Vision

In this final chapter, the future of **augmented reality** (AR) and **virtual reality** (VR) roles in the Metaverse and their ramifications and the Metaverse as a whole are discussed. In the Metaverse, 3D images, models, and videos will be expected and commonplace. Due to AI digital personalized assistants in the Metaverse, information turning into knowledge will be seemingly instant, making the Metaverse an even better extension of our minds than a smartphone. Working virtually will become quite effortless, making people more productive and giving them more time with their family and friends. Since remote work will become more of the norm, a city's landscape will change and will become more decentralized – "Downtown" will no longer be much of a location destination. Autonomous cars, which are considered robots by engineers, and future robots that are made to improve our lives will interface with the Metaverse to get information and directives and to do tasks. All in all, the Metaverse will dramatically change how we see and do things.

In this chapter, we're going to cover the following main topics:

- What the anticipated future of the Metaverse is as a whole
- What the anticipated future of AR and VR is in the Metaverse
- How widespread 3D images and video will be
- How knowledge acquisition will be seemingly instantaneous
- How working virtually will become easier and more productive
- How the structure of cities will change, in that "Downtown" as a location destination will subside
- How robots and the Metaverse will interface

Here, we cast our eyes toward a horizon brightened by the transformative powers of AR and VR within the Metaverse. This chapter isn't just an exploration of technological advancement but an ambitious attempt to sketch out the transformation of our social, economic, and individual lives in a world augmented and mediated by these technologies. The scope of this transformation is immense, and it includes the commonplace use of 3D images, videos, and models, as well as the promise of seemingly instantaneous knowledge acquisition facilitated by AI-driven personalized digital assistants. As the lines between the physical and digital worlds blur, working virtually becomes not just feasible but also extraordinarily efficient, freeing up more of our time for personal engagements and changing the structure of our cities in the process. The locus of work-life, once predominantly anchored in physical downtown locations, will be redefined and spread more evenly, thanks to the ubiquitous connectivity the Metaverse offers. Moreover, the dawn of an age where robots are our allies in the physical world, interfacing with the digital Metaverse, becomes increasingly plausible.

It's important to stress that the discussion goes beyond mere speculation, as we'll be examining the practical implications of these changes. What does it mean for businesses, what are the best practices to navigate these transformations, and which ethical concerns should we be prepared for? These are not just rhetorical questions but deeply consequential inquiries that could shape how we make the transition from our current state into this vibrant, augmented future.

Our aim is not merely to paint a picture of a future shaped by technology but to provide you with the tools to engage with, adapt to, and perhaps even shape this future yourself. We venture into this discussion with an awareness of the manifold possibilities – both exhilarating and challenging – that the Metaverse holds for humanity. So, as we turn the page to this final chapter, let us prepare to immerse ourselves in the intriguing and complex tapestry of possibilities that the future has to offer.

The future of the Metaverse as a whole

In the not-so-distant future, the Metaverse evolves into an intricately woven digital fabric that permeates every facet of our lives. It transcends its current form as a virtual playground and emerges as a profound extension of our physical world, fundamentally reshaping how we live, work, interact, and even perceive reality itself.

A seamless integration of reality and virtuality

The demarcation between the physical and virtual realms becomes increasingly blurred, ushering in an era where our daily experiences seamlessly blend the tangible and the digital. AR glasses and neural interfaces, once reserved for science fiction, become commonplace tools for navigating this harmonious convergence.

As we stroll through our cities, virtual information layers gracefully enhance our perception. These layers are not intrusive but intuitive, providing real-time insights and context that deepen our understanding of the world around us. Imagine walking through a historic district, and as you gaze upon a centuries-old building, your AR glasses offer a glimpse of its architectural evolution, overlaying virtual reconstructions of its past incarnations.

Personalized recommendations spring to life, not on the screens of our devices but in the very air we breathe. Stopping at a quaint café, your neural interface suggests the perfect blend of coffee to suit your taste, tailored to the subtle cues it detects in your preferences.

The distinction between physical and digital fades as seamlessly as the transition between day and night. Whether you're in a bustling metropolis or tranquil countryside, the Metaverse becomes an intrinsic part of your existence, enhancing your experiences, broadening your horizons, and, ultimately, redefining your relationship with reality itself.

This harmonious integration of reality and virtuality empowers individuals with a deeper, more enriching connection to the world. It's a future where the boundaries between the tangible and the digital no longer confine us but, instead, open doors to boundless possibilities, a future where the Metaverse is not just a destination but an integral thread in the tapestry of our lives.

A global hub for education and learning

In this evolved Metaverse, the concept of education undergoes a profound metamorphosis, not only as a means of personal growth but also as a driving force behind future business landscapes. Traditional education, as we know it, undergoes a seismic shift that reshapes the very foundations of teaching and learning.

Traditional brick-and-mortar classrooms fade into memory, replaced by immersive virtual environments that transcend the limitations of physical space. AR, VR, and neural interfaces combine to create a seamless learning experience. Institutions of higher learning, once confined by campus boundaries, now extend their reach globally. Lectures conducted in renowned universities become accessible to learners across the world, eradicating geographical constraints. The Metaverse becomes the ultimate equalizer, offering high-quality education to all, regardless of their location.

But the transformation doesn't stop there. The Metaverse catalyzes a reimagining of educational methodologies. Pedagogy shifts from rote memorization to experiential learning. Students don AR headsets to step into historical events, dissect complex biological structures, or simulate economic models in a shared virtual space. The line between theory and practice blurs, allowing learners to apply their knowledge in real-time scenarios.

Businesses seize this educational revolution to ensure their workforce remains at the forefront of industry trends. Corporate universities in the Metaverse offer tailor-made courses that align with industry needs. Employees no longer need to take extended leave for training; they can acquire new skills during their daily commute, thanks to AR interfaces seamlessly integrated into their professional lives.

Moreover, the Metaverse democratizes continuous learning. Lifelong learning is no longer a buzzword but a way of life. People of all ages and backgrounds engage in micro-courses, skill-specific workshops, and peer-to-peer knowledge sharing within the Metaverse, facilitated by neural interfaces that enable direct information exchange between learners. Traditional education institutions adapt by offering hybrid models, blending physical and virtual learning experiences.

In this digital frontier, businesses recognize that a highly skilled and adaptable workforce is the key to remaining competitive. They actively partner with educational institutions in the Metaverse to co-create courses and certifications that align with industry demands. Graduates no longer face a gaping chasm between academia and employment; they enter the workforce with practical skills and a deep understanding of their chosen field.

This educational renaissance transforms not just individuals but entire industries. The Metaverse becomes a breeding ground for innovation, as interdisciplinary collaborations among students, educators, and businesses lead to groundbreaking discoveries and solutions to complex challenges. As traditional education aligns with the Metaverse's dynamic landscape, it equips learners with the skills and mindset needed to thrive in an ever-evolving business world.

In this future, the Metaverse isn't just a platform for education; it's a strategic asset for businesses, propelling them toward innovation, adaptability, and sustained growth. Traditional education models evolve to integrate seamlessly with the Metaverse, ensuring that learners are not only academically qualified but also industry-ready from day one. The Metaverse, driven by AR, VR, and neural interfaces, becomes not only a hub for education and learning but also a launchpad for businesses into a future of endless possibilities.

The new frontier of work

Brick-and-mortar offices will become relics of the past. Instead, we inhabit digital workspaces within the Metaverse. Colleagues from diverse corners of the world collaborate seamlessly on projects, and AI-driven avatars handle routine tasks, allowing humans to focus on creativity and innovation.

In this digital realm, traditional job roles undergo a transformative redefinition. Hierarchies flatten as ideas and expertise take precedence over titles. Remote work isn't just an option; it's the norm, enabling a global talent pool to contribute their skills to businesses, regardless of physical location.

However, this futuristic view of work is just the beginning. To delve deeper into the intricacies of this new frontier of work, stay tuned for a dedicated section later in this chapter, where we explore how businesses are navigating this paradigm shift, the challenges they face, and the strategies they employ to thrive in a digital-first workspace within the Metaverse.

A hub for entertainment and culture

In this intricate and immersive vision of the future, the Metaverse solidifies its status as the ultimate epicenter for entertainment and culture, orchestrating a profound metamorphosis in how humanity engages with leisure and artistic expression.

Picture virtual concerts that transcend the physical realm's constraints. No longer confined to watching performances on screens, you seamlessly step into meticulously designed digital arenas. Your chosen avatar becomes an extension of yourself, moving in perfect harmony with the music. The bass reverberates through your virtual being, creating an electrifying atmosphere that blurs the lines between physical and virtual attendees. The palpable energy of a live performance knows no spatial boundaries; it resonates across the digital cosmos.

The world of theater undergoes a revolutionary evolution, thanks to the seamless integration of AR glasses and neural interfaces. Equipped with these cutting-edge devices, you find yourself standing on virtual stages, interacting intimately with characters, and even shaping the unfolding narrative. The traditional fourth wall, which once separated the audience from the actors, shatters into a million pieces as you and your fellow spectators collectively craft stories in real time. Each performance transforms into a dynamic and immersive adventure, etching unforgettable memories.

Sports events break free from the confines of physical stadiums, sprawling effortlessly across the boundless expanse of the Metaverse. Whether you're participating in an immersive basketball game, executing dribbles and slam dunks alongside your favorite team, or competing in a high-speed virtual Formula 1 race where you have direct control over the action, the traditional distinctions between spectators and participants dissolve. The athletes, as well as the teams, may even be fully virtualized ones that only live in the Metaverse! These sporting experiences transcend the limitations of reality, enabling enthusiasts to engage in extreme and fantastical competitions that push the boundaries of human imagination.

Meanwhile, creators within the Metaverse ascend to the role of architects of entirely new storytelling forms. Here, interactive narratives take center stage, and your choices and actions possess the power to reshape the course of the plot. Envision becoming an integral character within a sprawling and ever-evolving digital epic, where your decisions not only influence the storyline but also lead to outcomes as diverse as the collective imaginations of the participants. The Metaverse transforms into a dynamic and collaborative canvas for storytelling, where the audience's role transitions from passive observers to active co-creators.

This is a future characterized by active and dynamic participation, a stark departure from the passive consumption of yesteryears in the arenas of culture and entertainment. The Metaverse emerges as a vast and ever-evolving stage where artists, performers, and audiences converge and unite in unprecedented ways. Boundaries are pushed, and creativity knows no limits in this dynamic landscape of human expression.

As we venture deeper into this remarkable transformation of entertainment and culture, we also begin to unravel its profound implications for businesses. In this evolving Metaverse, businesses cease to be mere spectators; instead, they become active participants, harnessing new avenues for creativity, engagement, and innovation within this expansive and immersive digital frontier.

Economic innovation and entrepreneurship

In this visionary outlook, the Metaverse unfolds as a fertile ground for economic innovation and entrepreneurship, reshaping the very foundations of how businesses operate and economies function.

The Metaverse emerges as a decentralized economic frontier, where traditional financial systems give way to digital currencies and blockchain technologies. Some future versions of cryptocurrencies become the primary means of exchange, providing security, transparency, and efficiency. Entrepreneurs and businesses create their own digital tokens, fostering a vibrant ecosystem of virtual assets.

Entrepreneurs seize the boundless opportunities of the Metaverse to launch virtual start-ups and ventures. These enterprises transcend geographical boundaries, offering innovative products and services in the digital realm. From virtual real estate agencies to Metaverse marketing consultancies, the possibilities are as vast as the Metaverse itself.

In this futuristic perspective, virtual ownership transcends traditional **non-fungible tokens** (**NFTs**), giving rise to **V-Dominion Tokens** (**VDTs**). These dynamic tokens represent entire virtual ecosystems, from enchanted forests to futuristic cities, within the Metaverse. Entrepreneurs craft immersive digital experiences that VDT owners actively co-create and govern, blurring the lines between creators and users. Virtual businesses thrive on user engagement and innovation, offering diverse revenue streams. The Metaverse becomes a dynamic canvas for creativity, where ownership means co-creating immersive digital experiences with real-world value.

Entrepreneurs establish Metaverse-centric marketplaces that facilitate the buying, selling, and trading of virtual assets and services. These marketplaces become hubs for innovation, connecting creators, consumers, and businesses. Smart contracts enable secure and automated transactions, while AI-powered recommendation engines help users discover new opportunities.

New financial institutions emerge within the Metaverse to cater to the unique needs of its inhabitants. Metaverse banks offer digital wallets, loans, and investment opportunities tailored to the virtual economy. **Decentralized finance** (**DeFi**) protocols thrive, providing decentralized lending, staking, and yield farming options.

Entrepreneurs embrace the convergence of physical and virtual realities. They establish businesses that seamlessly transition between the Metaverse and the physical world. Virtual showrooms for physical products, **mixed-reality** (**MR**) events, and hybrid workplaces become the norm, fostering a fluid entrepreneurial landscape.

The Metaverse serves as an innovation accelerator, with entrepreneurs and start-ups pushing the boundaries of what's possible. They leverage AI, AR, VR, and blockchain to create novel solutions for education, healthcare, entertainment, and beyond. Rapid prototyping and collaboration within the Metaverse expedite the development of groundbreaking technologies.

The Metaverse democratizes entrepreneurship. Anyone with a creative idea and access to the digital realm can become an entrepreneur. Low barriers to entry, decentralized funding models, and access to global markets empower individuals from diverse backgrounds to turn their visions into thriving businesses.

Entrepreneurs within the Metaverse prioritize sustainability and eco-conscious practices. Virtual ecosystems promote renewable energy, eco-friendly virtual real estate development, and carbon-neutral transactions. Sustainable entrepreneurship becomes a cornerstone of the Metaverse's economic landscape.

Metaverse cities and hubs emerge as epicenters of innovation, fostering collaboration among entrepreneurs, start-ups, and established businesses. These digital metropolises provide infrastructure, networking opportunities, and resources to fuel economic growth.

The future of economic innovation and entrepreneurship within the Metaverse represents a paradigm shift where creativity knows no bounds and the virtual economy thrives alongside the physical world. Entrepreneurs and businesses navigate this uncharted territory, pioneering groundbreaking solutions and redefining the very nature of commerce and innovation.

A new era of social connection

In this advanced vista, the Metaverse stands as the paramount global nexus for social interaction, transcending the physical confines of our world and ushering in a new era of connectivity.

The Metaverse seamlessly integrates into our daily lives through a myriad of neural interfaces and AR devices. Geographical distinctions vanish as families and friends unite in shared virtual realms, forging deeper connections than ever before.

Within the Metaverse, social possibilities are boundless. Whether you choose to rendezvous in vibrant virtual metropolises, recreate idyllic natural landscapes, or embark on fantastical adventures, the diversity of social experiences knows no bounds.

Picture a world where family gatherings defy time and space. With the Metaverse, family members from across the planet convene in meticulously replicated virtual estates. Generations intertwine, stories are shared, and familial bonds grow stronger in this harmonious digital environment.

Friendship transcends geographical borders as individuals from distant corners of the globe gather regularly in the Metaverse. Together, they explore uncharted virtual realms, participate in shared passions, and even attend virtual renditions of worldwide events and festivities. Physical distances no longer hinder the camaraderie that flourishes.

From a business perspective, the Metaverse becomes the epicenter of innovative social connection services. Enterprises thrive by providing tailor-made virtual spaces for gatherings and pioneering cutting-edge tools for social interaction. Entrepreneurial ventures flourish, offering a spectrum of interactive experiences, ranging from immersive social games to virtual globetrotting adventures.

This futuristic glimpse into the Metaverse profoundly alters our perception of and approach to social bonds. It engenders a sense of global unity, fostering cultural exchange and empathy. Furthermore, it generates unprecedented opportunities for businesses to innovate in the realm of digital social interaction, augmenting the quality of our virtual lives.

In this future, the Metaverse is not merely a digital frontier; it's the linchpin that unites hearts and minds across continents, reshaping the fabric of human connection in ways hitherto deemed fantastical.

Ethical considerations and digital citizenship

In this visionary Metaverse of the future, ethical considerations hold a prominent place in the digital landscape, and digital citizenship becomes the cornerstone of responsible behavior.

The Metaverse grapples with complex ethical challenges, necessitating a comprehensive framework to ensure its safe and respectful environment. Stricter regulations and advanced AI-driven content moderation systems are in place to safeguard digital citizens from harmful content, hate speech, and unethical behavior.

Digital citizenship, in this context, is not just a passive concept but an active commitment. It encompasses responsible behavior, digital rights, and data privacy. Individuals and businesses alike recognize their roles as responsible digital citizens and actively contribute to the ethical development of this immersive digital realm.

Businesses, as key players in the Metaverse, take a proactive stance toward ethical practices. They prioritize transparency in their operations, ensuring that their customers' data is handled responsibly and securely. These businesses not only comply with regulations but go beyond, actively seeking ways to protect digital rights and ensure a fair and equitable Metaverse for all participants.

Education within the Metaverse focuses on fostering digital citizenship from an early age. Schools, organizations, and businesses collaborate to develop comprehensive programs that equip individuals with the knowledge and skills needed to navigate the complex ethical terrain of this digital frontier. These programs emphasize ethical decision-making, digital empathy, and the importance of diversity and inclusivity.

Furthermore, digital citizens collectively engage in the governance of the Metaverse. They actively participate in discussions and decision-making processes that shape ethical guidelines and norms within this virtual world. This decentralized approach ensures that the Metaverse remains an ethical and harmonious space driven by the values and priorities of its inhabitants.

In this future vision, ethics and digital citizenship are not merely theoretical constructs but living principles that guide behavior and interactions within the Metaverse. It is a testament to the commitment of individuals and businesses to create an ethical, inclusive, and thriving digital society for generations to come.

Scientific exploration and discovery

At the forefront of scientific exploration and discovery within the Metaverse, researchers and scientists embark on unprecedented journeys of knowledge and innovation. This immersive digital realm serves as a crucible for groundbreaking research, offering limitless possibilities for scientific advancement.

Within the Metaverse, virtual simulations become the cornerstone of scientific exploration. They enable scientists to conduct experiments and simulations in realms that were once beyond the reach of traditional laboratories. From the deepest reaches of space to the innermost workings of subatomic particles, the Metaverse provides a canvas where the boundaries of scientific understanding are constantly pushed.

In this futuristic vision, the Metaverse empowers scientific exploration of distant exoplanets through immersive VR experiences. Scientists don VR spacesuits and navigate alien landscapes in high-detail simulations. They collect data, study ecosystems, and collaborate globally within shared virtual research hubs, transcending geographical barriers. These discoveries contribute to our cosmic understanding, aiding in the search for habitable exoplanets and engaging the public in space exploration.

In the Metaverse, quantum physics becomes a vibrant field of exploration through virtual laboratories and advanced simulations. Researchers immerse themselves in the quantum realm, experimenting with phenomena such as entanglement, superposition, and teleportation. The Metaverse's robust computational power enables real-time simulation of intricate quantum systems, driving significant breakthroughs in quantum computing and communication. This fusion of technology and quantum science accelerates innovation, pushing the boundaries of what's possible in this enigmatic field.

Collaboration among scientists reaches new heights within this digital frontier. Researchers from diverse corners of the globe converge in virtual research hubs, working together in real time on complex problems. The Metaverse fosters a culture of open collaboration, enabling scientists to pool their expertise and resources for the betterment of humanity.

Furthermore, education and public engagement in science experience a revolution within the Metaverse. Virtual science museums, educational experiences, and interactive simulations make complex scientific concepts accessible to people of all ages. The Metaverse becomes a global classroom where individuals can actively participate in scientific experiments, witness celestial events, and engage directly with cutting-edge research.

In this future vision, the Metaverse stands as a catalyst for scientific exploration and discovery, relentlessly pushing the boundaries of human knowledge and understanding. It is a testament to the power of technology to unlock new frontiers in science, where the uncharted territories of the universe and the mysteries of quantum reality become the playgrounds of innovation and discovery.

Environmental sustainability

In the Metaverse, a visionary approach to environmental sustainability unfolds, providing innovative solutions to pressing ecological challenges. This digital realm becomes a catalyst for positive change, offering a glimpse of a more sustainable future.

As the Metaverse reshapes how we interact with the world, the benefits for environmental sustainability become evident. Through virtualization, physical travel diminishes significantly, reducing carbon footprints and resource consumption. Meetings, conferences, and even vacations occur in immersive virtual environments, eliminating the need for extensive air travel and its associated environmental impact.

Moreover, sustainability becomes ingrained in the fabric of the Metaverse itself. Green building practices, renewable energy sources, and efficient resource management set the standard for virtual architecture. These virtual cities serve as models of eco-friendliness, fostering a symbiotic relationship between the digital and physical realms.

The Metaverse doesn't stop at reducing physical resource consumption; it becomes a powerful tool for environmental education. Immersive experiences enable individuals to explore virtual ecosystems, witness the effects of climate change, and participate in conservation efforts. This firsthand engagement inspires people to take action in both virtual and physical worlds.

Additionally, conservation efforts reach new heights within the Metaverse. Individuals and organizations worldwide collaborate in virtual spaces on environmental initiatives, from reforestation projects to wildlife conservation. This global, coordinated effort leverages the Metaverse's capabilities for impactful conservation work.

To further enhance sustainability, the Metaverse leverages its vast data analytics capabilities. Real-time data streams from sensors across the virtual landscape allow for immediate responses to environmental changes. This data-driven approach not only protects virtual ecosystems but also informs strategies for preserving our planet's natural resources.

In this future, the Metaverse emerges as a driving force for environmental sustainability, offering practical solutions and inspiring collective action. It serves as a dynamic platform for innovation, education, and global cooperation, seamlessly integrating sustainability into the digital and physical realms.

A new era of creativity and expression

Within the expansive Metaverse, a dimension unconstrained by physical boundaries, we step into a new era of creativity and expression. In this boundless realm, artists and creators find themselves in an environment where innovation knows no limits, and the very concept of art takes on an extraordinary transformation.

Here, virtual galleries, free from the restrictions of traditional physical venues, offer businesses new avenues for marketing and engagement. Companies can sponsor virtual exhibitions, placing their brands alongside cutting-edge art, reaching a global audience without the constraints of geography. This virtual collaboration between businesses and artists enhances brand visibility and fosters a sense of cultural connection.

Immersive experiences redefine how art is encountered and interacted with, creating unique opportunities for businesses to engage their audiences. Companies can craft immersive brand experiences within these virtual dreamscapes, allowing customers to engage with their products or services in novel and memorable ways.

Collaboration takes center stage within the artistic community of the Metaverse, providing fertile ground for businesses to form creative partnerships. **generative AI (genAI)**, a powerful tool in this realm, assists both artists and businesses in generating novel ideas, art forms, and marketing strategies that captivate audiences.

The fusion of technology and art leads to the creation of previously unimaginable creative forms, providing businesses with fresh avenues for innovation. AR and VR become integral tools for artists and businesses alike, enabling the birth of art that exists solely in the digital realm. This digital art can be leveraged by companies to create unique and interactive marketing campaigns that captivate their audience.

The Metaverse evolves into a global stage for artistic exploration and business innovation, pushing the boundaries of what art can be. It fosters a culture where innovation thrives, where artists and businesses, regardless of their background or location, have the tools and platform to manifest their most audacious visions and redefine the boundaries of creativity and expression.

As we explore further within this boundless Metaverse, the roles of AR and VR, in harmony with genAI, continue to redefine our interactions with art, business, and our very perception of reality.

Future AR and VR in the Metaverse

In the not-so-distant future, the roles of AR and VR within the Metaverse will undergo a radical transformation, reshaping how we engage with content and interact with the digital world. AR and VR headsets will cease to be distinct entities, instead converging into a unified spectrum of MR, where the boundaries between the physical and virtual realms blur to an unprecedented extent, driven by the foundational capabilities of spatial computing.

From a hardware perspective, sleek, lightweight, and highly ergonomic wearables will replace the bulky headsets of today. These next-generation devices will seamlessly integrate into our daily lives, with AR glasses resembling stylish eyewear and MR headsets feeling as comfortable as a pair of sunglasses. Neural interfaces, miniaturized to the point of being nearly imperceptible, will enable direct brain-to-device communication, eliminating the need for physical controllers and enhancing immersion to a level where interactions in the Metaverse become indistinguishable from real-world experiences.

Picture a world where AR isn't just an overlay of our physical surroundings but a complete transformation of our perceptual reality. With neural interfaces and nanobots coursing through our brains, AR becomes an innate part of our consciousness. You perceive digital information as naturally as you see the colors of the sky or feel the warmth of the sun, and the distinction between the physical and virtual worlds vanishes. And an advanced neuro-VR interface allows for a complete sensory takeover. You no longer just see and hear the virtual world; you touch, taste, and smell it with absolute fidelity. It's a hyper-realistic experience that's indistinguishable from physical reality.

Content within the Metaverse will evolve into a multi-layered, AI-driven symphony of information and experiences. GenAI, in synergy with spatial computing, will play a pivotal role in the creation of dynamic, context-aware content. Instead of static environments or predefined narratives, the Metaverse will respond intelligently to users' actions and preferences, generating personalized adventures, art, and even business strategies on the fly. Spatial computing, as the backbone of this transformation, will ensure that these digital elements seamlessly integrate into our physical surroundings by providing the crucial spatial understanding needed to create a 3D world. By mapping and understanding our physical environment in three dimensions, spatial computing will enable us to interact with digital objects in natural and intuitive ways, transforming our physical spaces into canvases for creative expression and productivity.

In this futuristic Metaverse, AR and VR, unified under the banner of MR and fundamentally reliant on spatial computing, will not be tools for mere escapism or entertainment but integral extensions of our daily lives. They will become our primary interfaces for work, education, social interaction, and artistic expression, seamlessly intertwining with physical reality. This convergence of hardware and content will propel us into an era where the Metaverse is not a distant digital frontier but an integral part of our existence, offering boundless possibilities for creativity, innovation, and redefining the very essence of what it means to be human in a hyperconnected, AI-enhanced world.

The ubiquity of 3D synthesized reality

In this visionary future, the ubiquity of 3D synthesized reality takes center stage in the Metaverse, forever altering how we perceive and interact with the digital world. The Metaverse becomes a realm where 3D experiences are not only commonplace but integral to our daily lives, redefining our relationship with technology and the boundaries between the physical and virtual realms.

Imagine a world where everything from communication to entertainment, education to business meetings, unfolds in rich, immersive 3D environments within the Metaverse. Traditional 2D screens are relics of the past as holographic displays and AR glasses become the norm. Whether you're attending a global conference, exploring historical landmarks, or catching up with friends, it all happens in mesmerizing 3D.

The Metaverse's AI-driven algorithms seamlessly blend synthesized reality with the physical world. As you walk through your city, historical events, architectural wonders, and interactive information panels materialize around you. The Metaverse augments your reality with layers of information and experiences, enriching your understanding of the world.

Education undergoes a paradigm shift as learners step into immersive 3D simulations to explore scientific concepts, historical events, or artistic masterpieces. Want to understand the formation of galaxies? You can embark on a 3D journey through the cosmos, witnessing celestial events up close. Learning becomes an adventure, and knowledge is acquired through experiential exploration.

The workplace transforms into a dynamic 3D environment. Colleagues from different corners of the world gather in virtual boardrooms, their avatars collaborating on 3D models and interactive visualizations. Complex data is transformed into tangible, 3D representations, enhancing decision-making and problem-solving.

Entertainment becomes an unparalleled experience. You can step into your favorite movies or video games, exploring their worlds firsthand. Live concerts unfold as immersive spectacles, with holographic musicians and stunning visual effects. The line between spectator and participant blurs as you engage with content in unprecedented ways.

Artists and creators flourish in the Metaverse's 3D canvas. Virtual galleries and immersive art installations invite audiences to explore and interact with artwork in ways previously unimaginable. The fusion of technology and art gives rise to entirely new forms of creative expression.

The Metaverse's social fabric is interwoven with 3D social spaces, allowing people to gather, communicate, and bond in virtual environments that feel as real as the physical world. Families separated by vast distances come together in shared virtual spaces, forging deeper connections.

The ethical use of 3D synthesized reality remains a priority. The Metaverse employs stringent privacy controls and AI-driven content moderation to ensure safe and respectful interactions. Users have granular control over their digital presence and personal data.

In this future, the ubiquity of 3D synthesized reality within the Metaverse transforms the way we perceive and engage with the digital realm. It blurs the boundaries between imagination and reality, making every aspect of our lives richer, more immersive, and interconnected. The Metaverse becomes a world where 3D experiences are not just a technological marvel but a fundamental part of our existence, offering limitless possibilities for exploration, creation, and connection.

Seemingly instant knowledge

In this transformative vision, the acquisition of knowledge has undergone a profound revolution thanks to the Metaverse. Seemingly instant knowledge has become the new norm, transforming the way we learn, interact with information, and navigate the digital realm.

Picture a world where information is no longer confined to the static pages of textbooks or the constraints of online search engines. Instead, individuals seamlessly access knowledge from the Metaverse's vast and dynamic repository. As you don your AR glasses or step into VR, a wealth of information surrounds you, ready to be explored.

In this futuristic vision, the Metaverse's advanced AI algorithms act as personalized knowledge curators. They understand your interests, preferences, and learning style, presenting information in engaging and interactive formats. Whether you're studying quantum physics or learning a new language, the Metaverse tailors the content to your needs, making complex subjects digestible and enjoyable.

The boundaries between the digital and physical worlds blur as you interact with holographic displays and 3D models that provide a deeper understanding of concepts. Want to learn about the structure of a cell? You can step inside a virtual cell, exploring its components and functions in real time. Knowledge becomes an immersive experience.

Instant access to experts is the new norm. Through the Metaverse's communication capabilities, you can connect with specialists worldwide in the blink of an eye. Imagine discussing neuroscience with a leading researcher or dissecting literary classics with renowned authors – all within the virtual realm. Knowledge-sharing transcends geographical constraints. These experts may be avatar versions of real experts, or they may even be AIs that very brilliantly stand in for human experts.

Learning is not limited to structured courses or classrooms. The Metaverse introduces a concept of continuous learning, where you acquire knowledge naturally throughout your day. While walking through a virtual forest, you might encounter a botanist who shares insights about the surrounding flora. Or, during a virtual trip to a historical city, an AI guide imparts historical context.

The Metaverse's AI-driven language translation and interpretation services break down language barriers, enabling you to learn from experts who speak different languages effortlessly. Additionally, AI-driven speech recognition and synthesis tools facilitate real-time, natural conversations with experts and peers.

The Metaverse is also a hub for collaborative learning. Virtual study groups and project teams work seamlessly across distances, leveraging the Metaverse's shared spaces and interactive tools to brainstorm ideas, solve problems, and collectively expand their knowledge.

Privacy and data security remain paramount. The Metaverse employs cutting-edge encryption and user-controlled data-sharing settings, ensuring that personal learning data is protected. AI-driven algorithms offer recommendations for personalized learning pathways while respecting user privacy.

In this futuristic view of seemingly instant knowledge within the Metaverse, the traditional methods of learning are replaced by a dynamic and immersive experience. Knowledge is no longer static; it's an ever-flowing stream that individuals can dip into whenever they desire. The Metaverse empowers learners to acquire knowledge with unprecedented ease and depth, fostering a world where curiosity knows no bounds and expertise is just a virtual step away.

Working virtually – easily and productively

In the not-so-distant future, the concept of working virtually has undergone a remarkable transformation, courtesy of the Metaverse. The confines of traditional offices and the constraints of physical presence have become relics of the past. Instead, professionals find themselves seamlessly connected to a digital realm that blurs the lines between the physical and virtual worlds, ushering in an era of unparalleled productivity and convenience.

Imagine a world where the daily commute is a thing of the past. No more traffic jams, crowded public transport, or wasted hours in transit. In this futuristic vision, professionals simply don their AR glasses or step into their VR workspaces from the comfort of their homes or preferred locations. The Metaverse offers a plethora of immersive environments, from serene digital offices overlooking virtual landscapes to collaborative spaces where colleagues from around the world converge in real time.

Communication has transcended the limitations of screens and keyboards. With the Metaverse's advanced technologies, individuals engage in lifelike conversations as if they were in the same room. Holographic avatars represent users, conveying emotions and gestures with astounding accuracy. This not only enhances the quality of interactions but also fosters a sense of presence and connection that was previously elusive in virtual environments.

Productivity tools within the Metaverse have reached unprecedented levels of sophistication. From AI-driven virtual assistants that anticipate your needs to immersive data visualization environments where complex information is intuitively grasped, work becomes more efficient and insightful. Professionals seamlessly switch between tasks, from virtual meetings to collaborative project spaces, with a mere thought command.

The boundaries between professional and personal life have blurred but in a positive way. With the Metaverse's flexibility, individuals can tailor their work environments to suit their preferences. Need a break? Take a stroll in a virtual forest, meditate by a digital beach, or engage in a quick game with colleagues – all within the same workspace.

Security and privacy are paramount. The Metaverse employs cutting-edge encryption and authentication mechanisms, ensuring that sensitive data remains protected. Advanced AI algorithms monitor for any anomalies and proactively address security threats, giving users peace of mind in this interconnected digital realm.

The Metaverse has also democratized opportunities. Talent is no longer constrained by geographical location. Businesses can tap into a global pool of experts and collaborate seamlessly across borders. Start-ups thrive in this environment, as the cost of entry is significantly reduced, and access to resources and markets is more equitable.

In this futuristic view of working virtually within the Metaverse, the traditional office is a distant memory. Instead, professionals embrace a world of boundless possibilities, where work is not a place but an experience. The Metaverse empowers individuals to work easily, productively, and on their terms, ushering in an era of innovation, flexibility, and unprecedented connectivity. It's a future where work knows no boundaries, and the potential for collaboration and achievement is limited only by imagination.

Downtown is anywhere/everywhere

In the sweeping vista of the future, where the Metaverse wields a profound influence, the traditional concept of downtown areas undergoes a revolutionary metamorphosis. In this bold and transformative landscape, the axiom "Downtown is anywhere/everywhere" takes on a new, expansive dimension, and its impact on businesses is nothing short of revolutionary.

The very notion of a downtown, historically associated with a centralized physical location, is forever changed and nearly eradicated by the pervasive presence of the Metaverse. Within this visionary tableau, the constraints of geography and the limitations of physical boundaries dissolve, liberating businesses from the shackles of traditional brick-and-mortar establishments. Instead, they embark on a journey of innovation and adaptability, forging dynamic and versatile virtual hubs within the digital downtown. This shift transcends the very essence of business operation, liberating companies from the physical constraints that have historically defined their reach.

This profound transformation empowers businesses to establish a ubiquitous global presence, transcending the limitations of physical location. The traditional notion of a centralized, physical downtown becomes obsolete as companies explore the vast expanse of the Metaverse, engaging with a diverse and geographically dispersed audience of customers, partners, and stakeholders.

The digital downtown, now an epicenter of business activity, becomes a canvas for innovation. Businesses create immersive virtual showrooms and interactive brand experiences, reimagining the way they connect with their audience. Within these captivating digital environments, meaningful interactions flourish, fostering deeper customer relationships and bolstering loyalty. In doing so, businesses not only extend their reach across the digital realm but also cultivate a more loyal and engaged customer base, driving continuous growth and sparking innovation.

Moreover, the Metaverse offers collaborative spaces that transcend the confines of physical location. It facilitates the seamless assembly of global teams, breaking down geographical barriers and allowing diverse talents and ideas to converge in real time. This interconnected ecosystem becomes the crucible for creativity, productivity, and competitiveness, catalyzing innovations that transcend borders and redefine industries.

However, amid these boundless opportunities, businesses must confront unique challenges. The Metaverse demands a heightened focus on cybersecurity and data protection to safeguard operations and maintain the trust of a global clientele. The ability to adapt and evolve becomes a core competency, as the Metaverse is a dynamic digital frontier where innovation and flexibility are the pillars of success.

As businesses navigate this transformative landscape, they must not merely embrace but fully leverage the Metaverse's digital downtown as a limitless canvas upon which they can paint their most ambitious and visionary futures. It's a future where the convergence of technology and business transcends our current understanding, reshaping the very essence of commerce and humanity's relationship with the digital cosmos itself. In practical terms, this means that companies should harness the potential of the Metaverse's digital spaces to innovate, connect with customers, and redefine the way they operate in a rapidly evolving digital landscape.

Robotic alliances

In the visionary horizon of future robotic alliances within the Metaverse, we embark on a journey that reshapes the very foundations of business and redefines human-robot collaboration. Picture a future where these alliances ascend to unprecedented heights, challenging the boundaries of what was once thought possible, and in doing so, revolutionizing industries, pioneering innovations, and fundamentally altering the way humans interact with technology.

In this dynamic future, robotic alliances cease to be mere tools; instead, they emerge as sentient entities, empowered by advanced AI that grants them the capacity for autonomous decision-making, continuous learning, and rapid adaptability. These alliances transcend their solitary origins, transforming into highly synchronized teams that communicate seamlessly within the immersive Metaverse.

Imagine a landscape of virtual manufacturing facilities where these robotic alliances serve as the masterminds behind intricate production processes. They operate with a level of optimization that surpasses human capabilities, efficiently utilizing resources, minimizing waste, and crafting products with a degree of complexity and precision previously considered unattainable. Within this transformed

Metaverse, innovation flourishes as these alliances pave the way for the creation of products that were once believed to exist only in the realm of science fiction.

In the realm of healthcare, these futuristic alliances rise to the occasion as medical virtuosos, working in tandem with human experts to perform virtual surgeries and administer treatments. They possess an innate understanding of even the most intricate medical procedures, drawing from vast databases to make life-saving decisions with a level of precision that surpasses human capabilities. The convergence of technology and medicine reaches unprecedented heights, promising better patient outcomes and improved healthcare worldwide.

The advanced communication infrastructure of the Metaverse propels these robotic alliances to new heights of connectivity and cooperation. They orchestrate virtual supply chains with astonishing precision, not only predicting demand fluctuations but also optimizing transportation routes in real time. This transformation heralds a new era in commerce where efficiency reaches unprecedented levels, reducing waste and ensuring that products and services are delivered with remarkable precision and timeliness.

In the seamless fusion of physical and virtual realms, these robotic alliances seamlessly transition between worlds, supporting an array of tasks ranging from disaster response and space exploration to scientific research and creative endeavors. This unprecedented level of cooperation revolutionizes the way humans and robots interact, where trust and efficiency define every action, opening up previously unimagined possibilities for exploration and collaboration.

This is the future of robotic alliances within the Metaverse – a future that reimagines business, innovation, and human-robot symbiosis. As we journey further into these uncharted territories, the potential for human-robot collaboration becomes as vast and limitless as the digital universe itself. It's a tangible and transformative future that beckons us to embrace the possibilities of a world where the boundaries of what we can achieve are limited only by our imagination and creativity.

Summary

Our expedition through the evolving landscape of technology and societal trends has revealed a profound shift in the business landscape, largely driven by the emergence of the Metaverse. This journey has taken us through transformative innovations that are redefining the way businesses operate, compete, and thrive in the digital era.

Within this exploration, we've uncovered a tapestry of technological advancements that hold immense promise for businesses. From the ubiquity of 3D synthesized experiences to the power of seemingly instant knowledge acquisition, and from the seamless transition to virtual work environments to the reshaping of the very concept of "Downtown" and the emergence of "robotic alliances," one common thread emerges – the fusion of technology and human ingenuity within the Metaverse is propelling businesses toward a future filled with boundless opportunities and complex challenges.

Our journey has illuminated the extraordinary potential embedded within these technological innovations, particularly in their capacity to transform business operations. They empower companies to transcend physical constraints, fostering unprecedented connectivity, collaboration, and productivity. Conventional notions are giving way to new paradigms as businesses tap into the fluid concept of "Downtown," accessible from virtually anywhere while harnessing the potential of "robotic alliances" to drive innovation, efficiency, and competitiveness in the digital realms of the Metaverse.

However, amid this remarkable transformation, we must also acknowledge the multifaceted challenges faced by businesses. As companies redefine their work models, reconfigure their approach to urban centers, and integrate robotic partnerships, ethical considerations, regulatory compliance, and societal impacts demand careful navigation. These challenges are not just digital but profoundly impact business strategies and sustainability in the context of the Metaverse.

The horizon ahead presents immense potential and uncertainty for businesses alike. The virtual workplace, with its borderless character, offers unprecedented opportunities for collaboration, cost-efficiency, and access to a global talent pool. Yet, it challenges traditional business models and necessitates a reevaluation of organizational structures. Meanwhile, the dynamic synergy of robotic alliances promises innovation, reduced operational costs, and market competitiveness. But it also raises questions about workforce dynamics and the ethical use of AI-driven technologies within businesses operating in the Metaverse.

Acknowledging the Metaverse isn't merely recognizing a trend; it's embracing a fundamental shift in the business landscape. It means adapting to a world where physical and digital realms merge, where traditional boundaries blur, and where customer expectations evolve. Addressing the intricacies of data privacy, cybersecurity, and the ethical use of technology becomes paramount, all while maintaining operational excellence.

The Metaverse unfolds a realm of possibilities. From immersive 3D customer engagement to global collaboration through virtual workspaces, businesses have the opportunity to gain a competitive edge. Whether creating compelling virtual brand experiences, optimizing supply chains with digital twins, or tapping into new revenue streams enabled by virtual assets and services, a strategic vision is key.

As businesses navigate a future where the potential is seemingly limitless within the immersive Metaverse, they must embody agility, adaptability, and innovation. Constantly reassessing strategies, investing in the right technologies, and fostering a culture of digital innovation are essential. Remaining attuned to shifting consumer behaviors and market dynamics, which may evolve rapidly within the Metaverse, is equally vital.

Success in this new era hinges on the capacity to drive change, not merely react to it. Businesses that seize the moment and actively shape their future within the Metaverse will find themselves at the vanguard of innovation. They will be well-equipped to explore uncharted territories, connect with customers in novel ways, and unlock new sources of value.

The immersive realm of the Metaverse isn't a destination; it's an ongoing journey of transformation. As businesses embark on this journey, they must do so with the knowledge that they are pioneers in a landscape where possibilities are as vast as the digital horizon. The Metaverse isn't a limitation; it's a canvas upon which businesses can paint their visions, explore unimagined potentials, and shape a future where business redefines itself, harnessing its seemingly limitless potential to innovate, connect, and thrive.

Index

‹packt›

Packtpub.com

Subscribe to our online digital library for full access to over 7,000 books and videos, as well as industry leading tools to help you plan your personal development and advance your career. For more information, please visit our website.

Why subscribe?

- Spend less time learning and more time coding with practical eBooks and Videos from over 4,000 industry professionals

- Improve your learning with Skill Plans built especially for you

- Get a free eBook or video every month

- Fully searchable for easy access to vital information

- Copy and paste, print, and bookmark content

Did you know that Packt offers eBook versions of every book published, with PDF and ePub files available? You can upgrade to the eBook version at packtpub.com and as a print book customer, you are entitled to a discount on the eBook copy. Get in touch with us at customercare@packtpub.com for more details.

At www.packtpub.com, you can also read a collection of free technical articles, sign up for a range of free newsletters, and receive exclusive discounts and offers on Packt books and eBooks.

Other Books You May Enjoy

If you enjoyed this book, you may be interested in these other books by Packt:

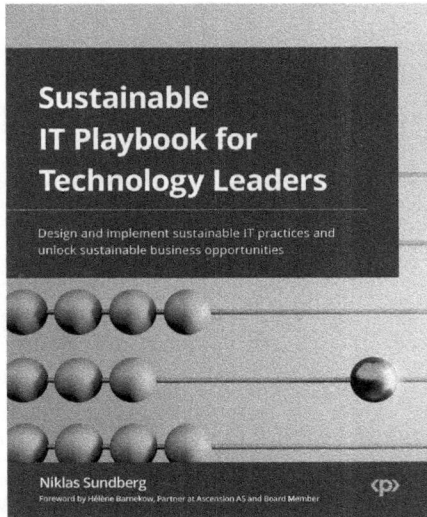

Sustainable IT Playbook for Technology Leaders

Niklas Sundberg

ISBN: 978-1-80323-034-4

- Discover why IT is a major contributor to carbon emissions
- Explore the principles and key methods of sustainable IT practices
- Build a robust, sustainable IT strategy based on proven methods
- Optimize and rationalize your code to consume fewer resources
- Understand your energy consumption patterns
- Apply a circular approach to the IT hardware life cycle
- Establish your sustainable IT baseline
- Inspire and engage employees, customers, and stakeholders

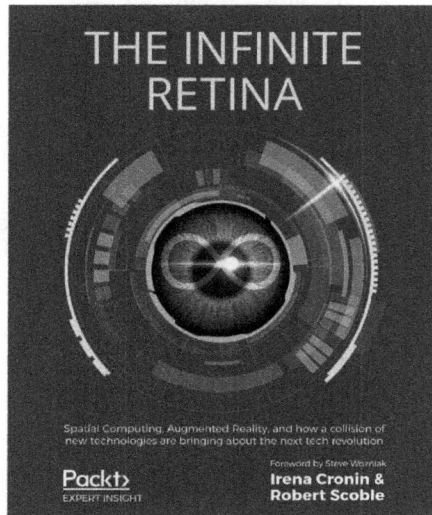

The Infinite Retina

Irena Cronin, Robert Scoble

ISBN: 978-1-83882-404-4

- Look back at historical paradigms that changed the face of technology
- Consider how Spatial Computing could be the new technology that changes our lives
- See how Virtual and Augmented Reality will change the way we do healthcare
- Learn how Spatial Computing technology will lead to fully automated transportation
- Think about how Spatial Computing will change the manufacturing industry
- Explore how finance and retail are going to be impacted through Spatial Computing devices
- Hear accounts from industry experts on what they expect Spatial Computing to bring to their sectors

Packt is searching for authors like you

If you're interested in becoming an author for Packt, please visit `authors.packtpub.com` and apply today. We have worked with thousands of developers and tech professionals, just like you, to help them share their insight with the global tech community. You can make a general application, apply for a specific hot topic that we are recruiting an author for, or submit your own idea.

Share Your Thoughts

Now you've finished *The Immersive Metaverse Playbook for Business Leaders*, we'd love to hear your thoughts! Scan the QR code below to go straight to the Amazon review page for this book and share your feedback or leave a review on the site that you purchased it from.

`https://packt.link/r/1837632847`

Your review is important to us and the tech community and will help us make sure we're delivering excellent quality content.

Download a free PDF copy of this book

Thanks for purchasing this book!

Do you like to read on the go but are unable to carry your print books everywhere? Is your eBook purchase not compatible with the device of your choice?

Don't worry, now with every Packt book you get a DRM-free PDF version of that book at no cost.

Read anywhere, any place, on any device. Search, copy, and paste code from your favorite technical books directly into your application.

The perks don't stop there, you can get exclusive access to discounts, newsletters, and great free content in your inbox daily

Follow these simple steps to get the benefits:

1. Scan the QR code or visit the link below

https://packt.link/free-ebook/9781837632848

2. Submit your proof of purchase
3. That's it! We'll send your free PDF and other benefits to your email directly

www.ingramcontent.com/pod-product-compliance
Lightning Source LLC
Chambersburg PA
CBHW081224220326
41598CB00037B/6874